HC 3/23/90

POLAR AND ARCTIC LOWS

Edited by

Paul F. Twitchell
Office of Naval Research
Washington, D.C., USA

Erik A. Rasmussen
University of Copenhagen
Copenhagen, Denmark

and

Kenneth L. Davidson
Naval Postgraduate School
Monterey, California, USA

A. DEEPAK Publishing   **1989**
A Division of Science and Technology Corporation
Hampton, Virginia USA

Studies in
Geophysical Optics and Remote Sensing
Series Editor: *Adarsh Deepak*
Published Volumes and Volumes in Preparation

Proceedings of the

Fourth International Workshop on Polar/Arctic Lows, and additional contributed papers

Sponsored by the

Office of Naval Research

POLAR AND ARCTIC LOWS

A. DEEPAK Publishing
A Division of Science and Technology Corporation
101 Research Drive
Hampton, Virginia 23666–1340   USA

Library of Congress Cataloging-in-Publication Data

Polar and Arctic Lows.
      "Contains most of the technical papers presented at the Fourth
International Conference on Polar/Arctic Lows held in Madison,
Wisconsin, 30–31 March 1988"—Pref.
      Includes bibliographical references.
      1. Polar vortex—Congresses.   2. Fronts (Meteorology)—Arctic
regions—Congresses.   3. Baroclinicity—Congresses.
4. Mesometeorology—Congresses.   I. Twitchell, Paul F.
II. Rasmussen, Erik A.   III. Davidson, Kenneth L.   IV. International
Conference on Polar/Arctic Lows (4th : 1988 : Madison, Wis.)
QC994.75.P64   1989   551.55'12'0998   89–23495
ISBN 0–937194–19–0

Printed in the United States of America

FOREWORD

It is only rarely that entirely new phenomena are discovered in the atmosphere. The probability of making such discoveries should be steadily decreasing since the atmosphere is under almost constant surveillance and has been watched by polar orbiting and geostationary satellites for quite a number of years. Nevertheless, the polar low is a new discovery made certain by satellite observations.

Before the era of satellites there were suspicions that depressions other than the well-known polar front low existed in the high latitudes. As a young forecaster I was warned by my supervisor that small dangerous systems occasionally existed in a northerly flow between Scotland and Norway, but this was a matter of experience only, and no physical explanation was given. Such smaller scale depressions were also mentioned in textbooks on forecasting, but the first real treatment of polar lows in a scientific journal is apparently the paper by Harrold and Browning in 1969 using radar observations. They considered the polar low as a baroclinic disturbance.

The polar low has similarities to a baroclinic disturbance, but there are very significant differences such as the scale, the characteristic time, and many aspects of the structure. Nevertheless, baroclinic instability of a special kind may play a role in creating a vorticity maximum (or a rather weak depression) in which polar lows may form. Polar lows have also similarities to tropical storms, and the CISK mechanism has been proposed as important for polar low formation; but once again, there are also differences. As yet, no agreement has been reached on the relative roles played by these two mechanisms.

In addition to the purely scientific interests in the polar lows they are very important in operational meteorology where the problem of *forecasting them is particularly difficult due to their size and relatively short life time. As operations of various kinds continue to increase in the arctic and polar regions the solution of the forecasting problem becomes increasingly urgent, and the polar lows are indeed excellent test cases for any mesoscale or fine mesh model. Such models should contain realistic parameterizations of the transfer of heat, momentum, and moisture between the underlying surface and the atmosphere, as well as a realistic convection prescription — two notoriously difficult physical processes to incorporate in a numerical prediction model. Just as important as the model is the definition of the initial state, which is difficult in the rather data sparse high latitude regions.*

The present book contains the papers from a special conference on polar lows and also papers on operational procedures. They give a picture of the present understanding of the many aspects of polar lows. It is my hope that this book will increase interest in the study of these important atmospheric disturbances.

Aksel Wiin-Nielsen

CONTENTS

CHAPTER 3 — MODELING / SIMULATIONS

4 — CASE STUDIES

CHAPTER 5 — OBSERVATIONAL STUDIES

PREFACE

This volume contains most of the technical papers presented at the Fourth International Workshop on Polar/Arctic Lows held in Madison, Wisconsin, 30–31 March 1988 and contributed papers on the topic not formally presented at the meeting. The workshop was held in conjunction with the American Meteorological Society's Second Conference on Polar Meteorology and Oceanography held 29–30 March 1988. This volume also contains additional technical papers on storms investigated in the Norwegian program as well as in other arctic regions and expands the topic to similar phenomena occurring near the Antarctic continent. Therefore, this volume is more than the proceedings of the meeting in Madison; it is a reference text on the topic, reporting on progress in the understanding of the origin, evolutions, and characteristics of these intense mesoscale winter storms. The progress in the understanding of the physics of the winter storms has led to numerical models that have prediction capability. The editors believe the topical area of polar/arctic lows has progressed to a level of understanding requiring a reference volume.

Inasmuch as this book was designed to be a reference the papers have been grouped according to their area of interest, presenting **reviews** *or describing* **theory, modeling/simulations, case studies,** *and* **observations/climatologies.** *For each of the five chapters in this book there is an introduction prepared by the editors, and for the entire volume a foreword by Aksel Wiin-Nielsen.*

This volume is the result of several independent but cooperative international efforts that began in 1983. These include the Norwegian program, the U.S. Navy's support of meteorology studies in the East Greenland Sea marginal ice zone in 1983 (MIZEX), and the 1984 United States National Oceanic and Atmospheric Administration Arctic

Cyclone Expedition (ACE). In 1984 the U.S. Navy also began the process to fund a research program specifically on arctic lows.

The first symposium on the present series of polar/arctic low meetings was held at the University of Copenhagen in August 1984. Many of the technical papers presented at the meeting were published in a special issue of Tellus in October 1985 (vol. 37A, no. 5). Interest by the U.S. Navy in these often violent storms that develop within hours with wind speed of hurricane force led to a workshop held on 9-10 May 1985 in Boulder, Colorado. Progress from this meeting was reported by Kellogg and Twitchell in the Bulletin of the American Meteorological Society in February 1986 (vol. 67, no. 2). The results of the three-year Norwegian research program on polar lows in the Norwegian, Greenland, and Barents Seas were reported at the International Conference on Polar Lows held near Oslo, Norway, 20-23 May 1986. Many of the technical presentations were published in another special issue of Tellus in August 1987 (vol. 39A, no. 4).

Polar/Arctic Lows are most frequently found in the seas surrounding Scandinavia and it is therefore natural that Norwegian and British meteorologists first studied these phenomena. In 1983 the Polar Lows Project was started in Norway. Jack Nordo, late scientist at the Norwegian Meteorological Institute, played an important role in the initiation of the project. In recognition of this contribution, this book is dedicated to him.

Paul F. Twitchell
Erik A. Rasmussen
Kenneth L. Davidson

FRONTISPIECE

A polar low as recorded on HRPT data received at Dundee University, Dundee Scotland. The data tape was provided through the courtesy of Professor Erik A. Rasmussen of the University of Copenhagen. Image processing, image enhancement, and photography were done by L.S. Fedor, NOAA/ERL/Wave Propagation Laboratory.

The full image (not shown) is from NOAA-9 on orbit #11379, 27 February 1987 from 04:10:59 to 04:17:35 UTC, AVHRR channel 4 (10.8 µm). The coldest infrared temperature in the full image is −63°C from cloud tops at the tropopause over the east coast of Greenland (at 75°N) and the warmest infrared temperature is +6°C from the open ocean northwest of Trondheim, Norway.

In the figure shown, a portion of the main image, centered on the major polar low at 71.5°N 23°E has been expanded by a factor of 2 in each direction. The minor polar low to the east is at 72°N 36°E: a distance of 450 km from the major system. Infrared temperatures for the major system are: −43°C to −38°C in the lower right-hand quadrant and −3°C to +2°C in the eye. Temperatures of the roll clouds west of this system are −18°C while the temperature is −23°C for Bear Island. The coldest temperatures over the sea ice north of Bear Island are −28°C.

Image processing was accomplished on a Digital Equipment Corporation Microvax II with a Ramtek RM-9460 display system.

CHAPTER 1 – REVIEWS

Introduction

One of the most basic features of polar and arctic lows is their small horizontal scale and the fact that they normally develop over the sea, often in remote regions. It is thus no surprise that the interest in polar lows more or less started with the advent of meteorological satellites that provide observations on the mesoscale even in normally data sparse regions.

Numerous satellite images have revealed that a variety of mesoscale disturbances, including polar lows, exist in the cold air masses poleward of the polar front. This knowledge is still comparatively new and almost nothing can be found about polar lows in meteorological text books. Although not a text book, the present book can be considered as a reference text defining the state of art of our knowledge of polar lows here at the end of the eighties.

As mentioned in the preface, most of the papers in the various chapters were presented at the Fourth International Workshop on Polar/Arctic Lows in Madison, Wisconsin, U.S.A., March 1988. Almost all of these papers were reports of recent research results from scientists involved in polar low research for several years. Two review papers have been included in the introductory chapter to provide the reader with the background of past and present research on the topic. In these papers a substantial part of the last 10 to 15 years' studies of polar and arctic lows has been summarized in a nonmathematical way. Although this summary is quite new, and as such fairly up to date, it should be noted that some of the ideas mentioned in the two review papers are challenged already in the following chapters of the book. We consider this as a positive aspect that indicates that research of polar

1

and arctic lows is vigorous and active. As an example of "old ideas" that are being challenged we might mention the explanation of the development of reverse-shear polar lows where the ideas of Reed and Duncan based on quasi-geostrophic theory are confronted with Moore and Peltier's more complex théory (see Chapter 2). Another interesting point of conflict is the role of conditional instability of the second kind (CISK) versus air-sea interaction instability (ASII). This issue is far from settled as the readers will realize by comparing, for example, the papers in this chapter with the papers by Craig and Cho, van Delden, and Emmanuel and Rotunno in Chapter 2.

If the two following papers could be summarized in one word (which of course cannot be done) probably the best word would be "variety." Fifteen to twenty years ago most meteorologists tended to believe that polar lows were either small baroclinic waves or small tropical storm-like systems. However, time and much work have shown that nature is much more complicated than originally thought, and that a whole spectrum of polar lows exist as amply demonstrated in this book.

POLAR LOWS[1]

Steven Businger
North Carolina State University
Raleigh, North Carolina, U.S.A.

Richard J. Reed
University of Washington
Seattle, Washington, U.S.A.

ABSTRACT

This paper constitutes a review of recent advances in our understanding of cyclogenesis in polar air masses. The review is primarily comprised of a survey of the observed features of polar lows as documented in a number of case studies presented in the recent literature. The review is organized on the basis of a combination of observational and physical considerations and is aimed at diagnosing common types of developments. Theoretical ideas concerning the origins of polar lows and results of numerical modeling experiments aimed at simulating their development are also summarized. Finally, a discussion of approaches to the operational problem of forecasting polar lows is given.

1. INTRODUCTION

Research and operational meteorologists have long been aware of the existence of subsynoptic-scale cyclogenesis in the cold air masses over open water. These disturbances command special attention for several reasons. They often evolve quickly and can result in adverse weather conditions that affect the safety of operations at sea. Due to the violent and unforeseen nature of polar lows, Norwegian and Icelandic history is filled with tragic accounts of accidents

[1]This material was based upon work supported by the National Science Foundation under grants ATM-8318857, ATM-8806276, and ATM-8421396-01.

POLAR AND ARCTIC LOWS
Paul F. Twitchell, Erik A. Rasmussen,
and Kenneth L. Davidson (Eds.)

3

involving small fishing boats. The scale of the disturbances is often too small for numerical weather prediction models to adequately simulate. Until only recently, atmospheric scientists were forced to rely on sparse observations from ships and islands to investigate the evolution of polar lows. While the introduction of satellite imagery has aided researchers, it has also revealed an unexpected profusion of cyclonic disturbances in polar air masses behind or poleward of the polar front (World Meteorological Organization, 1973; Anderson et al., 1969). The purpose of this paper is to provide a review of recent advances in our understanding of cold-air cyclogenesis in a framework that the authors' hope will prove useful to the research community in dealing with this ubiquitous phenomenon.

The term polar low or polar trough was originally used by British meteorologists to describe cold air depressions that affect the British Isles (Meteorological Office, 1962). Since some controversy surrounds, or at least in the past has surrounded, the use of the term "polar low," it is important to define at the outset what the authors mean by the term. We will use it here in a broad sense to denote any type of small synoptic- or subsynoptic-scale cyclone that forms in a cold air mass poleward of major jet streams or frontal zones and whose main cloud mass is largely of convective origin. We use the term polar low in a generic sense to include all phenomena that fit the above description. Some justification for this usage is provided by reexamination of cases that have gone under the rubric of polar lows in the early literature. Only by employing a broad definition of the term polar low can these varied cases be grouped together. A further discussion of the classification of polar lows is deferred to Section 2.

In view of the paucity of observational data available to early investigators it is only recently that polar lows have become a focus of research (Harley, 1960; Pedgley, 1968; Lyall, 1972; Harrold and Browning, 1969; Mansfield, 1974; Rasmussen, 1977, 1979; Reed, 1979). The lack of observations has contributed to the difficulty of synthesizing coherent physical models describing the triggering mechanisms and internal energetics of polar lows and to the slow progress that operational forecast models have made in predicting their development and movement.

In recent years significant research efforts have been made, directed toward improving our knowledge of polar lows. Through a combination of field experiments [Arctic Cyclone Expedition (ACE) in February of 1984 (Shapiro and Fedor, 1986); the Alaska Storms Experiment in February and March of 1985] and mesoscale modeling work [Seaman (1983); Sardie and Warner (1985); and

Grønås et al. (1986a, 1986b)], progress has been made in our understanding of these potentially destructive storms (Kellogg and Twitchell, 1986; Rasmussen and Lystad, 1987).

The framework for the overview of polar lows presented in this paper is derived from an observational as well as a physical perspective. In Section 2 general characteristics of the systems and the larger environment in which they form are described. Three types of polar lows are distinguished: (i) short-wave/jet-streak type, (ii) arctic-front type, and (iii) cold-low type. Discussions of case studies that are most representative of each of the three types are presented in Sections 3, 4, and 5, respectively. Multitype cases that represent combinations of the above types are discussed in Section 6. Theoretical and numerical modeling studies relevant to polar lows are briefly summarized in Section 7. A discussion that focuses on the problem of forecasting polar lows is presented in Section 8. Finally, summary and conclusions are given in Section 9.

2. GENERAL CHARACTERISTICS OF POLAR LOWS

Polar lows span a continuum of scales, from a few hundred kilometers to more than 1000 km in diameter, and a wide range of intensities, from moderate breezes to hurricane force winds (Reed, 1979; Locatelli ct al., 1982; Rasmussen, 1981, 1983; Mayengon, 1984; Forbes and Lottes, 1985; Shapiro et al., 1987).

Under favorable conditions, polar lows can intensify rapidly (e.g., Seaman et al., 1981; Rasmussen, 1985c; Shapiro et al., 1987; Rabbe, 1987), producing strong surface winds and locally heavy precipitation. Occasionally, they spawn severe thunderstorms. For example, seven tornados were reported in one storm that struck the California coast on 8 November 1982 (Reed and Blier, 1986b). Heavy snowfalls over Britain and the Netherlands have also been attributed to polar lows (Harrold and Browning, 1969; Seaman et al., 1981). Locatelli et al. (1982) noted that a significant number of the cyclones that affect the Pacific Northwest are polar lows, while Monteverdi (1976) found that as much as 50% of the annual precipitation in San Francisco is produced by nonfrontal disturbances occurring in the polar air mass and related only to positive vorticity advection (PVA) aloft.

Polar lows occur most often over the oceans in winter, originating in regions of enhanced convection and developing a comma- or spiral-shaped cloud pattern as they mature. The comma cloud shape is the result of the superposition of a positive vorticity center and a moderate background current. If the magnitude of the background flow is small, the vorticity will produce a spiral cloud pattern (Figure 1). (The polar low seen in Figure 1 will be the subject of a forthcoming paper by the authors). As an illustration of the seasonal variation of polar lows, the number of systems affecting the Norwegian coast or ships near the coast are shown in Figure 2a. The maximum frequency occurs in the period between October and April. A 9-year climatology of polar low occurrence in the Bering

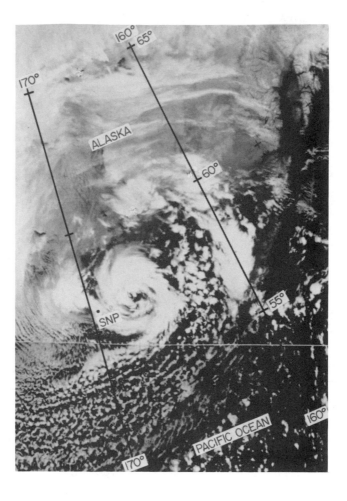

Figure 1. NOAA-5 infrared satellite photograph of a polar low and cloud streets over the Bering Sea at 2100 UTC 8 March 1977. SNP indicates the location of the rawinsonde station at St. Paul Island (R. Anderson, personal communication).

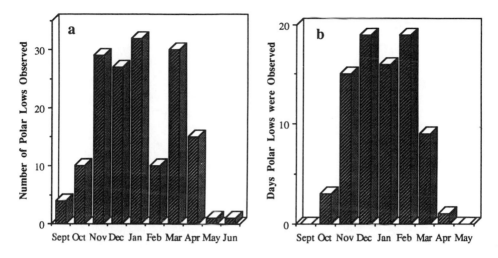

Figure 2. (a) Frequency distribution of polar lows near Norway for the period 1971–1982 (from Lystad, 1986), and (b) frequency distribution in the Bering Sea and Gulf of Alaska for the period 1975–1983 (from Businger, 1987).

Sea and Gulf of Alaska (Figure 2b) reveals a similar seasonal dependence and an even sharper winter peak.

The airstreams within which polar lows form are invariably characterized by cyclonic flow or shear (Reed, 1979; Mullen, 1979) by near neutral or unstable lapse rates in the boundary layer with conditionally unstable lapse rates extending locally as high as the middle to upper troposphere (e.g., Rasmussen, 1977; Mullen, 1979; Businger, 1987) and by substantial heat and moisture fluxes from the underlying ocean (e.g., Reed and Blier, 1986b; Shapiro et al., 1987). Composite charts (Figure 3) of surface pressure and 1000- to 500-mb thickness for 42 cases of polar lows, reproduced from Businger (1985), clearly show that large cold pools of cyclonically rotating air provide the most favorable environment for polar low outbreaks in the Norwegian and Barents Seas. Similar results were found for polar low outbreaks in the northern Pacific Ocean region (Businger, 1987).

Types of Polar Low Development. Some controversy continues as to whether there are several distinct types of polar lows with different underlying physical

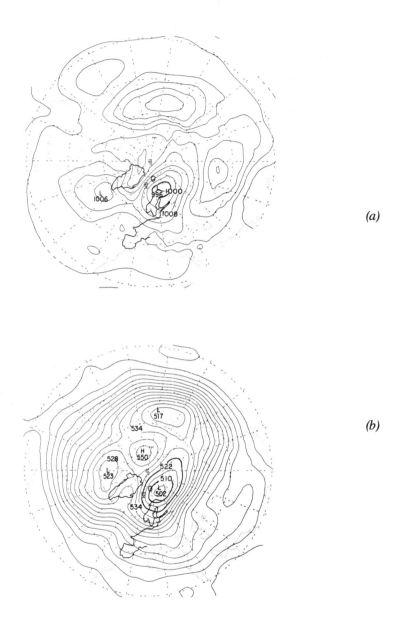

(a)

(b)

Figure 3. Composite charts of (a) surface pressure and (b) 1000- to 500-mb thickness for 52 cases (1971–1982) of polar lows in the Norwegian Sea (from Businger, 1985).

instability mechanisms, or whether these disturbances are basically varied manifestations of the same underlying physical mechanisms. The comma-shaped cloud system generally occurs in proximity to a trailing front or the jet stream axis and the spiral-shaped system tends to be located further back in the polar or arctic air mass (Locatelli et al., 1982). Rasmussen (1983) in a review article on mesoscale disturbances in cold air masses refers to the former type as "comma clouds" and the latter as "true" or "real" polar lows (sometimes also referred to as the "arctic instability lows"). Locatelli et al. refer to both types as "vortices in polar air streams."

As yet no widely accepted method exists for classifying polar lows, although schemes have been proposed based on their appearance in satellite imagery (Carleton, 1985; Forbes and Lottes, 1985) and on the synoptic situation (Lystad, Hoem, and Rabbe in Lystad, 1986; Rasmussen and Lystad, 1987; Reed, 1987). To facilitate the organization of this review, we will differentiate three elementary types of polar low development based on associated distinctive synoptic patterns. These types are also physically distinct in the degree and distribution of baroclinicity, static stability, and surface fluxes of latent and sensible heat. The three types are: (1) short-wave/jet-streak type, characterized by a secondary vorticity maxima and PVA aloft deep, moderate baroclinicity, and modest surface fluxes; (2) arctic front type, associated with ice boundaries and characterized by shallow baroclinicity and strong surface fluxes; and (3) cold-low type, characterized by weak baroclinicity, strong surface fluxes, and deep convection. Of course it is possible for a combination of the above types to exist. In some cases extreme winds (30 to 35 m s^{-1}) are associated with small, hurricane-like vortices embedded within a larger polar low or synoptic-scale low. It is not clear at this stage whether such systems should be regarded as a distinct class of polar low or as an embedded substructure within a larger system. In any case they are of sufficient importance to warrant separate discussion in Section 6.

3. SHORT-WAVE/JET-STREAK TYPE

At the larger end of the spectrum of polar lows is the comma cloud. This type of polar low is characterized by a large mesoscale to small synoptic-scale, comma-shaped cloud pattern that develops in regions of enhanced tropospheric baroclinicity, often just poleward of pre-existing frontal boundaries. The term "comma cloud" ascribed to this type of polar low is an abbreviation of "comma-shaped cloud pattern," and was introduced by satellite meteorologists to denote a characteristic cloud signature seen in cold air masses (Anderson et al., 1969;

World Meteorological Organization, 1973). Care must be exercised when using the term comma cloud, however, since comma-shaped cloud systems occur on many scales, and the same terminology has been used to describe the cloud pattern accompanying midlatitude frontal cyclones (Carlson, 1980; Carr and Millard, 1985).

Examples of comma clouds can be found in Reed (1979), Mullen (1983), Reed and Blier (1986a, 1986b), and Businger and Walter (1988). Figure 4 shows a satellite image of a well-developed comma cloud of small scale located to the rear of a synoptic-scale frontal cyclone.

Comma clouds form in regions of enhanced positive vorticity advection at midtropospheric levels (Anderson et al., 1969; Reed, 1979). Typically the comma cloud starts as a region of enhanced convection in the PVA region ahead of an upper level short-wave trough (or from an alternative point of view in the left front exit region of a jet streak (Uccellini and Kocin, 1987)). A surface trough (often marked TROF by U.S. analysts) lies near the rear of the comma tail. In the more strongly developed cases a low pressure center lies beneath the comma

Figure 4. GOES visible satellite image of small comma cloud (see arrow) located to the west of a synoptic-scale frontal cyclone. The length of the arrow is ~ 400 km. The west coast of the United States is visible on the right-hand side of the image (from Reed, 1979).

head and the trough may assume frontal characteristics in the later stages of development (Locatelli et al., 1982; Mullen, 1983; Reed and Blier, 1986a, 1986b). Weak to moderate baroclinicity exists within the cold air mass throughout the depth of the troposphere (Mullen, 1979). A schematic representation appears in Figure 5a. Using geostationary imagery, Zick (1983) showed that in some instances comma clouds over the Atlantic Ocean form from pre-existing vorticity centers that migrate from the front side to the rear side of large-scale troughs. The schematic in Figure 5b depicts the evolution of the cloud shield associated with a developing comma cloud as the latter moves down the trailing side of a 500-mb trough.

Comma clouds can occur anywhere in the extratropical Pacific or Atlantic Oceans, but based on the authors' experience, are most likely to form in the western oceans where the prevailing winds often transport cold, continental air across warm oceans currents. Convection develops in the cold air masses in response to heating and moistening of the boundary layer. Inspection of satellite

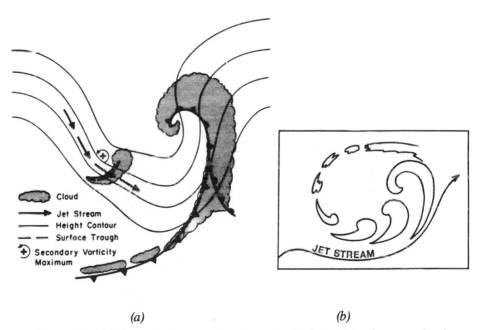

(a)   (b)

Figure 5. (a) Schematic diagram showing typical relationship of comma cloud to major frontal cyclone and upper level jet. (b) Schematic depiction of the principal life cycle of a comma cloud from the incipient stage to the mature vortex (from Zick, 1983).

imagery reveals that comma clouds typically form in regions of enhanced convection (Figure 6), and that the convection becomes organized into banded structures as the comma clouds mature. Recent aircraft observations taken in a region of enhanced convection (shown in Figure 6) over the Gulf of Alaska (Businger and Walter, 1988), show that when the mean wind profile is dominated by cold advection and the air is conditionally unstable through most of the troposphere, the rainbands tend to orient with their long axes parallel to the wind shear vector and do not propagate with respect to the mean flow. For a discussion of physical mechanisms for the maintenance of rainbands in comma clouds the reader is referred to Parsons and Hobbs (1983), and Businger and Hobbs (1987).

Figure 6. NOAA-6 infrared-satellite image of enhanced convection (A) and incipient comma cloud (B), at 1800 UTC 12 March 1985 (from Businger and Walter, 1988).

Interactions Between Comma Clouds and the Polar Front. When a comma cloud and its accompanying region of PVA approach a polar front, a wave often develops on the front. In reviewing the literature on such interaction, it is evident that a variety of interactions occur (Anderson et al., 1969; Reed, 1979; Carleton, 1985). In cases where the separation is large, the wave may remain independent of the comma cloud and develop into a typical midlatitude cyclone. In cases of closer proximity, the comma cloud often merges with the frontal wave in a process termed "instant occlusion," in which frontogenesis associated with the comma cloud provides the occluded front and the frontal wave provides the warm and cold fronts to the storm (Anderson et al., 1969). After the instant occlusion these storms appear identical to classical occluded systems in the cloud signatures seen on satellite imagery.

Since comma clouds generally occur over oceans, few case studies of their interaction with polar fronts have been studied with the benefit of mesoscale observational data. Mesoscale analysis by Locatelli et al. (1982) of three comma clouds that affected the Washington coast showed that the advection and precipitation patterns associated with these comma clouds remained independent of the advection and precipitation patterns associated with the polar fronts, even though they appeared to merge when viewing standard operational data. Browning and Hill (1985) used a mesoscale data set over the north Atlantic Ocean and Great Britain to describe the circulation pattern in a case where a polar trough interacted with a polar front. The authors developed a simple conceptual model that relates the principle cloud features to ascending air flows or conveyor belts. McGinnigle et al. (1988) analyzed several cases of instant occlusion over the north Atlantic Ocean using synoptic and mesoscale data and numerical model diagnostics. They suggest an alternative analysis scheme to the tradiational instant occlusion in which the instant occlusion is replaced by a warm front, and the secondary cold front associated with the polar trough becomes the more significant cold air boundary. The authors suggest a model of the principal airflows during the interaction and merging of the two cloud features and provide guidelines for the prediction of the development sequence.

Rapid cyclogenesis has been observed in some cases of interaction between comma clouds and polar-front cloud bands. Mullen (1983) documented an extremely rapid central pressure fall (22 mb in 12 hr) in a storm that occurred over the Pacific Ocean. In a paper by Businger and Walter (1988), satellite imagery revealed that four separate comma cloud systems evolved from a region of enhanced convection associated with a cold core 500-mb trough and a region of enhanced PVA (see Figure 6). The development of a wave cyclone was triggered when the comma clouds and the associated upper level short-wave trough approached a pre-existing polar front cloud band. Satellite imagery showed one of the comma clouds (labeled A in Figure 7a) to be absorbed by the cyclone, concurrent with a cyclone central pressure drop of more than 25 mb in 12 hrs. The satellite image for 0200 UTC on 14 March showing the interaction of the mature wave cyclone and the comma cloud is given in Figure 7a. The corresponding surface pressure analysis is given in Figure 7b.

4. ARCTIC–FRONT TYPE

Arctic fronts, separating modified and unmodified boundary-layer air, appear to play a significant role in the formation of some polar lows, even in the absence of significant PVA. Despite the nomenclature used in this section, it should be noted that arctic fronts are not usually analyzed on operational weather maps over ocean regions where polar lows occur. This may in part be due to the lack of observational data in their vicinity, and in part to varying degrees of sharpness of the associated baroclinicity. Favored regions of formation of arctic fronts are the Greenland, Norwegian, and Barents Seas, and the area south and east of Iceland (see papers by Shapiro and Fedor, and Shapiro, Hampel, and Fedor in this volume). Arctic fronts also commonly occur over the Bering Sea and the northern Gulf of Alaska (e.g., Businger, 1987). The regions referred to above are all located along the ice margin, where relatively warm, open water lies adjacent to ice fields or cold continents. In these regions, strong low-level baroclinicity exists due to differential heating of the boundary layer over open water and over ice covered surfaces. The low-level baroclinicity appears to play an important role in the formation of polar lows that occur in conjunction with outflows of surface air from ice and snow covered regions over open water. These outflows are characterized by the presence of cloud streets in the boundary layer (Anderson et al., 1969; Brown, 1980; Businger, 1985, 1987).

During the Arctic Cyclone Expedition (ACE) dropwindsonde and aircraft flight-level data were collected in traverses along a line perpendicular to an arctic front just west of Spitzbergen on 14 February 1984 (Shapiro and Fedor, in this volume). Cross sections constructed by Shapiro and Fedor using this data (their Figure 6b) show a relatively shallow arctic front extending to ~ 860 mb with a ~ 30 m s^{-1} jet stream above a significant low-level baroclinic zone. The frontal zone contained large relative and potential vorticity (Shapiro and Fedor's Figure 6b) that may have played an important role in the evolution of two arctic lows that formed on the southern edge of the arctic front.

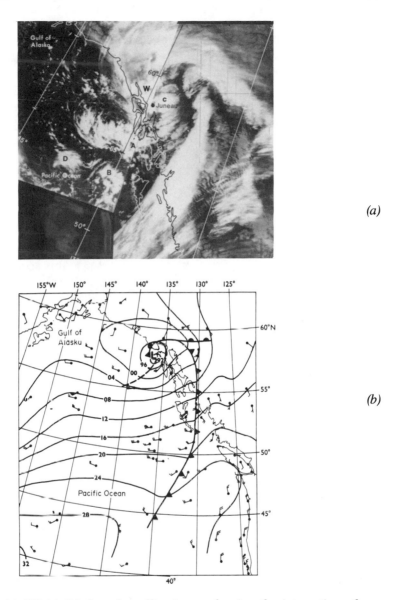

(a)

(b)

Figure 7. (a) NOAA-6 infrared-satellite image showing the interaction of developing comma cloud (A) and wave cyclone at 0200 UTC 14 March 1985. Lower case "c" indicates locations of enhanced convection along the occlusion cloud band (W). (B) and (D) indicate localized enhanced convection associated with mesoscale vortices in the cold air. (b) Surface pressure analysis for 0000 UTC 14 March 1985 (solid contours are isobars every 4 mb). The heavy dashed line indicates the position of the surface trough axis located to the rear of the comma cloud (from Businger and Walter, 1988).

The formation of boundary layer fronts through differential heating is not restricted to the west coast of Spitzbergen. Similar structures may form when the wind flow is approximately parallel to adjacent snow- or ice-covered and open-water surfaces. These fronts can maintain their identities when changing synoptic winds advect them out to sea, despite the modifying effect of the diabatic heating by the underlying open water. Arctic fronts can often be traced from their origin at the ice edge and propagate as far away as the coast of Norway (see papers by Shapiro and Fedor, and Shapiro, Hampel, and Fedor in this volume for further discussion).

A prominent class of polar low, which can be of the arctic front type, is the "reversed shear" disturbance commonly found over the seas to the west and north of Norway. Reversed shear refers to a situation in which the storm motion (or wind at the steering level) is in the direction opposite to the thermal wind, unlike the situation that prevails with the comma cloud type, which propagates in the direction of the thermal wind. Duncan (1978) was first to identify the reversed shear case and to elucidate the structure and dynamics of these disturbances.

A particularly striking example of reversed-shear arctic lows (Reed and Duncan, 1987), involving a succession or train of four disturbances that formed during a 2-day period, is shown in Figures 8–10. Figure 8 shows a NOAA-7 infrared satellite image of two of the lows (labeled 3 and 4) at 1255 UTC 31 January 1983. The disturbances are spaced about 500–600 km apart and are moving southwestward at about 5 m s^{-1}. Representative constant pressure charts for the period appear in Figure 9. It is evident that the lows were steered southwestward in conformity with the low-level winds and that the latter were strongest near the surface where the thermal gradient was also strongest. From the orientation of the isotherms (Figures 9a and 9b), it is apparent that the thermal wind was directed opposite to the low-level flow. Figure 10 depicts a cross section taken along the line AB in Figure 9. The storm track is located near the axis of maximum northeasterly winds at low levels. An increasingly deep layer with a near adiabatic lapse rate extends outward from the ice edge to the vicinity of the storm track. It is clear that the arctic lows formed within a shallow baroclinic zone of small static stability and that the thermal contrast was produced at least in part by differential surface heating. Air to the north flowed mainly over the ice and experienced little heating; air to the south flowed over relatively warm water and was strongly heated and moistened from below.

The reversed shear case is further illustrated in Figure 11 where it is contrasted with the forward shear type represented by the short wave/jet streak

Figure 8. NOAA-7 infrared satellite image of two polar lows (numbered 3 and 4) at 1255 UTC 31 January 1983 (from Rasmussen and Lystad, 1987).

systems. In both cases the upward motion and comma-shaped cloud pattern are located, in conformity with the Sutcliffe development principle (Sutcliffe and Forsdyke, 1950), where the thermal wind advects positive vorticity, i.e., down-shear of the trough. However, because of the opposite relationships between thermal winds and steering level winds, in one case the cloud system lies ahead of the trough and in the other case to its rear. In some cases (Grønås et al., 1986b) the reversed shear disturbances appear to initiate in a tongue of warm air that protrudes northward over the open sea to the west of Spitzbergen and Bear Island.

5. COLD–LOW TYPE

Sometimes small comma- or spiral-shaped cloud patterns of convective character are observed to flare up within the inner cores of old occlusions or

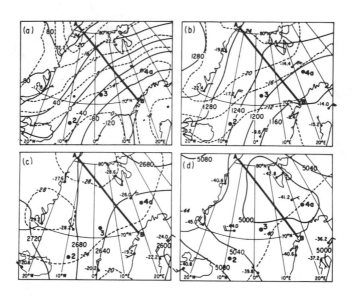

Figure 9. Charts for 1200 UTC 30 January 1983 at (a) 1000 mb, (b) 850 mb, (c) 700 mb, and (d) 500 mb. Solid lines contour at 40-m intervals; dashed lines isotherms at 4°C intervals; numbered dots indicate positions of polar lows (from Reed and Duncan, 1987).

cold lows, without any obvious association with upper level short waves or low-level baroclinic features. In particular, several interesting polar lows have recently been reported over the warm Mediterranean Sea (an otherwise unusual location for polar-low type developments), beneath cold-core upper level lows during late fall and winter in which baroclinic instability does not appear to play a significant role (Ernst and Matson, 1983; Mayengon, 1984; Billing et al., 1983; and Rasmussen and Zick, 1987). Although, observational data in these small storms was insufficient to give detailed accounts of their structure, synoptic data, ship reports, and satellite imagery suggest that they bear a resemblance to tropical cyclones. The common characteristics include very symmetric cloud signatures in satellite imagery, vigorous cumulonimbus surrounding a clear "eye," and a band of strong winds close to the eye. Evidence indicates that cyclonic circulations are most intense at sea level, gradually weakening with increasing height. Examples of two cases of this type of polar low that occurred over the Mediterranean Sea will now be discussed and illustrated.

Ernst and Matson (1983) described a case in which a hurricane-like feature formed in a decaying, old occlusion beneath an upper level cold trough. Visible

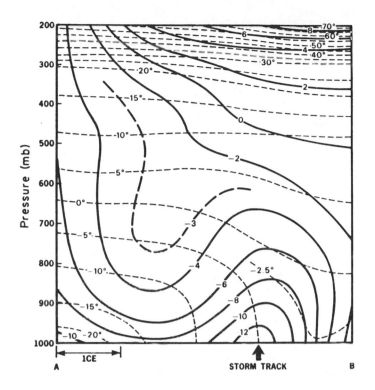

Figure 10. Cross section along line AB in Figure 9. Solid lines are isotachs (m s⁻¹); dashed lines are potential isotherms (°C) (from Reed and Duncan, 1987).

satellite imagery for 1247 UTC on 25 January 1982 (Figure 12) shows an eye-like feature and in-spiraling convective bands. A corresponding surface map (Figure 13a) shows only the dissipating occlusion. The upper level cold low for 0000 UTC on 25 January is depicted in Figure 13b. It is evident that an intense mesoscale vortex formed within the synoptic low as the latter decayed. A ship caught in the same storm on 26 January recorded a sustained wind of 25 m s⁻¹ indicating that the system well exceeded tropical storm intensity.

Another example of a hurricane-like development is described by Rasmussen and Zick (1987). In this case, as in the previous, the development occurred in the central part of a synoptic-scale cold core low of modest intensity, and deep convection was observed prior to the formation of the mesoscale vortex. The authors used METEOSAT imagery to observe the structure of the clouds and calculate the cloud level winds by tracking cloud elements as the disturbance

REVERSED SHEAR FLOW

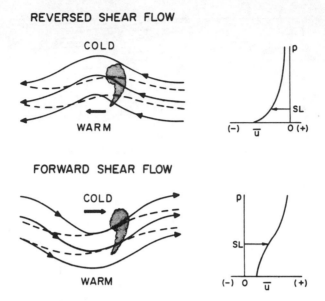

FORWARD SHEAR FLOW

Figure 11. Comparison of structures of disturbances in reversed-shear flow (top) and forward-shear flow (bottom). Solid lines, streamlines, and dashed lines—isotherms at the steering level; heavy arrows—phase propagation vector and steering level (SL) wind; stippling—comma cloud.

evolved. They found that deep convection preceded the formation of the vortex at the surface, and attribute the surface development to a rapid spin-up of surface vorticity, associated with the parent synoptic-scale low, by the intense convection. In its mature stage, the vortex was characterized by deep convection, a vertical axis and an upper level anticyclonic divergent outflow corresponding to a warm core structure. At this stage, the storm has a distinct similarity to its tropical counterpart to a casual observer (for example, see Figure 12).

6. MULTITYPE DEVELOPMENTS

The foregoing examples of short-wave and arctic-front types of polar lows highlight two quite distinct flow patterns that are associated with polar low development, one characterized by upper level PVA and the other by shallow baroclinicity. When an upper level short wave traverses the marginal ice zone, it is possible for a system to develop that combines both features. Two fascinating examples of multitype developments will now be described.

*Figure 12. NOAA-7 visible satellite image for 1247 UTC 25 January 1982.
The Island of Sicily is visible at the lower left-hand side of the image. The
location of the center of the cloud spiral is approximately 37.2°N 16.5°E.*

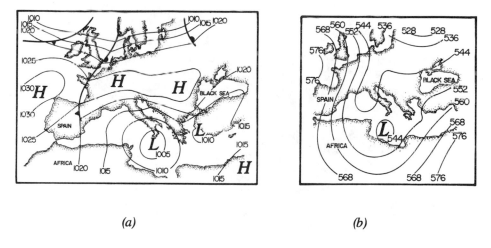

(a)   *(b)*

*Figure 13. (a) Surface analysis for 1200 UTC 25 January 1982, and
(b) 500-mb chart for 0000 UTC 25 January 1982.*

6.1 The Bear Island Case

The first case to be illustrated developed over the Barents Sea and has been documented by Rasmussen (1985a, 1985b). As pointed out by Rasmussen, there were two distinct stages in the development of this polar low. A satellite infrared image of the system taken at 0250 UTC 13 December 1982 is shown in Figure 14. At this time during the initial stage of development, a spiral-shaped cloud pattern was already present, though the surface low had yet to form. Antecedent conditions at 500 mb are displayed in Figure 15, and a sequence of surface maps commencing at about the time of the satellite picture appears in Figure 16.

The upper level chart for 1200 UTC 12 December (Figure 15b) shows a sharp trough and closed low west of Bear Island. The path of this short wave system is indicated by the broken line. It appears from the evolution at 500 mb that upper level vorticity advection played a role in the initial development. The surface chart for 0000 UTC 13 December (Figure 16a) shows only a trough in

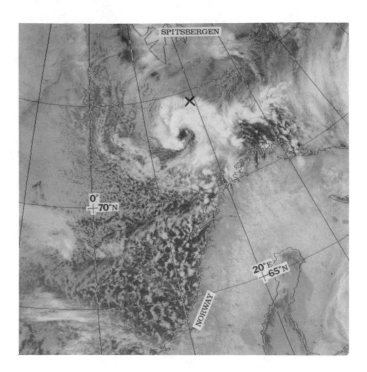

Figure 14. NOAA-7 infrared image 0250 UTC 13 December 1982. The position of Bear Island is marked by an "X" (from Rasmussen, 1985a).

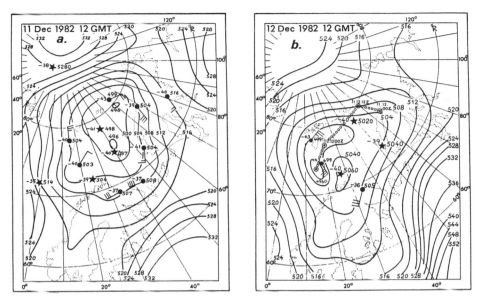

Figure 15. Five-hundred millibar charts for (a) 1200 UTC 11 December 1982, and (b) 1200 UTC 12 December 1982. Contours in decameters. Track of upper low is shown by broken line in (b) (from Rasmussen, 1985a).

the vicinity of the incipient low. This trough, located within the region of strong low-level thermal contrast, strengthened during the ensuing 12 hr, while a well-formed, comma- or spiral-shaped cloud remains on the satellite picture (not shown). This cloud pattern was associated with but not identical to the original spiral cloud seen in Figure 14. The later surface charts reveal that a dramatic second stage of development took place between 1200 UTC on the 13th and 0000 UTC on the 14th (Figures 16b and 16c). Rasmussen (1985a) ascribes this rapid deepening stage to the effect of organized deep convection (see paper by Rasmussen in this volume). A small, tight surface low is seen to cross weather-ship AMI located just off of the Norwegian Coast, bringing a pressure fall of 5.9 mb during a 3-hr period and winds of 20 m s^{-1} within a distance of 50–100 km of the low center. The surface pressure trace for weathership AMI during the low's passage (Figure 17) shows a rapid pressure fall between points A and B, followed by a more gradual rise between points B and C as the low moves away from AMI. It is perhaps of some significance that the hurricane-like core formed as the low passed over the warmest waters. Due to the upper level vorticity advection and enhanced low-level baroclinicity, this system can be classified as a combined type, but it had the added ingredient of a hurricane-like core.

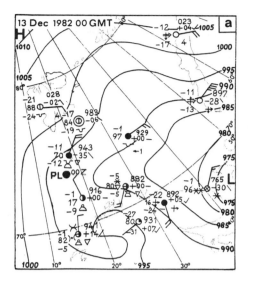

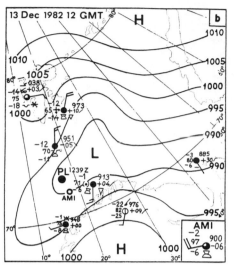

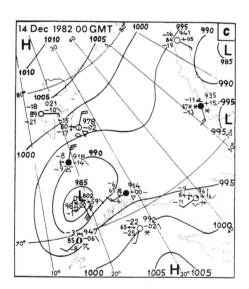

Figure 16. Surface maps for (a) 0000 UTC 13 December, (b) 1200 UTC 13 December 1982, and (c) 0000 UTC 14 December 1982 (contours every 5 mb). PL indicates position of the polar low (from Rasmussen, 1985a).

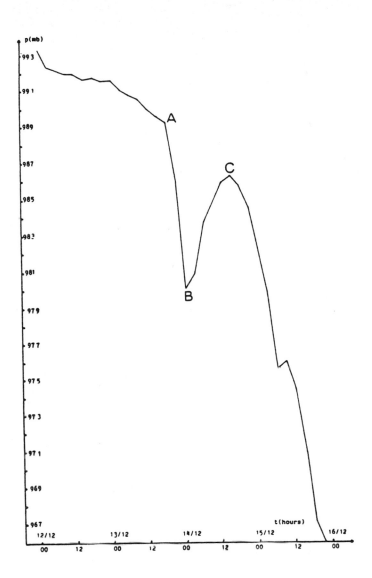

Figure 17. The surface pressure at AMI as function of time from 12 to 16 December 1982. The part of the curve between A and B shows a sudden deepening of the arctic low over AMI on 13 December, following a more gradual deepening as the low approached AMI (from Rasmussen, 1985b).

6.2 The ACE Case

The second example of a multitype polar low occurred during the ACE in 1984 when a research aircraft successfully penetrated a mature mesoscale vortex (Shapiro et al., 1987). This resulted in the best available documentation of the structure of a small, intense polar low to date. The 1000-mb and 700-mb operational analyses from the European Centre for Medium-Range Weather Forecasts (ECMWF) for this case are shown in Figure 18. The circled cross marks the position of the polar low. Lacking the flight data, the 1000-mb analysis gives no hint of the low. The surface pressure and 300-m winds determined from the aircraft observations (Figure 19) show a small cyclone with central pressure of about 980 mb and winds near the surface in the 25–35 m s^{-1} range at radial distances of 50–100 km present in the rear part of the synoptic-scale low. Aircraft soundings revealed that the low possessed a warm core. The satellite picture for the time in question (Figure 20) shows a comma-shaped cloud pattern with the suggestion of an eye-like feature in the head. The arctic low was the most intense of five mesoscale vortical circulations that were observed within the general cloud shield during a period of about 12 hr. Scales ranging from the synoptic scale down to the small mesoscale appear to be involved in this notable case.

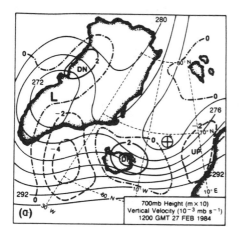
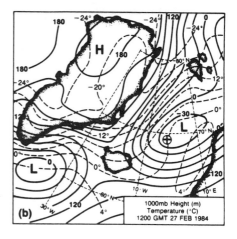

Figure 18. ECMWF analyses for 1200 UTC 27 February 1984. (a) 700-mb heights (thin solid lines in dm); and vertical velocity 10^{-3} mb s^{-1} (heavy dashed lines are positive, heavy dot-dash lines are negative). (b) 1000-mb heights (solid lines in m) and isotherms (dashed lines in °C). Circled cross marks position of the polar low (from Shapiro et al., 1987).

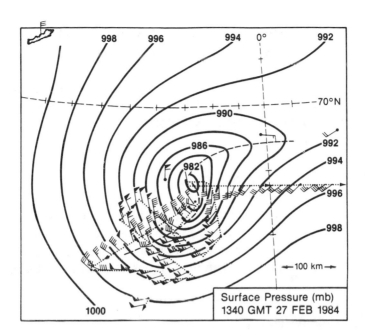

Figure 19. Surface pressure (mb) and 300-m winds at 1340 UTC 27 February 1984. Full barb 5 m s⁻¹; pennant 25 m s⁻¹ (from Shapiro et al., 1987).

Figure 20. NOAA-7 satellite infrared image for 1340 UTC 27 February 1987 (from Shapiro et al., 1987).

7. THEORETICAL STUDIES AND NUMERICAL PREDICTION EXPERIMENTS

In this section we will give a brief summary of some theoretical ideas and studies and numerical modeling experiments relevant to the polar low problem that have been presented in the recent literature.

7.1 Theoretical Ideas and Studies

Several mechanisms exist in current instability theory that in combination or alone might explain the formation and/or intensification of polar lows. They include baroclinic instability, conditional instability of the second kind (CISK), air-sea interaction instability, and barotropic instability, which are discussed below.

7.1.1 Baroclinic instability

The presence of significant baroclinicity through a considerable depth of the troposphere in the vicinity of polar lows, as revealed by a number of studies (Harrold and Browning, 1969; Mullen, 1979; Reed, 1979; Locatelli et al., 1982), as well as broad regions of clouds and precipitation, support the view that baroclinic instability can play a primary roll in the formation and intensification of polar lows. This was not the early view, however. Most early writings on polar lows attributed their origin to thermal instability within cold air masses flowing over warm seas (e.g., Meteorological Office, 1962). Harley (1960) appears to have been the first to raise the question of whether at least some polar lows are of baroclinic origin, citing the case of a wavelike, upper level frontal perturbation that progressed southeastward from Greenland and spawned a non-frontal surface low in the North Sea. It appears from his maps that the upper level system in question would today be termed a short wave or secondary vorticity maximum.

Interest in baroclinic instability as a factor in polar low formation was much stimulated by Harrold and Browning's well-known paper (1969). With the use of radar these authors showed that in the case of a polar low passing over England the precipitation occurred mainly in slantwise ascent as in a typical baroclinic disturbance. On the basis of a cursory inspection of several small comma cloud developments over the Pacific Ocean, Reed (1979) stated that these systems invariably form in the cyclonic-shear zone poleward of the jet stream and that this zone is marked by conditional instability and weak to moderate baroclinicity. Mullen (1979) obtained a composite picture of the large-scale environment of 22 small comma clouds that formed over the Pacific Ocean that reinforced the foregoing description and interpretation. Locatelli et al. (1982) showed that the mesoscale structures of several larger comma clouds exhibited wind, temperature, and precipitation patterns similar to those observed in larger extratropical cyclones that form on the polar front. Reed (1979) pointed out that relatively intense comma clouds can develop in polar air masses over land (e.g. see Wallace and Hobbs, 1977, p 110; Mullen, 1982) in the absence of significant sensible heat and moisture fluxes from the surface, which indicates that the potential energy released by baroclinic instability alone might be sufficient to account for the development of some polar lows.

A succession of theoretical papers dealing with baroclinic instability as a mechanism of polar low formation using numerical models have followed Harrold and Browning's work. Mansfield (1974) applied the Eady linear perturbation model to a shallow (1.6 km) rigidly bounded layer, using the observed mean state in the Harrold and Browning case. Despite the neglect of moist convection, he obtained realistic growth rates (e-folding times of 1–2 days) and realistic sizes (600–800 km) for the most unstable disturbances. Friction and sensible heat flux were found to inhibit the development. Though it may be questioned whether the basic state employed by Mansfield adequately represented the true state, his results provided the first quantitative evidence that baroclinicity could play an important role in at least some polar low developments.

Duncan (1977) employed a linear quasi-geostrophic inviscid model to find normal mode solutions for unstable disturbances with small static stability near the Earth's surface. Three observed cases of polar lows were used to specify the basic state wind and temperature distributions in vertical sections normal to the flow and the vertical profiles of static stability. Sensible and latent heating and friction were neglected. The presence of small static stability at low levels in conjunction with the existence of low-level baroclinicity (small Ri, Richardson number) resulted in the formation of shallow disturbances that, in two out of three cases investigated, resembled the observed disturbances in size, growth rate, and propagation speed. A numerical study by Staley and Gall (1977), employing a three-level model, called attention to the effect of the small Richardson number at low levels in producing small disturbances, without specifically mentioning polar lows. In a later paper Duncan (1978) applied his model to a case of baroclinic instability in a reversed-shear flow and elucidated the structure of the reversed-shear disturbance.

Blumen (1980) studied the nonlinear evolution of unstable two-dimensional Eady waves in a two-layer model characterized by smaller static stability in the lower layer than in the upper. Two types of unstable disturbances were found, one being a short wave solution confined to the lower layer. His results may be relevant to cases in which there is evidence of different wave scales at upper and lower levels.

Reed and Duncan (1987) have recently applied the dry baroclinic model of Duncan (1977) to the observed background state in the case of the train of four more or less evenly spaced polar lows described in Section 4. Their computations yielded a wavelength of maximum instability that was consistent with the observed wavelength of 500–600 km, suggesting a possible baroclinic origin for the wave train. However, these authors pointed out that some factor other than baroclinicity, presumably latent-heat release in deep convection, must have been an important factor in the development, since the observed lows moved at significantly slower speeds than predicted by the dry baroclinic model and the growth rates given by the model, though substantial, were not sufficiently large, especially in view of the neglect of friction.

The theoretical treatments discussed above have been based on normal mode solutions. By taking nonmodal initial conditions, corresponding to an upper level trough overtaking a surface depression, Farrell (1982, 1984) shows that transient growth ensues, in which the scale of the cyclogenesis is that of the upper level trough, and is independent of the most rapidly growing normal mode. As a short-wave trough approaches a surface depression, a tilt in the resulting vertical trough axis produces a favorable phasing of the wind and temperature distributions, and this results in rapid deepening of the initial surface pressure deficit (3–5 times), over short time scales (≤ 12 hr). Farrell's results particularly have application to interactions between polar lows and the polar front, in which rapid cyclogenesis is induced along the polar front by the 500-mb short wave trough associated with the polar low. Orlanski (1986), using a two-dimensional model, has applied the initial value approach to the problem of meso-alpha cyclone development on a mean baroclinic state. His results highlight the importance of small static stability at low levels and moist processes in promoting more vigorous development.

The above cited baroclinic studies using dry models all yield unrealistically shallow structures. That latent heat release can have a profound effect on the structure and growth of polar lows is evident from the work of Gall (1976). In studying baroclinic wave growth, he found that the effects of condensational heating are to increase amplitudes near 500 mb, where cloud signatures suggest that considerable rotation exists, and to increase the growth rate of short waves.

7.1.2 Conditional instability of the second kind

Paralleling the studies of the role of baroclinic instability in polar low development have been a number of studies that have examined quantitatively the role of diabatic processes. The underlying concept in most of these studies has been that of conditional instability of the second kind introduced by Charney and Eliassen (1964) to explain the growth of the hurricane depression. The CISK theory explains their growth as the result of a cooperative interaction between cumulus convection and the large-scale circulation. The CISK process occurs if the convection becomes sufficiently organized, so that a positive feedback develops between the cloud scale and the developing vortex scale, in which the vortex motion provides moisture convergence for the convective process and the cumulus scale provides latent heating that intensifies the large-scale disturbance.

Økland (1977) formulated an analytical model that examined the effects of surface fluxes and latent heat release in clouds in intensifying small cyclones in cold air masses over the ocean. His results showed that release of latent heat, although comparatively small under the specified conditions, may be a sufficient energy source to account for the spin-up. The results also indicated that sensible heat flux can modify the dynamics of the vortex considerably and that the static stability is an important parameter that must be below a certain limit for intensification to occur.

Using a three-layer, linear, quasi-geostrophic model, Rasmussen (1977, 1979) obtained normal mode solutions for CISK-driven disturbances under typical polar low conditions. His results showed that provided the heating in the upper layer of the model was sufficiently strong relative to that in the lower layer, a short wave cut off to the instability could be obtained. The maximum growth rate for disturbances larger than the cut-off size occurred at wave lengths comparable in size to polar lows. Computed growth rates of $2-3 \times 10^{-5}$ s^{-1} (e-folding times of approximately one-half day) were easily large enough to account for the rapidity of polar low development.

Bratseth (1985), employing a linear analytic model, found that CISK may be important in polar air if heating takes place in a shallow layer at low levels. The layer must be elevated above the surface in order for a short wave cut off to exist. Another finding of Bratseth was that the artificial unconditional heating used by Rasmussen (and others), i.e., a CISK formulation in which the heating in regions of ascent is matched by cooling in regions of descent, greatly overestimates the growth rates. Even so, it seems apparent that the CISK mechanism can be a vital factor in polar low development.

A somewhat different idea for the role of CISK in polar low development has been proposed in a later paper by Økland (1987). In this paper he develops a new CISK formulation that lacks a short wave cut off. He then proposes that the small dimension of the polar low is due to the presence of local regions of deep convection within the large-scale environment. For instance, baroclinic development may produce isolated regions of less stable air aloft. When heating causes the convective boundary layer to reach the base of the overlying less stable layer, convection grows rapidly to great height and precipitation and latent heat release ensue. In this paper Økland also shows that a small vortex may have a greater amplification rate if it develops inside a larger disturbance. His remarks obviously have relevance to the problem of the hurricane-like inner core of systems discussed earlier, and indeed Økland presents an example of such a system.

The combined effects of baroclinicity and CISK have been studied by Sardie and Warner (1983) with the use of a three-layer, two-dimensional, quasi-geostrophic model, which included both the effects of latent heating and baroclinicity. They found that in general both moist baroclinicity and CISK were of importance in polar low formation, though moist baroclinic processes alone were sufficient to account for the comma cloud type of development. Prior to Sardie and Warner's work most investigators emphasized either baroclinic instability or diabatic heating as the primary cause of polar low development. Lately there seems to be greater recognition that the interaction of both mechanisms needs to be taken into account.

7.1.3 Air-sea interaction instability

Although widely cited as an explanation for the development of hurricanes, CISK is a conceptual model that does not directly address the storm dynamics. Objections to the CISK model have recently been voiced by Emanuel (1986), who states: "the average generation of Convective Available Potential Energy (CAPE) in the tropical atmosphere by radiation and sea surface fluxes is apparently just enough to balance the dissipation of kinetic energy within cumulus clouds and, without augmentation, could not account for the greatly increased dissipation of kinetic energy in the hurricane boundary layer.'' Furthermore, numerical experiments (Ooyama, 1969; Rosenthal, 1971) clearly show the crucial importance of heat fluxes through the sea surface. As an alternative to CISK, Emanuel (1986) has proposed that tropical cyclones result from an air-sea interaction instability in which anomalous sea surface fluxes of sensible and latent heat induced by strong surface winds and falling pressure, lead to increased temperature anomalies, and thereby to further increases in surface winds and pressure deficit. He shows, using a simple nonlinear analytical model, and an axisymmetric numerical model, that this hypothesis is consistent with observations of tropical cyclones.

The possible significance of an air-sea interaction mechanism for cyclogenesis in polar air masses becomes apparent in light of the large fluxes of sensible and latent heat into the boundary layer of incipient polar lows measured during the ACE. Data indicate that combined fluxes of sensible and latent heat may have been as high as 1000 W m^{-2} over the Norwegian Sea during the development of a particularly intense arctic low (Shapiro et al., 1987), (see Section 6.1). This value is comparable to the heat fluxes observed in hurricanes. However, as Rasmussen and Lystad (1987) have pointed out, CAPE is large in the cold air outbreaks within which polar lows develop, and therefore the arguments applied to the tropical hurricane may not apply to the arctic systems.

Emanuel and Rotunno (1988) tested Emanuel's air-sea interaction mechanism for polar lows by using conditions observed in the atmosphere over the Norwegian Sea prior to a complex polar low development documented by Rasmussen (1985a, 1985b) (see Section 6.1). This case was chosen in part because of the non-advective warming seen at the core of the polar low. Results from Emanuel's calculations using a simple nonlinear analytical model, and an axisymmetric numerical model, show that the air-sea interaction hypothesis is consistent with observed arctic low development, but requires a pre-existing disturbance to act as a triggering mechanism before the air-sea interaction instability can operate. This two-stage development is consistent with evidence for two stages in the deepening of this disturbance. However, the rate of deepening calculated numerically is less than that observed, a discrepancy that requires further study.

7.1.4 Barotropic instability

Mullen (1979) showed that the necessary condition for barotropic instability generally is met in areas conducive to the development of comma clouds. However, as first discussed by Reed (1979), it appears very unlikely that a jet stream could be narrow enough to account for the mesoscale and small synoptic-scale wavelengths observed in polar lows, if barotropic processes alone are responsible. Reed's observation is consistent with Sardie and Warner (1985), who show that a jet maximum of 80 m s^{-1} with a half-width of 5° latitude would result in a growth rate of $\sim 1.6/10^{-6}$ s^{-1}. This very small rate implies that, although the necessary condition for barotropic instability is met, it cannot be an important contribution to disturbance growth. It is interesting to note, however, that disturbances developing in a flow that is both baroclinically and barotropically unstable tend to have maximum geopotential perturbation at the jet-stream level, while those that are barotropically damped tend to have a maximum near the surface (Mudrick, 1974).

In conclusion, although barotropic instability is present in many polar low developments, calculations indicate that it represents only a minor contribution to the rapid development observed in polar lows.

7.2 Numerical Prediction Experiments

We review here briefly the few prediction experiments that have been carried out using observed initial data and mesoscale models of sufficiently fine resolution to at least roughly represent systems of the dimensions of polar lows. Comments concerning the operational application of numerical prediction models to the problem of forecasting polar lows will be deferred to Section 8. Information concerning the models, the cases investigated, and the data sources is given in Table 1 for simulations conducted by Seaman (1983), Sardie and Warner (1985), Grønås et al. (1986a, 1986b, 1987), and Nordeng (1987). The two models utilized thus far, the Pennsylvania State University/National Center for Atmospheric Research (PSU/NCAR) and the Norwegian Meteorological Institute (DNMI) mesoscale models, are advanced models that represent the basic physical processes believed to be of importance in polar low development. The processes include surface and boundary layer fluxes and release of latent heat by resolvable motions (explicit convection) and by parameterized, subgrid-scale convection.

Short range (12 hr) forecasts obtained by Seaman were moderately successful, though they underestimated the intensity of the disturbance. Surface fluxes, convection, and forcing by an upper level short wave all proved necessary for the maintenance of the storm. The Atlantic low investigated by Sardie and Warner developed baroclinically at first, but latent heat release and surface fluxes were

TABLE 1. POLAR LOW SIMULATIONS USING FULL PHYSICS MESOSCALE MODELS AND OBSERVATIONAL DATA

Investigators	Model	Resolution	Layers	Cases	Area	Sources
Seaman (1983)	PSU/ NCAR	50 km	10	1	North Sea	NMC grid data
Sardie and Warner (1985)	PSU/ NCAR	80 km	10	1 1	Atlantic Pacific Ocean	NMC grid data, Manual analysis
Grønås, et al., (1986a, 1986b, 1987)	DNMI meso- scale	Nested grids 150, 50 and 25 km	10	7	Norwegian Sea	ECMWF grid
Nordeng (1987)	DNMI meso- scale	50 km	10	2	Norwegian Sea	ECMWF grid

required for its subsequent maintenance. The Pacific low was well predicted in the control experiment that employed parameterized convection. However, baroclinicity was identified as the major factor in the deepening, since latent heat release accounted for only 5 mb of 14 mb deepening. An experiment using explicit convection produced a significant over-intensification, a common feature of the explicit scheme, at least in earlier studies.

In all cases studied by Grønås et al. (1986a, 1986b, 1987) disturbances were predicted that could be associated with the observed polar lows. A failure of the forecasts was their inability to predict the strength of the disturbances when the scale was small. The low investigated by Shapiro et al. (1987) was a case in point. The failure may arise from a lack of sufficient resolution in the model to capture essential features of the initial state or to represent essential physical processes.

Nordeng (1987) has developed a parameterization method for slantwise convection in the DNMI mesoscale model. Although the potential of his parame- terization scheme is clear, the forecasts made with the revised model were still unable to reproduce the magnitude of the surface deepening observed in the polar low cases simulated, a shortcoming he attributes to the small scale of polar low developments and the modest 50-km grid-point resolution of the model.

8. FORECASTING POLAR LOWS

The numerical simulations described in the previous section, carried out with real data but in the research mode, demonstrate the ability of advanced mesoscale models to forecast polar lows with at least some degree of success in a variety of situations and locations. In this section we address the more relevant question of the performance of operational forecast models in predicting polar lows. After noting the capabilities and limitations of the operational models, we describe some forecast methods, developed by Norwegian meteorologists, that use model output in conjunction with other sources of information, particularly satellite data, as a means of dealing with the practical problem of forecasting polar lows. Due to the previously emphasized range of sizes and structures that characterizes these systems, the success of operational models depends greatly upon the type of low being considered. Accordingly, it is convenient to discuss model performance in terms of the types identified in Sections 3–6.

Considering first the short wave, jet streak type, we can assert on the basis of everyday experience that these systems, which lie at the larger end of the size spectrum, are usually handled with some measure of success by current operational models (e.g., the NMC spectral and NGM models). We are unaware of any verification studies that have been carried out to test this assertion, but we can point to specific examples, for instance, the two cases examined by Reed and Blier (1986a, 1986b) in which the systems can be identified in both the initial and the prognostic fields. These cases also furnish examples of the tendency of the models to underpredict this type of development (but not as badly as for some of the other types).

The comparative success of the operational models in predicting the upper level, short wave type stems from the ability of the analysis systems to capture the upper level vorticity maxima and jet streaks that are crucial to their development. Some of the vorticity maxima have long histories, being remnants of earlier synoptic-scale disturbances. Others seem to form when a lobe of high vorticity protrudes outward from a pool of cyclonic relative vorticity contained within a large upper level cold low. In either case, the characteristic field of enhanced convection, which later becomes organized in a comma shape pattern, appears in the PVA region ahead of the upper level vorticity maximum.

The arctic front type of polar low presents a much more difficult forecast problem. Due to its small size and the lack of a sufficiently dense observing network in the regions of interest, this type of system is beyond the capabilities of present day numerical prediction. A case in point is the reversed shear case illustrated in Section 4. Being only a few hundred kilometers in diameter, the polar lows in this case (Figure 8) were not resolved by the ECMWF analysis system (Figure 9a). Forecast maps are not available, but in view of the heavy weight assigned by the ECMWF system to forecast (first-guess) fields in regions of sparse data, it is evident from the analysis that the lows were not predicted. Indeed, in cases where the lows grow spontaneously from infinitesimal disturbances, as may have happened in this case, the systems are inherently unpredictable in the incipient stage.

Similar remarks apply to the cold-low type of polar low. For example, the time of formation, if not the exact location, of the small vortex over the Mediterranean Sea, discussed in Section 5 and illustrated in Figures 12 and 13, was essentially unpredictable. However, once formed, this system, because of its association with an earlier synoptic-scale low, could conceivably have been tracked with some degree of success by operational models despite their inability to resolve the true intensity of the inner vortex. The actual performance of the operational models in this case has not been ascertained.

In the case of the multitype developments described in Section 6, some predictive skill is possible, since they involve identifiable migratory upper level vorticity maxima as well as a favorable lower level environment that is at least partly resolvable by the model. The larger scale features of the Bear Island and ACE cases discussed in Section 6 (e.g., Figure 15a), which could be associated with upper level PVA, were predicted by the numerical forecasts. However, the second stage of development in these cases, featured by the appearance of small, intense vortices (e.g., Figure 15c), was essentially unpredictable because the initial part of the rapid-intensification phase could not be uniquely associated with the upper level forcing (each of the vortices was only one among a number of small vortices) and because at later stages the vortices could not be adequately resolved by the observing network and forecast system.

Given the limited usefulness of the operational models in the prognosis of polar lows, how should the forecaster deal with the prediction problem? This question was addressed by Norwegian meteorologists in connection with the Polar Lows Project that was conducted in the Norwegian and Barents Seas between January 1983 and December 1985 (Lystad, 1986). On the basis of their experience, the investigators divided the problem into two parts: the problem of monitoring, which involves the synthesis of all available data into the most probable representation of the atmospheric state, and the problem of making the actual forecast. There were three elements in the monitoring: first was the use of analyses based on model assimilated data (a fine mesh, 50-km-grid model was employed in the assimilation); second was attention to individual reports from land, ship and drifting buoy locations, and from radiosonde stations; and third, and most important, was interpretation of infrared images from polar-orbiting satellites. Due to the characteristic underprediction of the intensity of the lows by the model, subjective corrections were made to the numerical analyses when the other sources of information suggested the presence of a polar low.

Forecasting methods were developed and tested during the Polar Lows Project (Lystad, 1986) that treated both the problem of identifying areas of likely polar low development, without attempting to pinpoint the actual locations of the storms, and the problem of predicting the track of a low in cases where one was detected. The method for selecting areas with significant probability of polar low occurrence made use of three simple criteria: (1) cold air advection at the sea surface; (2) 850- to 500-mb thickness of less than 3960 m (adjusted according to the sea surface temperature); and (3) cyclonic or zero curvature of contours at 500- and 700-mb levels. In 2.5 months of operational testing this method was successful in identifying all cases of polar low occurrence. However, the method also forecast about the same number of cases in which polar lows failed to appear. Clearly fine-tuning of the method is required before it can be regarded as fully useful.

The best method for predicting the tracks of already identified lows proved to be construction of storm paths from model predicted 850- and 700-mb winds under the assumption that the systems are steered by the winds at these levels. Testing of the method using 850-mb winds revealed an average position error for 18-hr forecasts of 200 km compared with a mean path length of 800 km (Midtbø, 1986). Though developed for the Norwegian and Barents Sea areas, the foregoing forecast methods could with modification be applied to other regions.

What does the future hold for polar low prediction? Opportunities for better monitoring of the lows exist that could lead to a significant enhancement of predictive skill. In the case of satellite observations, scanning passive microwave radiometers that measure total water vapor content and, more importantly, precipitation rate in an atmospheric column (McMurdie and Katsaros, 1985) provide a new tool for better detection of polar lows. In addition, microwave scatterometers give promise of yielding high resolution surface wind fields. The high temporal frequency needed to observe rapidly developing storms, which cannot be obtained from polar orbiting satellites, may, at least in some areas, be achievable for geostationary orbit with use of an image processing technique developed by Zick (1986) that stretches the high latitude portions of the satellite picture. Obviously, deployment of an increased number of drifting buoys and of coastal radars are steps that can be taken to improve the observational base. Radar reflectivity data can be used to track the movement of storm centers and predict areas susceptible to high winds by charting the velocities of convective cells (Monk et al., 1987).

In view of the difficulty of obtaining high resolution three-dimensional fields of the basic variables for insertion in numerical models, it is unlikely that the very fine mesh limited area models now under development will have a dramatic impact on polar low prediction. While it is probable that they will result in better predictions of the larger size lows and will achieve some success in predicting smaller systems that are sparked by resolvable upper level vorticity advections, it is unlikely that they will improve the prediction of the small seemingly random vortices that have been described elsewhere in this review. Clearly the forecaster will continue to play a major role in polar low prediction for the foreseeable future. Applying his or her interpretive skills to buoy and ship data, satellite images, model-based analyses, and numerical forecasts offer the best route for advancing the monitoring and prediction of polar lows over the oceans.

9. SUMMARY AND CONCLUSIONS

This review paper highlights the diversity of mesoscale phenomena that exist in polar air masses, and the forecasting problems that they embody. We have looked at polar lows from the standpoints of observation, theory and numerical modeling, and forecasting. From the observational standpoint it is clear that a variety of systems fall under the general heading of polar lows. We have organized the review of case studies on the basis of a combination of observational and physical considerations. This classification scheme is aimed more at diagnosing common types of developments rather than as a classification for its own sake. Further work in this direction is still needed to resolve issues such as whether the small, hurricane-like vortices seen in some cases should be regarded as a distinct type of polar low or as a substructure embedded within a larger sub-synoptic system.

It is clear that progress towards understanding and forecasting these varied phenomena depends on continued research efforts. If the subject is to advance, far more observations of the type taken during the Arctic Cyclone Expedition are urgently required. Since these maritime phenomena cannot be resolved with the current observational networks, they require special observations that can only be obtained by aircraft. Extensive field studies are needed to obtain multi-level observations during the life cycles of a representative sample of polar lows to resolve the many remaining questions regarding the triggering mechanisms and energetics of these fascinating and sometimes destructive storms.

Theoretical investigators in the past have tended to divide into two schools of thought: one school emphasizing baroclinic effects and the other championing CISK or other heating mechanisms. Lately there has been widespread recognition that the interaction of both mechanisms are important. Accordingly, the problem has become more complex, and it has become increasingly difficult to advance theoretical understanding using only simple models, though these still have a place. Particularly promising for future advancement are the regional primitive-equation models that are now under development at a number of research centers and that are coming into operational use at some forecast centers. These mesoscale models offer the opportunity for a wide variety of experimentation on the polar low problem.

Polar lows, with their rapid development and small scale present a special challenge to the operational forecaster. The ability to forecast polar lows will benefit from advances on several fronts: improved observations (drifting buoys, coastal weather radars, satellite systems), the application of regional numerical weather prediction models with finer resolution to represent polar low scale developments, continued refinement of empirical techniques, and further basic research.

ACKNOWLEDGMENTS

The authors wish to thank Professor Erik Rasmussen and Dr. Melvin Shapiro for helpful conversations. The comments of anonymous reviewers were much appreciated.

REFERENCES

Anderson, R.K., J.P. Ashman, F. Bittner, G.R. Farr, E.W. Ferguson, V.J. Oliver, and A.H. Smith, 1969: *Application of Meteorological Satellite Data in Analysis and Forecasting*. ESSA Tech. Rep. NESC51, Government Printing Office, Washington, D.C. (NTIS AD-697033).

Billing, H., I. Haupt, and W. Tonn, 1983: Evolution of a hurricane-like cyclone in the Mediterranean Sea. *Beitr. Phys. Atmos.*, *56*, 508–510.

Blumen, W., 1980: On the evolution and interaction of short and long baroclinic waves of the Eady type. *J. Atmos. Sci.*, *37*, 1984–1993.

Bratseth, A.M., 1985: A note on CISK in polar air masses. *Tellus, 37A*, 403–406.

Brown, R. A., 1980: Longitudinal instabilities and secondary flows in the Planetary boundary layer: A review. *Rev. Geophys. and Space Phys.*, *18*, 683–697.

Browning, K.A., and F.F. Hill, 1985: Mesoscale of a polar trough interacting with a polar front. *Quart. J. Roy. Meteor. Soc.*, *111*, 445–462.

Businger, S., 1985: The synoptic climatology of polar low outbreaks. *Tellus, 37*, 419–432.

Businger, S., 1987: The synoptic climatology of polar low outbreaks over the Gulf of Alaska and Bering Sea. *Tellus, 39*, 307–325.

Businger, S., and P.V. Hobbs, 1987: Mesoscale and synoptic scale structures of two comma cloud systems over the Pacific Ocean. *Mon. Wea. Rev.*, *115*, 1909–1928.

Businger, S., and B. Walter, 1988: Comma cloud development and associated rapid cyclogenesis over the Gulf of Alaska: A case study using aircraft and operational data. *Mon. Wea. Rev.*, *116*, 1103–1123.

Carlson, T., 1980: Airflow through midlatitude cyclones and the comma cloud pattern. *Mon. Wea. Rev.*, *108*, 1498–1509.

Carleton, A.M., 1985: Satellite climatological aspects of the "polar low" and "instant occlusion." *Tellus, 37*, 433–450.

Carr, F.H., and J.P. Millard, 1985: A composite study of comma clouds and their association with severe weather over the Great Plains. *Mon. Wea. Rev.*, *113*, 370–387.

Charney, J.G., and A. Eliassen, 1964: On the growth of hurricane depression. *J. Atmos. Sci.*, *21*, 68–75.

Duncan, C.N., 1977: A numerical investigation of polar lows. *Quart. J. Roy. Meteor. Soc.*, *103*, 255–268.

Duncan, C.N., 1978: Baroclinic instability in a reversed shear flow. *Met. Mag.*, *107*, 17–23.

Emanuel, K.A., 1986: An air-sea interaction theory for tropical cyclones, Part I: Steady-state maintenance. *J. Atmos. Sci.*, *43*, 585–604.

Emanuel, K.A., and R. Rotunno, 1988: Polar lows as arctic hurricanes. *Tellus* (in press).

Ernst, J.A., and M. Matson, 1983: A Mediterranean tropical storm? *Weather, 38*, 332–337.

Farrell, B.F., 1982: The initial growth of disturbances in a baroclinic flow, *J. Atmos. Sci.*, *39*, 1663–1668.

Farrell, B., 1984: Modal and non-modal baroclinic waves. *J. Atmos. Sci.*, *41*, 668–673.

Forbes, G.S., and W.D. Lottes, 1985: Classification of mesoscale vortices in polar airstreams and the influence of the large-scale environment on their evolutions. *Tellus, 37*, 132–155.

Gall, R.J., 1976: The effects of latent heat release in growing baroclinic waves. *J. Atmos. Sci.*, *33*, 1686–1701.

Grønås, S., A. Foss, and M. Lystad, 1986a: *Numerical Simulations of Polar Lows in the Norwegian Sea, Part I: The Model and Simulations of the Polar Low 26–27 February 1984.* Tech. Rep. No. 5, Norwegian Meteorological Institute, Oslo, 51 pp.

Grønas, S., A. Foss, and M. Lystad, 1986b: *Numerical Simulations of Polar Lows in the Norwegian Sea, Part II: Simulations of Six Synoptic Situations.* Tech. Rep. No. 18, Norwegian Meteorological Institute, Oslo, 32 pp.

Grønas, S., A. Foss, and M. Lystad, 1987: Numerical simulations of polar lows in the Norwegian Sea. *Tellus, 39,* 334–353.

Harley, D.G., 1960: Frontal contour analysis of a "polar low." *Meteor. Mag., 89,* 146–147.

Harrold, T.W., and K.A. Browning, 1969: The polar low as a baroclinic disturbance. *Quart. J. Roy. Meteor. Soc., 95,* 710–723.

Kellogg, W.W., and P.F. Twitchell, 1986: Summary of the workshop on arctic lows 9–10 May 1985, Boulder, Colorado. *Bull. Amer. Meteor. Soc., 67,* 186–193.

Locatelli, J.D., P.V. Hobbs, and J.A. Werth, 1982: Mesoscale structures of vortices in polar air streams. *Mon. Wea. Rev., 110,* 1417–1433.

Lyall, I.T., 1972: The polar low over Britain. *Weather, 27,* 378–390.

Lystad, M. (Ed.), 1986: *Polar Lows in the Norwegian, Greenland, and Barents Seas.* Final Rep., Polar Lows Project, The Norwegian Meteorological Institute, Oslo, 196 pp.

Mansfield, D.A., 1974: Polar lows: The development of baroclinic disturbances in cold air outbreaks. *Quart. J. Roy. Meteor. Soc., 100,* 541–554.

Mayengon, R., 1984: Warm core cyclones in the Mediterranean. *Mariners Wea. Log., 28,* 6–9.

McGinnigle, J.B., M.V. Young, and M.J. Bader, 1988: The development of instant occlusions in the North Atlantic. Meteor. 0.15 Internal Report, Meteorological Office, London Road, Bracknell, 25 pp.

McMurdie, L.A., and K.B. Katsaros, 1985: Atmospheric water distribution in a midlatitude cyclone observed by the Seasat scanning multichannel microwave radiometer. *Mon. Wea. Rev., 113,* 584–598.

Meteorological Office, 1962: *A Course in Elementary Meteorology.* Met. 0.707, Her Majesty's Stationary Office, London, WC1V 6HB, 189 pp.

Midtbø, K.H., 1986: Polar low forecasting. *Proceedings of the International Conference on Polar Lows,* Olso 1986, The Norwegian Meteorological Institute, 363 pp.

Monteverdi, J.P., 1976: The single air mass disturbance and precipitation characteristics at San Fransisco. *Mon. Wea. Rev., 104,* 1289–1296.

Monk, G.A., L.A. Browning, and P. Konas, 1987: Forecasting applications of radar-derived precipitation echo velocities in the vicinity of polar lows. *Tellus, 39,* 426–433.

Mudrick, S.E., 1974: A numerical study of frontogenesis. *J. Atmos. Sci., 31,* 869–892.

Mullen, S.L., 1979: An investigation of small synoptic-scale cyclones in polar air streams. *Mon. Wea. Rev., 107,* 1636–1647.

Mullen, S.L., 1982: Cyclone development in the polar airstreams over the wintertime continent. *Mon. Wea. Rev., 110,* 1664–1676.

Mullen, S.L., 1983: Explosive cyclogenesis associated with cyclones in polar air streams. *Mon. Wea. Rev., 111,* 1537–1553.

Nordeng, T.E., 1987: The effect of vertical and slantwise convection on the simulation of polar lows. *Tellus, 39,* 354–375.

Økland, H., 1977: *On the Intensification of Small-Scale Cyclones Formed in Very Cold Air Masses Heated Over the Ocean.* Inst. Rep. Ser. No. 26, Institutt for Geofysikk, Universitet, Oslo, 25 pp.

Økland, H., 1987: Heating by organized convection as a source of polar low intensification. *Tellus, 39,* 397–407.

Ooyama, K., 1969: Numerical simulation of the life cycle of tropical cyclones. *J. Atmos. Sci., 26,* 3–40.

Orlanski, I., 1986: Localized baroclinicity: A source for meso-α cyclones. *J. Atmos. Sci., 43,* 2857–2885.

Parsons, D.B., and P.V. Hobbs, 1983: The mesoscale and microscale structure and organization of clouds and precipitation in midlatitude cyclones XI: Comparisons between observational and theoretical aspects of rainbands. *J. Atmos. Sci., 40,* 2377–2397.

Pedgley, D.E., 1968: A mesoscale snow system. *Weather, 23,* 469–476.

Rabbe, A., 1987: A polar low over the Norwegian Sea, 29 February–1 March 1984. *Tellus, 39,* 326–333.

Rasmussen, E., 1977: *The Polar Low as a CISK Phenomenon.* Rep. No. 6, Inst. Teoret. Meteor., Kobenhavns Universitet, Copenhagen, 77 pp.

Rasmussen, E., 1979: The polar low as an extratropical CISK disturbance. *Quart. J. Roy. Meteor. Soc., 105,* 531–549.

Rasmussen, E., 1981: An investigation of a polar low with a spiral cloud structure. *J. Atmos. Sci., 38,* 1785–1792.

Rasmussen, E., 1983: A review of mesoscale disturbances in cold air masses. In D.K. Lilly and T. Gal-chen (Eds.), *Mesoscale Meteorology-Theories, Observations and Models,* Reidel, Boston, 247–283.

Rasmussen, E., 1985a: A case study of a polar low development over the Barents Sea. *Tellus, 37A,* 407–418.

Rasmussen, E., 1985b: *A Case Study of a Polar Low Development Over the Barents Sea.* Tech. Rep., Polar Lows Project, The Norwegian Meteorological Institute, Oslo, 42 pp.

Rasmussen, E., 1985c: Paskestormen et baroklink polart lavtryk. *Vejret, 4–7 Argang,* 3–17.

Rasmussen, E., and M. Lystad, 1987: The Norwegian polar lows project: A summary of the international conference on polar lows. *Bull. Amer. Meteor. Soc., 68,* 801–816.

Rasmussen, E., and C. Zick, 1987: A subsynoptic vortex over the Mediterranean Sea with some resemblance to polar lows. *Tellus, 39,* 408–425.

Reed, R.J., 1979: Cyclogenesis in polar air streams. *Mon. Wea. Rev., 107,* 38–52.

Reed, R.J., and W. Blier, 1986a: A case study of comma cloud development in the Eastern Pacific. *Mon. Wea. Rev., 114,* 1681–1695.

Reed, R.J., and W. Blier, 1986b: A further case study of comma cloud development in the Eastern Pacific. *Mon. Wea. Rev., 114,* 1696–1708.

Reed, R.J., 1987: *Proceedings of Polar Lows Seminar on the Nature of Prediction of Extratropical Weather Systems, 7–11 September 1987.* Available from ECMWF, Shinfield Park, Reading, RG2 9AX, U. K., 213–236.

Reed, R.J., and C.N. Duncan, 1987: Baroclinic instability as a mechanism for the serial development of polar lows: A case study. *Tellus, 39,* 376–384.

Rosenthal, S.L., 1971: The response of a tropical cyclone model to variations in boundary layer parameters, initial conditions, lateral boundary conditions, and domain size. *Mon. Wea. Rev., 99,* 767–777.

Sardie, J.M., and T.T. Warner, 1983: On the mechanism for the development of polar lows. *J. Atmos. Sci., 40,* 869–881.

Sardie, J.M., and T.T. Warner, 1985: A numerical study of the development mechanisms of polar lows. *Tellus, 37,* 460–477.

Seaman, N.L., H. Otten, and R.A. Anthes, 1981: A rapidly developing polar low in the North Sea of January 2, 1979. *First International Conference on Meteorology and Air/Sea Interaction of the Coastal Zone, May 10–14, 1982,* The Hague, Netherlands.

Seaman, N.L., 1983: Simulation of a mesoscale polar low in the North Sea. *Sixth Conference on Numerical Weather Prediction,* American Meteorological Society, Boston, 24–30.

Shapiro, M.A., and L.S. Fedor, 1986: *The Arctic Cyclone Expedition, 1984: Research and Aircraft Observations of Fronts and Polar Lows Over the Norwegian and Barents Sea, Part 1. Polar Lows Project,* Tech. Rep. No. 20. The Norwegian Meteorological Institute, Oslo, Norway, 56 pp.

Shapiro, M.A., L.S. Fedor, and T. Hampel, 1987: Research aircraft measurements within a polar low over the Norwegian Sea. *Tellus, 37,* 272–306.

Staley, D.O., and R.L. Gall, 1977: On the wavelength of maximum baroclinic instability. *J. Atmos. Sci., 34,* 1679–1688.

Sutcliffe, R.C., and A.G. Forsdyke, 1950: The theory and use of upper air thickness patterns in forecasting. *Quart. J. Roy. Meteor. Soc., 76,* 189–217.

Uccellini, L.W., and P.J. Kocin, 1987: The interaction of jet streak circulations during heavy snow events along the east coast of the United States. *Wea. and Forecasting, 1,* 289–308.

Wallace, J.M., and P.V. Hobbs, 1977: *Atmospheric Science: An Introductory Survey,* Academic Press, 417 pp.

World Meteorological Organization, 1973: The use of satellite pictures in weather analysis and forecasting. Tech. Note. No. 124, WMO No. 333. (Available from the Secretariat of the World Meteorological Organization, Geneva, Switzerland), 275 pp.

Zick, C., 1983: Method and results of an analysis of comma cloud developments by means of vorticity fields from upper tropospheric satellite wind data. *Meteor. Rdsch.*, *36*, 69–84.

Zick, C., 1986: Two Meteosat film scenes showing polar lows over the Norwegian Sea. *Proceedings of the International Conference on Polar Lows*, Olso, 1986, The Norwegian Meteorological Institute, 363 pp.

A COMPARATIVE STUDY OF TROPICAL CYCLONES AND POLAR LOWS

Erik A. Rasmussen
Geophysical Institute
University of Copenhagen
Copenhagen, Denmark

ABSTRACT

The similarity and differences between tropical cyclones and polar lows are discussed. The discussion is based on the polar low developments from 13 to 16 December 1982 and on a polar low-like cyclone over the Mediterranean from 27 September to 2 October 1983. Physical parameters important for both types of phenomena as, for example, the sea surface temperature, the degree of convective instability, and low-level vorticity are considered. Examples are presented showing the surface wind distribution around polar lows, and the potential for formation of eyes in connection with polar lows is discussed. It is concluded that although polar lows develop in regions very different from the genesis area of tropical cyclones, nevertheless there are significant similarities between (some) polar lows and tropical cyclones.

1. INTRODUCTION

In a paper on one of the first successful numerical model simulations of tropical cyclone developments, Ooyama (1969) pointed out (p. 38) that: "Results also demonstrate that the supply of heat and moisture directly from a warm ocean is a crucial requirement for growth and maintenance of a tropical cyclone, *a fact that characterizes the tropical cyclone as truly a creature of the tropical oceans." (Italics added.)* In this paper we want to show that a very similar statement can be made for polar lows, i.e., "the supply of heat and moisture directly from a (relatively) warm ocean is a crucial requirement for growth and maintenance of some polar lows, a fact that characterizes these polar lows truly as creatures of the arctic oceans."

POLAR AND ARCTIC LOWS
Paul F. Twitchell, Erik A. Rasmussen,
and Kenneth L. Davidson (Eds.)

The question whether polar lows have any similarities to tropical cyclones has been rather controversial (in one of the author's early papers the editor stressed that all comparisons with tropical cyclones should be left out). There is no doubt, however, that certain features *are* common for tropical cyclones and (some) polar lows. This point will be considered in detail in Section 3 of this paper following a short overview in Section 2.

2. OVERVIEW

Figure 1 shows a polar low. Satellite images such as Figure 1 show convincingly that polar lows exist as phenomena in their own right. Most of our knowledge of polar lows is of a relatively recent origin, and only 10 to 15 years ago most meteorologists were unaware of these phenomena. Going through the meteorological literature almost nothing can be found about polar lows prior to Harrold and Browning's 1969 paper on "The polar low as a baroclinic disturbance." What can be found shows that before 1969 the polar low was considered as a mainly nonfrontal, convectively driven system.

Figure 1. NOAA—9 infrared satellite image 0418 GMT 27 February 1987 showing an exceptional symmetric, "ideal" polar low just north of North Cape. Arrows marked S and B indicate respectively the southern point of Svalbard (Spitzbergen) and Bear Island.

Harrold and Browning's paper marks the starting point of a debate about the nature and structure of the polar low, a debate that is far from finished. In their paper they conclude that the polar low is basically a baroclinic disturbance of short wavelength. This point of view was challenged by Rasmussen (1977, 1979) and Økland (1977), who both supported the older point of view that the polar low was driven by deep convection and that baroclinic instability played a minor, if any, role at all. Rasmussen and Økland both made use of the conditional instability of the second kind (CISK) concept, and in the following years much of the debate was concentrated on whether polar lows were of baroclinic nature, or of the CISK type. One of the reasons why the CISK mechanism was only reluctantly accepted as a possible explanation for the formation of polar lows was probably the widely accepted point of view that tropical cyclones only develop over regions with sea surface temperatures around or higher than 27 °C. How, then, could the same basic mechanism possibly work over the Norwegian and even the Barents Sea? However, as demonstrated by Rasmussen and Økland, even in these hostile environments far north, sufficient energy in the form of latent *and* sensible heat was available (over the sea) to drive disturbances such as polar lows. An important point stressed by both authors is that sensible heat transport from the relatively warm sea surface plays a much more important role for polar low developments than it does for tropical cyclones.

Reed (1979) identified the polar low with the so-called comma cloud. Comma clouds develop typically in a baroclinic region with a pronounced upper level flow, and poleward of, but relatively near to the polar front. Numerous examples of comma clouds have been described in the literature by Reed and others, and observations and theoretical arguments alike show that comma clouds are basically baroclinic disturbances. They are of a relatively large horizontal scale, ~ 1000 km, and it seems possible to explain their basic dynamics from quasi-geostrophic theory. Comma clouds form mostly over the sea and often in regions of low static stability that ensure a strong response for a given dynamic forcing. Convection may add to the baroclinic development and possibly, as pointed out by Reed (1979), enhance some developments through a CISK mechanism. More recently comma clouds have also been studied by means of numerical models and they probably are some of the best understood cold air mass phenomena.

The question whether comma clouds are "real" or "true" polar lows has caused much debate, and the answer depends of course on the definition of a polar low. An unambigious definition, however, has never been generally accepted. Reed (1988) suggests a "broad sense definition" according to which the term polar low "denotes any type of small synoptic or subsynoptic cyclone, of an essential nonfrontal nature, that forms in a cold air mass poleward of major jet streams or frontal zones and whose main cloud mass is largely of convective origin." Other definitions as for example by van Delden (1985) "that a polar low is a warm core (like the tropical cyclone) vortex consisting of deep Cb-clouds" limits the number of polar lows much more and is difficult to use in practice. For practical purposes even Reed's definition seems too limited. Small-scale cyclones with many features similar to large extratropical cyclones quite often develop for example in the North Sea region and northwest of the British Isles. These "baroclinic polar lows" may start as "convective polar lows," which fulfil Reed's definition, but later on they change into small baroclinic waves showing the typical cloud structures of a baroclinic wave. An example of a baroclinic polar low is shown in Figure 2.

Between the "ideal" types of convective polar lows (Figure 1) and baroclinic polar lows (Figure 2) we find numerous hybrid types where both convection and baroclinic instability are important. This can be inferred from satellite images and has also been shown theoretically (Craig and Cho, 1988, 1989) and by numerical model studies (Sardie and Warner, 1985).

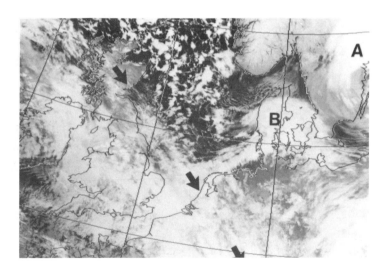

Figure 2. NOAA-7 infrared satellite image 1404 GMT 7 December 1981 showing baroclinic polar low (B) over Denmark. The thick arrows indicate the jet stream associated with the main baroclinic zone, situated further south.

It is not surprising that polar lows appear in so many forms. Some of them form very close to the ice edge, i.e., close to a semi-permanent secondary shallow baroclinic zone, and some, such as most comma clouds, much farther to the south (in this work only Northern Hemisphere phenomena are considered) close to the main baroclinic zone (the polar front). These factors as well as many others mean that a variety of forcing mechanisms are effective, which again leads to a whole spectrum of polar lows. A useful, practical definition should reflect this, and Reed's definition might be improved if changed to: "a polar low is a small-scale synoptic or subsynoptic cyclone that forms in the cold air mass poleward of the main baroclinic zone and/or major secondary fronts. It will often be of a convective nature but baroclinic effects may be important."

The proposed definition includes besides "real" polar lows as the one shown in Figure 1, comma clouds, baroclinic polar lows, and hybrid types. Excluded are the small but otherwise normal wave cyclones that quite often form on secondary frontal zones north of the major frontal zone.

3. SIMILARITIES AND DIFFERENCES BETWEEN TROPICAL CYCLONES AND POLAR LOWS

Since 1977 the formation and development of polar lows of a convective nature have been associated with CISK on numerous occasions. This implies that these polar lows in some way should have a structure similar to tropical cyclones. Some common features indeed have been found. For example, deep convection is essential for both types of phenomena, and warm cores and/or "eyes" have been observed in polar lows. A more detailed comparison has not been attempted so far, probably because, even now, relatively little is known about the detailed structure of polar lows.

Numerous papers, textbooks, etc., have been written on tropical cyclones. However, in the following we will limit the references to mainly two works, i.e., Ooyama's paper (Ooyama, 1969), in the following referred to as "Ooyama," and Anthes' book on tropical cyclones (Anthes, 1982), in the following referred to as "Anthes."

Physical factors important for the development of tropical cyclones include the release of latent heat in cumulus convection, the oceanic sources of moisture, and the conservation of angular momentum in maintaining the hurricane. Physical parameters related to these processes include (see Anthes, p. 49):

* sea surface temperature
* degree of convective instability
* low-level absolute vorticity, and
* vertical shear of the horizontal wind.

The parameters mentioned above will be discussed in the following in a polar low context. The discussion will be based mainly upon the December 1982 polar low episode discussed in detail by Rasmussen (1985a, 1985b, and 1988). The reader is referred to these papers for details and only a very short description of the synoptic situation and related matters will be given here.

During the late afternoon on 13 December 1982 a vortex-like polar low (in the following referred to as "polar low *A*," "polar vortex *A*" or simply "vortex *A*") formed in a pre-existing cyclonic disturbance a little west of the Norwegian weathership AMI (71.5 °N 19 °E). The fully developed vortex at midnight between 13 and 14 December 1982 is shown in Figure 3. A satellite

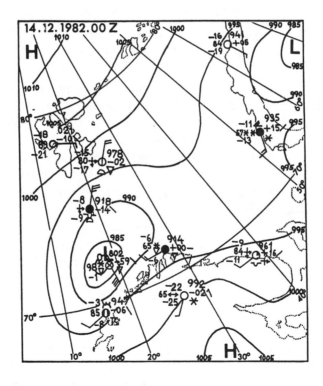

Figure 3. Surface map 0000 GMT 14 December 1982.

image received a few hours later at 0419 GMT 14 December 1982 is shown in Figure 8. The vortex subsequently drifted in an easterly direction and its center passed almost directly over the AMI. Figure 4a shows the surface pressure at the AMI, and the estimated surface pressure at the center of polar low A from 12 to 14 December. Figure 4b shows the surface mean wind velocity and pressure at the AMI during the passage of the vortex. Finally Figure 4c shows some important thermodynamic parameters. This figure is basically the same as Figure 8a in Rasmussen (1985) except that surface values of the dew point temperatures and the potential wet bulb temperatures, θ_w, have been added.

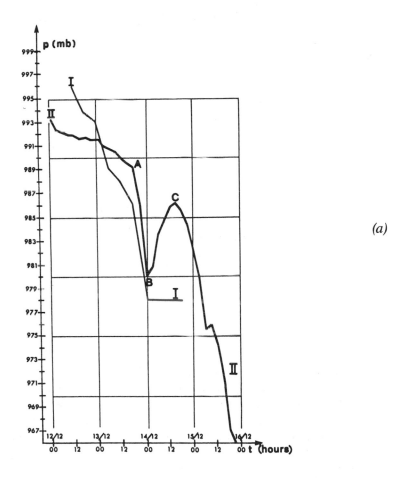

(a)

Figure 4. (a) Surface pressure as function of time at center of low/vortex A (curve I), and at weathership AMI (71.5°N 19°E), (curve II).

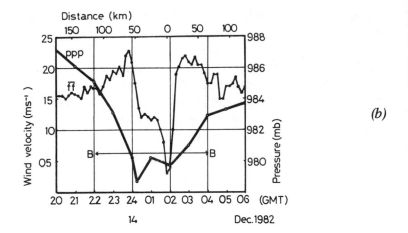

(b)

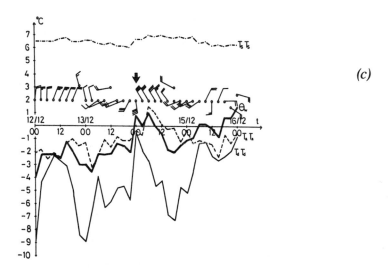

(c)

Figure 4. (Concluded) (b) Mean wind velocity $\overline{ff}$ in m s^{-1} (10 min mean) and surface pressure (ppp) from weathership AMI from 2000 GMT 13 December to 0600 GMT 14 December 1982. A length scale is shown on top of the figure. The distance B-B indicated on the figure is a measure of the horizontal scale of the system. (c) Surface winds in knots (long barbs indicate 10 knots) and north "upwards." Surface air temperature (T_aT_a, thin dashed line) and dew-point temperature (T_dT_d, thin solid line), surface wet bulb potential temperature (θ_w, thick solid line), and sea surface temperature (T_sT_s, dash-dotted line), all from weathership AMI. The arrow shows the time at which the center of vortex A passes the AMI.

3.1 The Degree of Convective Instability

During the passage of polar low A around midnight between 13 and 14 December the value of θ_w increased significantly at the AMI as seen from Figure 4c. This increase was due partly to an increase in the temperature and partly to an increase in the dew point. The relatively small decrease in the surface pressure, p_s, had little effect. Ooyama mentions that for tropical cyclones a sharp decrease in p_s may raise θ_e (and θ_w) significantly, and in this way "boost" the surface air shortly before it ascends into the warm core. On the other hand Ooyama's numerical experiments show, that "this effect does not seem to be of crucial importance for development of a tropical cyclone" (Ooyama, p. 25).

The increase in the surface value of θ_w is extremely important for the potential for deep convection in the vortex. This is shown in Figure 5, which

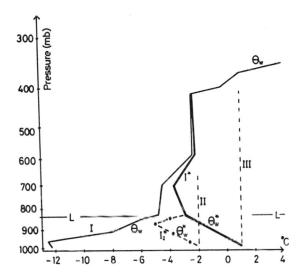

Figure 5. Stratification of fresh arctic air (curve I) shown by the wet-bulb poten-tial temperature θ_w. The thick curve marked I * shows θ_w^* corresponding to the modified Bear Island ascent obtained by replacing the temperatures below the layer L in 830 mb with a dry adiabatic layer. See text for details and for definition of θ_w^*. The dash-dotted curve I_2^* from the surface up to level L shows θ_w^* in case of intermediate modification of the boundary layer air (see text for further details and Figure 6). The vertical lines II and III describe pseudoadiabatic ascent from the surface with θ_w equal to respectively $-2°C$ and $1°C$.

has been drawn in a way analogous to Figure 2 in Ooyama. Curve I shows θ_w for the Bear Island radiosonde ascent 1200 GMT 13 December 1982, and Curve I^* θ_w^* corresponding to a "modified" Bear Island–radiosonde ascent. The modified ascent is derived from the 1200-GMT sounding by replacing the temperatures below 830 mb by a dry adiabatic layer (Figure 6). The modification is due to strong sensible heat fluxes and the modified sounding is representative for the vertical stratification just prior to the formation of the intense vortex. Emanuel and Rotunno (1989) use a similar modified sounding for their "Hot, Dry Case." As in Ooyama θ_w^* corresponds to the wet bulb potential temperature of hypothetically saturated air of the same temperature at each level (Ooyama uses the equivalent potential temperature θ_e instead of θ_w). In the modified sounding (curve I^*) the surface temperature is around 1°C, a value that was actually measured at the AMI during the passage of the center (see Figure 4c). Outside the center a more representative value of the surface temperature is $-2°$C. Another modified sounding representative for these conditions away from the center may be constructed by replacing the temperatures in the lowest layers of the original Bear Island sounding by a dry adiabatic layer from the $-2°$C surface value up to ~ 850 mb (see Figure 6). In this way a small inversion will remain from 850 to 830 mb θ_w^* from the surface to 830 mb for this situation is shown as curve I_2^* on Figure 5.

A parcel lifted pseudoadiabatically from the boundary layer will describe a straight vertical line on the diagram, and the parcel will be warmer than the ambient air where the line lies to the right of the appropriate θ_w^*-curve. It is seen immediately that the possibility for deep unstable moist convection in this atmosphere, just as in the tropics, depends critically on high values of θ_w in the boundary layer. Boundary layer air with a $\theta_w = -2°$C, which is representative for the conditions just prior to and just after the passage of vortex A (see Figure 4c), will *not* be able to penetrate very deep into the troposphere. Depending on the degree of entrainment, surface air with $\theta_w > -2°$C can penetrate into the upper troposphere. Assuming arbitrarily that deep convection may proceed when θ_w reaches a value between $-1°$C and 0°C we see from Figure 4b and 4c that this region is restricted to a relatively small region around the center of the vortex. Figure 4c and Figure 5 demonstrate the importance of "air of good quality," i.e., air with a high θ_w in the boundary layer, in order for deep, penetrating convection to occur.

As mentioned in Rasmussen (1985b) the high temperatures in the disturbed region must be the result of an enhanced sensible heat flux from the warm surface.

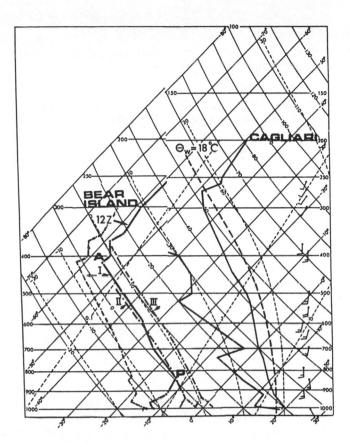

Figure 6. Radiosonde ascent (temperature and dewpoint) Bear Island 1200 GMT 13 December 1982 (curve I). The "modified ascent" consists of the Bear Island sounding above point P at 830 mb and a dry adiabatic layer (dashed) below. Curve II corresponds to saturated (pseudoadiabatic) ascent of surface air away from the strongly disturbed region near the center of the vortex (see Figure 4c), while curve III shows ascent of a particle with the high θ_w value measured at weathership AMI during the passage of the center. Also shown (on the right side of the diagram) is the Cagliari sounding from 0000 GMT 29 September 1983.

Correspondingly the increase in the dew point temperatures reflects the result of increased moisture fluxes. The potential wet bulb temperature, θ_w, which represents the temperature as well as the humidity, increases because of the increased surface fluxes due to the increasing wind velocity. This in turn effects the intensity of the convection and the release of latent heat, and through this the intensity of the system. An intensification of the system means stronger surface winds, and a positive feedback system has been established. Several researchers (including the author) have proposed that (some) polar lows are the result of CISK. In addition, other mechanisms such as baroclinic instability and the positive feedback mechanism mentioned above might contribute to the developments. Recently, however, Emanuel and Rotunno (1989) have argued against CISK as a mechanism for polar low development (as well as for tropical cyclones) pointing out that in the real world very little convectional available potential energy (CAPE) will be available as envisaged in the CISK theory. Instead they have argued that a new type of instability called air-sea interaction instability (ASII) alone explains the growth of a hurricane-like vortex, without a reservoir of available potential energy. According to ASII enhanced surface winds associated with an earlier formed disturbance lead to enhanced surface fluxes of sensible and latent heat, which are then redistributed aloft by convection. This leads to the formation of a warm core, i.e., an intensification of the system, and the positive feedback loop is closed.

The data from the AMI (Figure 4c) show that $\theta_w \simeq -2\,°C$ before (as well as after) the passage of polar low A around midnight between 13 and 14 December. Figure 6 shows that surface air with this value of θ_w represents little or (in the real world with entrainment and the effects of condensed water) probably *no CAPE at all*. Nevertheless an almost explosive development takes place in the 6 hr from 1800 GMT to 2400 GMT 13 December 1982. Since this development cannot be explained by the little or even nonexisting CAPE it seems to be a piece of evidence in favour of Emanuel and Rotunno's ASII theory.

Another positive feedback process, although somewhat weaker, is associated with variations in the sea surface temperature. If we consider the oceanic response to tropical cyclones then "the combination of upwelling and vertical mixing typically produces decreases in the surface ocean temperature of 1–3 °C and may occasionally produce decreases as large as 5 °C" (Anthes, p. 131). In the polar low case in December 1982 the sea surface temperature rises about 1°C (Figure 4c) during the passage of the low. Whether or not this is typical and important for polar low developments remains to be investigated.

As tropical cyclones polar lows may approach a steady state when the vertical lapse rate becomes neutral or near neutral to moist convection. Vortex A considered here is observed to exist in a nearly steady state for about one day after the spin-up late on 13 December. According to Ooyama the transition from the deepening stage to the steady state is best understood by considering the nondimensional thermodynamic variable, η. The parameter η is associated with the upward diabatic heat flux Q at the top of the boundary layer through

$$Q = \eta w , \tag{1}$$

where w is the vertical velocity. Through energy considerations η can be shown to be given (in an approximate form) by

$$\eta = 1 + [(\theta_e)_o - (\theta_e^*)_2] / [(\theta_e^*)_2 - (\theta_e)_1] \tag{2}$$

where indices 0, 1, and 2 indicate respectively the boundary layer, a middle layer, and the layer corresponding to the top of the convection.

Inserting appropriate values for the equivalent potential temperatures we find that $\eta \simeq 2.5$ prior to the onset of deep convection on 13 December 1982. This is, also compared to tropical standards, a fairly big value. Following Ooyama (p. 17) a typical value for tropical systems is $\eta = 2.0$. Only 6 hr later around midnight (0000 GMT 14 December) when the stratification in the core region is approximately moist adiabatic, η has decreased to only $\simeq 1$. Ooyama points out (p. 18) that "the fact that all tropical cyclones eventually stop intensifying must be explained by taking into account the reduction of η due to the establishment of the warm core." Another way to put it is to say, that when the core cannot get any warmer then the intensification must stop.

There are no radiosonde ascents available from the AMI to verify directly that the lapse rate actually changes to nearly moist adiabatic during the rapidly intensifying stage. This is shown indirectly, however, through the massive amount of deep convective clouds in the central region of the vortex (Figure 8 and Section 3.4) and through the observed surface pressure decrease (see Section 3.4).

3.2 Boundary Layer Transfers of Heat and Moisture

It is generally accepted that the supply of latent heat and moisture from the warm ocean is crucial for the growth and maintenance of tropical cyclones. While there is some doubt about the role of the oceans as a source of sensible heat for tropical cyclones, these fluxes are very important for polar lows. The importance of an in situ sensible heat flux was pointed out already by Rasmussen (1977, 1979) and Økland (1977), and is also verified by the present case.

Let us assume that the sensible and latent heat transports, H_s and H_e, respectively, are given by the usual bulk formulas

$$H_s = \rho_s c_p C_D (T_w - T_a) \, V_s \tag{3}$$

$$H_e = \rho_s L C_D (q_w - q_a) \, V_s \tag{4}$$

where the symbols have their conventional meaning.

Recent results (Boyle et al., 1987) indicate that the normally used bulk method underestimates stress by as much as a factor of three in storm conditions. In strongly disturbed regions the value of C_D therefore may be as high as 4×10^{-3}. Using this value for C_D together with values of the parameters appropriate for the development of vortex A, i.e., a temperature difference $T_w - T_a = 6°C$, a specific humidity difference $q_w - q_a = 2 \times 10^{-3}$, and a wind velocity of 20 m s^{-1} we find that

$$H_s \simeq 600 \text{ Wm}^{-2} \text{ and } H_e \simeq 500 \text{ Wm}^{-2} \, .$$

These values are in good agreement with the results obtained by Shapiro et al. (1987), during the Arctic Cyclone Expedition, 1984, and correspond to a Bowen ratio $\simeq 1$, i.e., much larger than the value 0.1 typical for tropical oceans. The total heat flux $H = H_s + H_e$ is very large, around 1100 Wm^{-2}. Note that the specific humidity *difference* 2 g kg^{-1} is close to the value 2.4 g kg^{-1} used by Anthes (p. 68) for *tropical* conditions and that the temperature difference on 6°C is much larger than the difference on 1°C used by Anthes in his estimate of surface heat fluxes in hurricanes (p. 69).

It is straightforward to compute the time Δt required for a heat flux of that magnitude to eliminate the area between the modified Bear Island sounding (see Section 3.1) and curve *III* (see Figure 6) corresponding to pseudoadiabatic ascent from the surface of air with properties as measured at the AMI during the passage of the center. Such saturated ascent conditions will be typical for the inner region of a vortex accompanied by deep convection. Assuming a heat flux of 1000 Wm^{-2} and provided *all* the heat is retained in the convective region, we find that Δt is $\simeq 4$ hr, i.e., a value in reasonable agreement with the 6 hr between the start of the spin-up process around 1800 GMT 13 December 1982 and the time when the fully developed vortex could be observed for the first time at the synoptic surface maps at 0000 GMT 14 December 1982.

In his discussion of tropical cyclones Ooyama (p. 11) points out that "the major contribution to the total convergence of water vapour in a vertical column should come from the convergence in the boundary layer," and that "it is unlikely that convective activity in a column would be supported mainly by the evaporative supply of water vapour from the part of the sea surface directly below the column." A somewhat similar point of view is taken by Palmen and Newton (1969, p. 498) in writing: "the essential source of total energy is derived from the lateral influx of water vapour in the moist surface layer." These authors, however, add "that also the additional flux of latent and sensible heat from the sea in the core region represents a heat source not to be neglected." Finally we may quote Anthes (p. 73): "The convergence rate is 2–5 times the evaporation rate, depending on the portion of the storm considered." In this respect polar lows seem to differ from tropical cyclones. This may be illustrated by comparing the latent and sensible heat fluxes computed above with the convergence of latent heat in the boundary layer into the central region of the low. Convergence of latent heat in the free atmosphere may be neglected. Assuming a depth of the boundary layer of 1000 m, and a rather high-cross contour low-level wind component of 5 m s^{-1} we can compute the latent heat flux into the vortex at a radius of 200 km by means of the data shown in Figure 4c. According to the figure the air outside the active core region is rather dry with a surface dewpoint $\sim -6\,°C$. The specific humidity corresponding to $-6\,°C$ is used as representative for the whole (well mixed) boundary layer. By dividing the total flux across the 200 km radius cylinder with the corresponding horizontal area we obtain a mean "convergence flux" expressed in Wm^{-2}. Compared to the in situ latent heat flux of 500 Wm^{-2}, the contribution from moist air import according to this computation is of approximately the same magnitude namely ~ 400 Wm^{-2}. The total energy available for the formation of polar lows is thus much smaller than that available for tropical cyclones.

Based on the approximations stated above the sensible heat flux, the latent heat flux, and the "convergent heat flux" are of the same magnitude, ~ 500 Wm^{-2} for the present case. In a study of two polar low developments near the British Isles, Rasmussen (1979) found a somewhat similar result. The contributions from moisture import and evaporation were nearly equal, whereas the contribution from the sensible heat flux in this case where the development occurred over a more southerly region was only half the value of the two others.

If we consider the development of vortex A from 1800 GMT 13 December to 0000 GMT 14 December 1982, we find (Figure 4a) a surface pressure fall of the order 10 mb in 6 hr. This corresponds to a fairly large pressure tendency even compared to tropical systems, which, according to Ooyama (p. 21), in their rapidly deepening stage may deepen 20–30 mb in 24 hr.

The fact that the contributions from the three types of energy supply discussed are of equal magnitude may be important for the rapid development of this type of polar lows. Only about one third of the energy has to be transported into the active region by advection in the boundary layer. The rest is already there and directly available.

3.3 The Surface Wind Distribution

The tangential wind speed profile of polar low A was shown in Figure 4b. The slight asymmetry of the profiles indicate the existence of at least two mesoscale vortices in the central region of the vortex. This is confirmed by satellite images. The wind field associated with a subsequent disturbance, C^*, which formed the following day on 15 December, showed more symmetry as seen from Figure 7. Both profiles resemble wind profiles from tropical storms (Anthes,

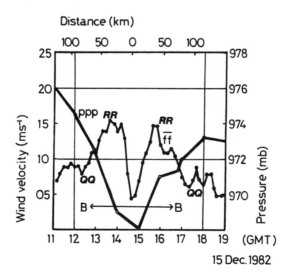

Figure 7. Mean surface wind velocity $\bar{ff}$ in m s^{-1} (10 min mean) and surface pressure (ppp) during passage of vortex C^* at Bear Island on 15 December 1982. A length scale is shown on top of the figure. The distance B-B marked on the figure is a measure of the horizontal scale of the system.

Figure 2.8). Inside the radius of maximum wind, R_o, the wind velocity may be approximated as a solid rotation, i.e.,

$$v(r) = v(R_o) \frac{r}{r_o} \qquad 0 \leq r \leq R_o \,. \qquad (5)$$

Outside R_o the radial variation of the tangential wind is often described by

$$v(r) = v(R_o) \, (R_o \,/\, r)^x \,, \qquad R_o < r \,. \qquad (6)$$

In addition to being somewhat asymmetric, vortex A is embedded in a relatively strong background flow and a value of x cannot be given. Vortex C^* is more symmetric and the background flow is weaker. A value for x representative for the points between RR and QQ in Figure 7 gives (for both branches) $x \approx 1.1$. This value is larger than the values 0.5 to 0.7 typical for tropical storms (Anthes, p. 22) and corresponds to negative relative vorticity assuming a circular symmetric cyclone and the wind field given by Eq. (6).

With a solid core rotation as given by Eq. (5) the vorticity in the core is given by $\zeta_o = 2v(R_o)/R_o$. Using the values for vortex A shown in Figure 4b we find that $\zeta_o \approx 2 \times 10^{-3}$ s^{-1}, i.e., a rather high value (according to Anthes, p. 53: "the absolute vorticity in a moderate hurricane has a typical value of 2×10^{-4} s^{-1})".

3.4 The Polar Low Eye

One of the most characteristic features of a tropical cyclone is the formation of an eye and the associated very low surface pressures and very high winds in the eye wall region. Without an eye tropical storms would be limited to much more modest surface pressures (Anthes p. 30). With this background it is an important question if polar lows produce eyes and associated weather phenomena similar to their tropical counterparts. The evidence so far is a little conflicting. Satellite images often show eyes or eye-like features such as that shown in Figure 1, and tangential wind profiles as shown in Figure 4b and Figure 7 likewise suggest the existense of an eye. So far there is no evidence, however, at least to the author's knowledge, of exceptionally low surface pressures and associated high winds in the central region of any polar lows. The apparent paradox can be solved in that way, that eyes do form as shown by the satellite images, but that they are much weaker than their tropical counterparts. Emanuel (1986) and Emanuel and Rotunno (1989) have pointed out that polar lows have a *potential* to become *very* intense with *very* deep central pressures. However, such very intense systems have not yet been observed. Since the formation of an eye "represents a crucial step in the transformation of a weak, disorganized storm to an intense hurricane" (Anthes, p. 32), the fact that polar lows never, or at least very seldom, seem to be able to form "strong" eyes may be one of the explanations why they never seem to reach intensities similar to tropical cyclones.

The high temperatures in the eyes of tropical cyclones are generally associated with high level air, partly originating from the eye wall, which is sinking and warming. The surface pressure in the eye may in this way become considerably lower than that associated with a column of (saturated) air of uniform θ_e equal to that of the surface air.

During the polar low episode in mid-December 1982 where eyes actually were observed, we can evaluate the potential for eye formation by considering the radiosonde ascents, as for example the one from Bear Island 1200 GMT 13 December 1982 shown in Figure 6. During the formation of vortex *A* later on that day individual updrafts approximately would follow the moist adiabatic shown as curve *III*. Descending upper level air taking part in an eventual eye formation would probably be slightly cooler than the temperature corresponding to curve *III*. If for example a descending air particle near 400 mb starts with a temperature corresponding to that of the environment (point *A*) it will be substantially *cooler* than the updraft except in the lower layers. In this case there is *no potential* for the formation of an eye by subsidence of high-level warm air. In fact no eye-like feature was observed on the satellite images in connection with the formation of vortex *A*.

The warmest possible eye at any time would form if high θ_e surface-air after ascent in the eye wall would start to subside in the (interior) eye region. Even in this case, however, only a relative small increase in the mean temperature of the column would occur. This is so because of the low moisture content of the cold air that results in relatively small differences between moist and dry adiabatic ascent/descent except in the lowest layers. Assume for example a downward vertical velocity of 100 mb day^{-1}, which is a reasonable value judging from Nordeng's (1986) numerical simulation of a polar low (see Nordeng's Figure 6a). In that case (see Figure 6) only air below a level around 700 mb will undergo a significant increase in temperature. Since a vertical velocity of that magnitude probably will not be effective all the way down to the surface, and since the subsiding air probably will have a lower value of θ_e than the newly ascending surface air, the increase in mean temperature, and the corresponding surface pressure decrease in the eye are bound to be small.

Although polar lows only develop weak if any eyes at all, they do develop a warm core. Anthes (pp. 74–75) discusses three different physical processes that may be responsible for tropospheric warming necessary to produce a hurricane from a weak disturbance. Especially the third of these mechanisms, "the replacement of an unsaturated column of air with a cloud with a higher mean temperature" seems relevant for the formation of some polar lows like polar low A discussed in this work. The satellite images show that deep convective clouds are present in and close to low A (Figure 8). Assuming that the lapse rate in these clouds is given by curve III on Figure 6, a simple calculation, based on the temperature increase between the modified Bear Island radiosonde ascent representing the conditions prior to the onset of the deep convection and curve III

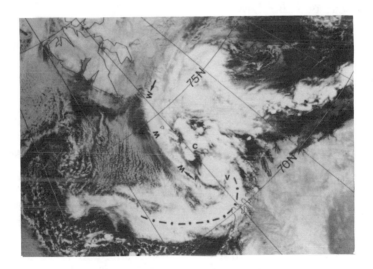

Figure 8. NOAA-7 infrared satellite image 0419 GMT 14 December 1982. The approximate position of the center of vortex A has been indicated by "V." A local arctic front at the leading edge of a shallow cold air outbreak west of Bear Island is shown by the thick dash-dotted line. The shallow cold air mass can easily be identified by the system of well-developed cloud streets. The western boundary of the warm, modified air mass east of the shallow cold air mass has been indicated W-W-W.

gives a surface pressure fall of the order 10 mb (Rasmussen, 1988). This (hydrostatic) pressure fall is in good agreement with the observed one. Pressure disturbances of this magnitude are typical for intense polar lows and in general therefore it does not seem necessary to invoke a special "eye-dynamics" to explain the observed minimum pressures.

3.5 Low-Level Vorticity and Vertical Shear

Among the factors related to tropical cyclone genesis are the low-level vorticity and the vertical shear of the horizontal wind. Both factors are of decisive importance also for polar low formations in the Bear Island region where polar lows typically develop in regions with little or no vertical wind shear and within a reservoir of low-level relative vorticity such as a (relatively) large-scale trough or cyclonic circulation. For detailed discussions see Rasmussen (1985 and 1988).

3.6 The Decay of Polar Lows

Hurricanes are known to weaken rapidly when they move over land. Anthes (pp. 61–62) lists three major physical effects for this. First, and probably most important according to Anthes, is the dramatic reduction in evaporation as the storm leaves the ocean. A second effect is that the land is cooler than the ocean, and the third is the increase in surface roughness.

Polar lows of the convective type considered here are known to decay rapidly after landfall or "icefall." As an illustration of this the polar low shown in Figure 1 should be compared with the satellite image of the same polar low only a few hours later at 0831 GMT (Figure 9a). The system at this time is situated a little inland and has already started to lose the well-organized structure seen in Figure 1. Nevertheless this particular low could be followed for a relatively long time (≈ 10 hr) after landfall. Around noon the disturbance could be seen as a small comma-like feature over the northern part of Sweden (Figure 9b), accompanied by a weak cyclonic circulation at the surface and scattered light snow. The last time the system could be traced on the satellite images was about 6 hr later. The reason for the rapid decay of polar lows after landfall is probably a mixture of all three effects mentioned above.

Numerous examples of rapid decay of polar lows when they move over land (or ice) can be found, and the behaviour illustrates the similarities between tropical

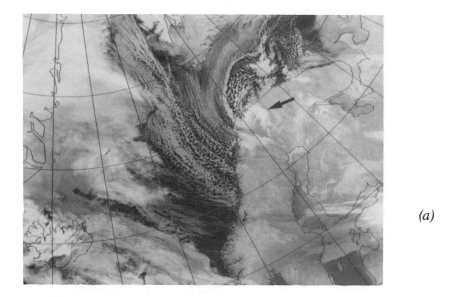

(a)

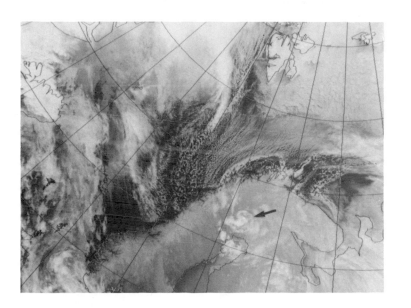

(b)

Figure 9. (a) NOAA-10 infrared satellite image, 0831 GMT 27 February 1987, showing the polar low on Figure 1 shortly after landfall. (b) NOAA-9 infrared satellite image 1226 GMT 27 February 1987 showing decaying polar low (marked by arrow) over northern Sweden.

cyclones and convective polar lows. Nevertheless, polar lows of a more baroclinic structure are *not* influenced in the same way by landfall for obvious reasons. On the contrary, these systems may sometimes even intensify when they cross a coastline.

3.7 The Arctic Instability Low

Polar low A, discussed in the preceding sections, formed as the result of a spin-up process of a previously formed low. Later on, on 15 and 16 December, another vortex C^* of very small horizontal scale, < 100 km, formed in the same region (Figure 10). Although of slightly smaller horizontal extent C^* showed many similarities to polar low A.

This is illustrated by comparing the surface wind and pressure data shown in Figures 4b and 7, respectively. Although the fully developed disturbances had many similarities their initial mode of formation was different. As the result of the rapid formation of vortex A a sharp internal air mass boundary was formed

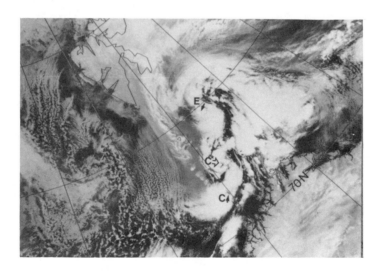

Figure 10. NOAA-7 infrared satellite image 0406 GMT 15 December 1982. Arrows marked C, C* and E show a decaying vortex C, the newly formed vortex C* and a cumulonimbus cluster that later on develops into a vortex E. The shallow cold air mass west of 20°E is clearly indicated by the system of cloud streets and the region of a shallow overcast.

separating a shallow, cold northerly surge of fresh arctic air from modified warmer air to the east (see Figure 10). This internal air mass boundary or front could be identified on the satellite images from around midnight between 13 and 14 December 1988 and until 16 December. At least two small-scale vortices formed along this shallow front, and the center of one of them, C^*, crossed Bear Island (see Figure 11). The Bear Island surface wind measurements show a not very strong, but very symmetric disturbance. The inner core region with a tangential wind distribution corresponding to a solid rotation has a vorticity around 10^{-3} s^{-1}, i.e., only a little less than that associated with vortex A. The diameter of this region with solid rotation is on the order of 50 km according to the surface observations, while the diameter of the cloud-free region is only about half of that (Figure 11).

The region around Bear Island has a rather unique climatology and geography. This may explain why vortices like C^* shown in Figures 10 and 11 have not so far, or at least very seldom, been observed elsewhere.

The rapid formation of C^* seems to be the result of a pre-existing shear zone along the shallow internal air mass boundary mentioned above, and deep convection. Anthes (p. 60) points out that giant thunderstorms or clusters of thunderstorms sometimes may persist for 6 hr or longer, and that convergence

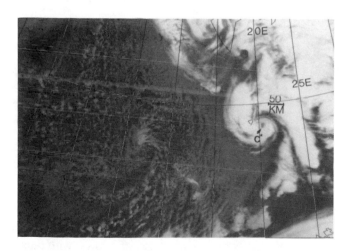

Figure 11. NOAA-7 infrared satellite image 1400 GMT 15 December 1982. Short arrow shows center of vortex C^* east of Bear Island, which is marked by B. The long arrow indicates direction of upper level anticyclonic outflow from C^*.

under such a region of intense convection reaches a magnitude of 10^{-4} s^{-1}. (An even larger value of the horizontal convergence, $D \simeq -5. \ 10^{-4}$ s^{-1} was measured during the passage of a convective band at Risø (near Copenhagen) in December 1983 (Nielsen, 1984)). If we, following Anthes, assume a constant horizontal divergence D of -10^{-4} s^{-1} near the surface under a complex of deep cumulonimbus clouds we can estimate the approximate time necessary for the formation of vortex C^*. Based on the observations shown in this paper and in Rasmussen (1988) the shear vorticity, ζ, across the shallow front is $\sim 2 \times 10^{-4}$ s^{-1}, corresponding to an initial absolute vorticity $\zeta + f \simeq (2 \times 10^{-4} + 1.4 \times 10^{-4})$s^{-1}. Knowing the final absolute vorticity $\simeq 10^{-3}$ and the (constant) divergence D, we may integrate the vorticity equation

$$\frac{d}{dt} (\zeta + f) \simeq -(\zeta + f) D$$

to obtain the time T required to change the vorticity from its initial to its final value. With the figures given above we find that $T \simeq 3$ hr. With a contribution from the planetary vorticity alone ($\zeta = 0$), the same process (with the same convergence) would require almost 6 hr. No vortices were observed inside the modified air mass away from the shallow air mass boundary.

The term polar low is generally used in a generic sense to include several types of phenomena. To distinguish between polar lows in general with a scale of a few hundred kilometers and disturbances like C^* that form in highly unstable airmasses and have a much smaller scale (≤ 100 km), it is suggested to denote the latter "arctic instability lows," a term, which sometimes but not very often, has been used as a synonym for polar lows.

4. THE SUBTROPICAL LINK

Cyclones similar to hurricanes are known sometimes to form over the Mediterranean Sea (Ernst and Matson, 1983; Billing et al., 1983). These small-scale cyclones may have a striking similarity to polar lows, with regard to horizontal scale, intensity, structure, and mode of formation.

One of the mechanisms responsible for the development of polar lows is convection. Conversely, more-or-less shallow baroclinic zones are almost invariably present in the regions where polar lows form, and it is difficult to assess the relative importance of convection versus baroclinic instability.

In a case studied by Rasmussen and Zick (1987) baroclinic instability did not seem to play any role at all for which reason this development, even if it took place over the Mediterranean, might be considered as an "ideal" development of a polar low of convective nature.

The analysis of the Mediterranean disturbance was carried out mainly by means of satellite data in form of cloud track wind data. In this section we will concentrate upon the similarity between the Mediterranean system and polar lows and show some new additional interesting data that were not available before.

All data available indicate, that the Mediterranean vortex, which could be followed for 6 days, was the result of a rapid spin-up caused by deep convection, of pre-existing vorticity associated with a synoptic scale of low modest intensity. This way of formation is *very similar* to that of vortex A briefly described in Section 3. Figure 12 shows the vortex a few hours after its formation, and Figure 13 shows the vortex the next morning on 28 September a little southeast

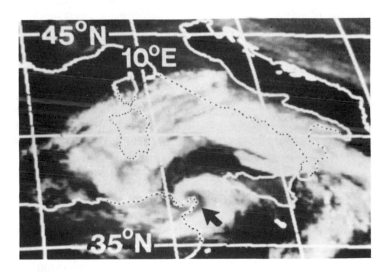

Figure 12. METEOSAT infrared satellite image, 1800 GMT 27 September 1983, showing the initial formation of a subsynoptic vortex (at the arrow) over the sea close to Carthage in Tunis, North Africa (copyright by ESA-EUMESTAT).

Figure 13. NOAA-8, visible channel satellite image, 0807 GMT 28 September 1983, showing dissolving large-scale cloud spiral with a subsynoptic vortex in the center (shown by arrow).

of Sardinia. The "active part" of the disturbance is associated with the tight cloud spiral with the eye. The cloud bands with the spiral structure around the central vortex are remains from the original synoptic scale cyclone (and quickly disappear). By comparing the central vortex in Figure 13 with the polar low shown in Figure 1 we note at once the striking similarity with regard to horizontal scale and general appearence. The vortex at this time is cyclonic through its whole depth as seen in Figures 14a and 14b.

On the following day, 29 September 1983, when the vortex is situated to the west of Sardinia and Corsica it has developed a dense cirrus shield (Figure 15). At the same time it has changed to a warm core system with anticyclonic, divergent flow aloft (see Figures 14c and 14d). The temperature of the cloud tops of the most active cumulonimbus in the vortex at this time is − 54 °C, which corresponds to a height around 250 mb. One day later the vortex makes landfall and the center of the vortex passes almost directly over Ajaccio on the west coast of Corsica.

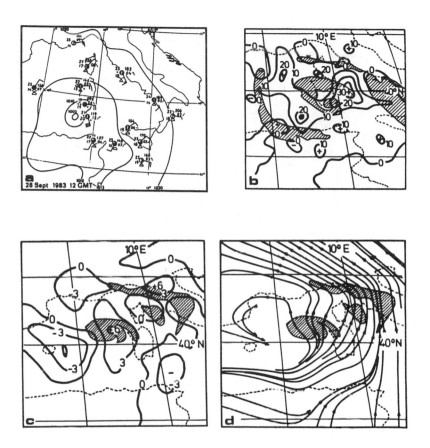

Figure 14. (a) Volume I Surface map 1200 GMT 28 September 1983 show-ing subsynoptic vortex at southern Sardinia. (b) Relative vorticity in units of $10^{-5} s^{-1}$ at high level (~300 mb) 1200 GMT 28 September 1983. The analysis is based on cloud track wind data. High-level clouds are hatched, and land contours stippled. (c) High-level (~300 mb) divergence in units of $10^{-5} s^{-1}$ 1200 GMT 29 September 1983 based on cloud track wind data. High-level clouds are hatched, and land contours stippled. (d) High-level (~300 mb) streamlines 1200 GMT 29 September 1983. High-level clouds are hatched, and land contours stippled.

Observations from Ajaccio before, during, and after the passage are shown in Figures 16 and 17.

Anemometer data (not shown) and the barograph recordings (Figure 16) show that the center passes a little south of Ajaccio at 1200 GMT. Even if the

Figure 15. NOAA-7 visible channel satellite image, 1418 GMT 29 September 1983, showing the vortex at the time when it is best developed. The white arrow shows rows of cumulus clouds converging cyclonically into the center at low levels. The black arrow indicates direction of anticyclonic outflow at upper levels. Corsica and Sardinia are seen to the east of the vortex.

center does not pass directly over Ajaccio the barograph curve (Figure 16a) shows a remarkable pressure drop of nearly 10 mb. Figure 16b shows barograph records from seven meteorological stations along the Corsican coastline. The barograph curves clearly show that *the central part of the vortex where the strong pressure falls are concentrated has a very small horizontal scale on the order of 50 km.*

In addition to the surface observations rawindsonde observations are available from Ajaccio for every 6 hr as shown in Figure 17. At 0600 GMT 30 September 1983 when the center of the vortex is situated over the Mediterranean 75–100 km west of Corsica, the upper winds at Ajaccio veer from a southerly direction in the lower layers to southwest at 5 km reflecting the influence of the approaching core. At 1200 GMT the vortex is very close to Ajaccio and the strong southeasterly low-level winds rapidly drop off to almost zero already at the height of 3000 m. (Unfortunately no radiosonde temperature and humidity data are available from Ajaccio at 1200 GMT 30 September).

At 0000 GMT 29 September the center of the vortex is over the Mediterranean, about 200 km to the northwest of the radiosonde station at Cagliari at the southern tip of Sardinia. The radiosonde ascent from Cagliari, at this time

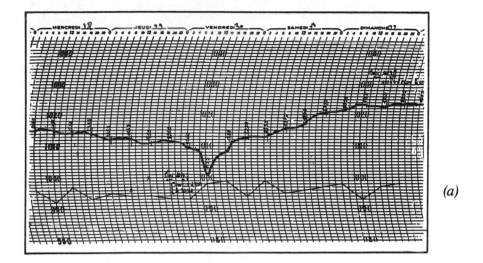

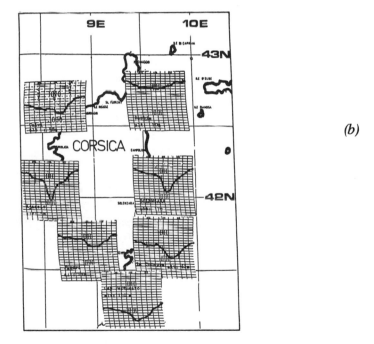

Figure 16. Barograph data from Corsica. (a) Barograph record from Ajaccio from 28 September to 2 October 1983 showing passage of subsynoptic vortex at 30 September 1983. (b) Barograph records from several Corsican stations on 30 September 1983.

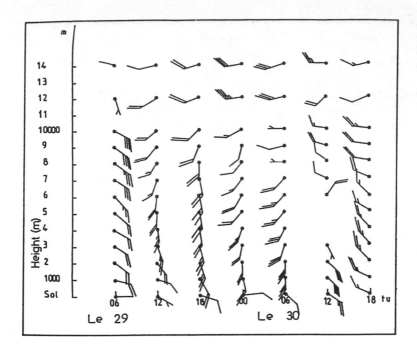

Figure 17. Six hourly rawindsonde observations from Ajaccio from 29 and 30 September 1983.

is plotted on Figure 6 together with the Bear Island ascent from December 1982. The two radiosondes illustrate in a striking way the difference between the air masses in which surprisingly similar small-scale vortices develop. Apart from the more obvious differences there is a more subtle one, however, namely the potential for CAPE. As discussed in Section 3, the conditions prior to the formation of vortex A on 13 December 1982 were characterized by a modified sounding with a nearly dry adiabatic layer from the surface up to around 850 mb, in which case very little CAPE will be available (Figure 6). To reach a situation with more CAPE a super-adiabatic layer would have to be formed in the lowest, say 100 mb. This is clearly not realistic. Conversely, the Cagliari ascent shows a substantial amount of CAPE defined by the positive area between the measured temperature and the $\theta_w = 18\,°C$-curve. The role of CAPE should not, therefore, completely be ruled out as a part of the explanation of the Mediterranean vortex. If we consider the estimated center surface pressures in the Mediterranean vortex shown by Rasmussen and Zick and reproduced here as Figure 18, they are characterized by two rapid pressure falls followed by rapid rises and a relatively

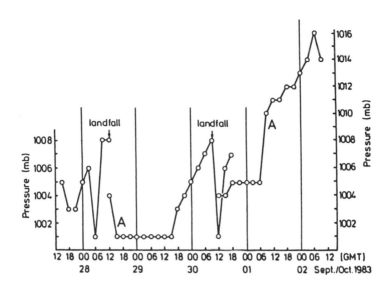

Figure 18. The center pressure of the Mediterranean vortex as function of time.

long period with a low center pressure when the system is over the open sea. It is likely that the rapid pressure falls are associated with sudden "burst-like" releases of CAPE, and in fact both incidents are characterized by deep convection (see Rasmussen and Zick for details).

Even in the long quasi-steady period from 1500 GMT 28 September 1983 to 1500 GMT 29 September when the surface pressure is (apparently) steady at its lowest value, it is *not* below 1000 mb. Emanuel (1986) in his Figure 2 shows the minimum sustainable central pressure of a (tropical or tropical-like) cyclone as a function of the sea surface temperature and the mean temperature of the lower stratosphere (where the outflow takes place). His figure is for 20° latitude and a radius of influence of the cyclone of 500 km, but (as pointed out by Chris Bretherton, personal communication) the latter parameters have only a weak influence on the central pressure. From noon 28 September to noon on the following day, the vortex moves over a region with sea surface temperatures between 22 °C and 23 °C. Using this value together with an outflow value of $\sim -52\,°C$ we find a minimum pressure around 980 mb, i.e., a value around 20 mb lower than that observed. In their discussion of the December 1982 polar low Emanuel and Rotunno (1989) argue that: "The rapidly moving polar low may simply not remain over warm water long enough to achieve its potential intensity." In the Mediterranean case, however, the vortex remains over the sea for more than 4 days, only interrupted by the short passage over Corsica, and still it does not achieve anything like the minimum pressure envisaged by the ASII theory.

5. CONCLUDING REMARKS

Although polar lows develop in regions very different from the genesis area of tropical cyclones over the warm tropical seas, there is increasing evidence that there are *significant similarities between (some) polar lows and tropical cyclones.*

The similarities include a tendency for an axisymmetric structure, sometimes around a vertical eye of small horizontal diameter. Polar lows never for reasons discussed above, reach the intensities of tropical cyclones, but their surface wind field and pressure distribution are qualitatively alike.

The remarks above are only valid for a certain group of polar lows associated with deep, vigorous convection. Polar lows form in regions where many forcing mechanisms operate simultaneously with the result that a whole *spectrum* of polar lows, ranging from the convective type discussed in the preceding to types resembling small-scale baroclinic waves, can be observed.

Polar lows normally are associated with horizontal scales of a few hundred kilometers. In the present paper it has been demonstrated, that polar lows may form on an even smaller horizontal scale, < 100 km, not very different from the scale of large thunderstorms. To distinguish between the "normal" polar lows of a scale of ~ 400–500 km and these very small-scale systems, it is suggested to use the term arctic instability lows for the latter. This term has sometimes, but not very often been used as a synonym for polar lows, and is well suited for the particular systems developing in a highly unstable air mass modified by strong air-sea interaction processes.

ACKNOWLEDGMENTS

The author wants to thank the Department of Electrical Engineering and Electronics, University of Dundee for providing the excellent satellite images shown in this paper. I also want to thank M. Rene Mayengon who kindly made available the data from the Mediterranean storm shown in Figures 16, 17, and

18, as well as all my colleagues at home and abroad who directly or indirectly have contributed to this paper.

REFERENCES

Anthes, R.A., 1982: Tropical cyclones. Their evolution, structure and effects. *Am. Meteor. Soc.*, 208 pp.

Billing, H., I. Haupt, and W. Tonn, 1983: Evolution of a hurricane-like cyclone in the Mediterranean Sea. *Beitr. Phys. Atmos.*, *56*, 508–510.

Boyle, P.J., K.L. Davidson, and D.E. Spiel, 1987: Characteristics of over-water surface stress during STREX. *Dynamics of Atmos. and Oceans*, *10*, 343–358.

Craig, G., and H.-R. Cho, 1988: Cumulus convection and CISK in the extratropical atmosphere, Part I: Polar lows and comma clouds. *J. Atmos. Sci.* (in press).

Craig, G., and H.-R. Cho, 1989. Baroclinic instability and CISK as the driving mechanisms for polar lows and comma clouds. *J. Atmos. Sci.*, xx–yy.

Emanuel, K.A., 1986: A two stage air-sea interaction theory for polar lows. Preprint, *The International Conference on Polar Lows*, Oslo, Norway, 20–23 May 1986, 187–200, 364 pp.

Emanuel, K.A., and R. Rotunno, 1989: Polar lows as arctic hurricanes. *Tellus*, *41A*, 1–17.

Ernst, J.A., and M. Matson, 1983: A Mediterranean tropical storm? *Weather*, *38*, 332–337.

Harrold, T.W., and K.A. Browning, 1969: The polar low as a baroclinic disturbance. *Quart. J. Roy. Meteor. Soc.*, *95*, 710–723.

Nielsen, W.N., 1984: Vejret julenat 1983. VEJRET, 6, 40–47 (in Danish).

Nordeng, T.E., 1986: Various condensation schemes and their impact on the simulation of polar lows. Preprint, *The International Conference on Polar Lows*, Oslo, Norway, 20–23 May 1986, 221–233, 364 pp.

Økland, H., 1977: *On The Intensification of Small-Scale Cyclones Formed in Very Cold Air Masses Heated by the Ocean.* Institute Rep. Series, No. 26, University of Oslo, Institutt for Geofysikk.

Ooyama, K., 1969: Numerical simulation of the life of tropical cyclones. *Atmos. Sci.*, *26*, 3–40.

Palmen, E., and C.W. Newton, 1969: Atmospheric circulation systems: Their structure and physical interpretation. Int. Geophys. Ser., 13, Academic Press, 603 pp.

Rasmussen, E., 1977: *The Polar Low as a CISK Phenomenon.* Rep. No. 6. University of Copenhagen, Institute for Theoretical Meteorology.

Rasmussen, E., 1979: The polar low as an extratropical CISK disturbance. *Quart. J. Roy. Meteor. Soc.*, *105*, 531–549.

Rasmussen, E., 1985a: *A Polar Low Development Over the Barents Sea.* Polar Lows Project Tech. Rep. No. 7, The Norwegian Meteorological Institute, Oslo, Norway, 28 pp.

Rasmussen, E., 1985b: A case study of a polar low development over the Barents Sea. *Tellus, 37A,* 407–418.

Rasmussen, E., 1988: On polar lows, arctic instability lows and arctic cyclones, An observational study. (Manuscript submitted to *Tellus.*)

Rasmussen, E., and C. Zick, 1987: A subsynoptic vortex over the Mediterranean with some resemblance to polar lows. *Tellus, 39A,* 408–425.

Reed, R.J., 1979: Cyclogenesis in polar airstreams. *Mon. Wea. Rev., 107,* 38–52.

Reed, R.J., 1988: Polar lows. In the nature and prediction of extratropical weather systems, 7–11 September 1987. *Seminar proceedings, Vol. 1, ECMWF,* 213–236, 280 pp.

Sardie, J.M., and T.T. Warner, 1985: A numerical study of the development mechanisms of polar lows. *Tellus, 37A,* 460–477.

Shapiro, M.A., L.S. Fedor, and T. Hampel, 1987: Research aircraft measurements of a polar low over the Norwegian Sea. *Tellus, 39A,* 272–306.

Van Delden, A., 1985: *Convection in Polar Outbreaks and Related Phenomena.* University of Utrecht, Netherlands, 163 pp.

CHAPTER 2 – THEORY

Introduction

One of the first theoretical works on polar lows was probably the paper by Mansfield from 1974 in which he, using Eady's model, discussed two polar lows described earlier by Harrold and Browning. Mansfield concluded as they did that polar lows were basically shallow baroclinic waves. This point of view was challenged some years later by Økland and Rasmussen who introduced the idea of CISK as an explanation of polar low developments. At that time it was not realized that different types of polar lows could be found, and the existence of two different theoretical models for the explanation of the "new" phenomenon gave rise to some years of intense discussion.

In the first paper "On the Precursors of Polar Lows" Wiin-Nielsen discusses the early baroclinic stage of some polar low developments in the Bear Island region by means of a three-level baroclinic model. The basic state of the model consists of a thermal stratification and thermal winds that closely correspond to the conditions along the ice edge in the Bear Island region, where developments of small-scale disturbances, probably of a baroclinic nature, have been observed. Wiin-Nielsen concludes that the baroclinic instability process can account for the initial formation of a wave that may form the necessary precursor for the later development of a convectively driven system.

Van Delden in his paper about CISK and the growth of polar lows by diabatic heating considers the adjustment processes during a CISK-forced polar low development. Recently Emanuel and Rotunno (see their paper "Polar Lows as Arctic Hurricanes") have argued against CISK as a mechanism for tropical and polar low cyclogenesis. Van Delden in his work argues against some of Emanuel and Rotunno's *ideas and defends the original CISK concept at least as an explanation for the intensification during the later stages of convective polar low developments. In accordance with other authors, van Delden points out that the development of a polar low takes place through various stages, and that "CISK cannot serve as an explanation of the genesis and initial growth."*

Craig and Cho start in their paper by observing that most investigations of the physical processes responsible for polar low formation in the past have been centered around one of two mechanisms, i.e., either CISK or baroclinic instability. They proceed to explore the combined effect of heating and a baroclinic environment for the development of a small-scale disturbance. They find that baroclinic and convective processes may interact, and provided that the diabatic heating rate from the convection is not too large the resulting sufficient amount of convective diabatic heating the disturbance will resemble a pure CISK phenomenon. For intermediate values of the heating, hybrid or transitional disturbances will result. The model results are in good agreement with the observations since satellite images often show systems that best can be characterized as hybrid types.

Moore and Peltier's paper "On the Development of Polar Low Wavetrains," reviews their theory of nonseparable baroclinic instability and its application to frontal cyclogenesis. Using the theory they investigate the stability of a realistic atmospheric frontal structure to arbitrary three-dimensional perturbations. The quasi-geostrophic approximation is not invoked, permitting for example the static stability to vary in the horizontal. One of the main results of this new theory is the demonstration of the existence of a new short wave branch of unstable normal modes not allowed by the quasi- or semi-geostrophic theories. Moore and Peltier re-analyze a reverse-shear polar low case considered earlier by Reed and Duncan using a quasi-geostrophic model and conclude that their "results are far more robust and in closer accord with the observations."

One of the controversial questions around the end of the seventies and the early eighties was whether (some) polar lows were small-scale extratropical hurricanes. In the paper "Polar Lows as Arctic Hurricanes," Emanuel and Rotunno consider this basic question by means of a recently developed theoretical model and by numerical simulations with an axisymmetric nonhydrostatic model. The authors argue against CISK as a mechanism for the formation of polar lows (as well as a mechanism for tropical cyclogenesis) and continue by proposing an air-sea interaction instability (ASII) theory. One of the important conclusions of this study is that the polar low environment appears to be stable to small amplitude axisymmetric perturbations, and that disturbances of a substantial amplitude seem to be necessary to initiate intensification by air-sea interaction.

Assuming that tropical cyclones and some polar lows are driven by the same basic mechanism, Hans Økland in his paper "On the Genesis of Polar Lows" considers the intriguing question why such convectively driven systems form in the tropics and in arctic regions but not in middle latitudes. Økland points out that deep convection seems to be a characteristic feature of especially small and intense polar lows. He stresses also the importance of upper level tropospheric troughs or small-scale vortices, which he believes to be "the nucleus for the polar low development rather than the cyclonic vorticity associated with the (surface) trough."

This chapter is closed by a discussion of a very fundamental question that has puzzled many meteorologists: Why and how do some polar lows acquire a spiral cloud configuration as that seen on the cover [frontispiece] of this book. William Raymond in his discussion of the problem performs an instability study by means of an axisymmetric barotropic model in a vertical plane and a normal mode approach. Based on a series of numerical experiments Raymond concludes that "comma vorticity configurations can be generated as a consequence of vortex stretching" and that some of the growing modes take on the familiar comma-spiral pattern.

ON THE PRECURSORS OF POLAR LOWS

A. Wiin-Nielsen
Geophysical Institute
University of Copenhagen
Copenhagen, Denmark

ABSTRACT

The investigation explores the possibility that the precursor for the formation of polar lows, i.e., a baroclinic wave with maximum amplitude in the lower layers of the atmosphere and with the scale of a few hundred kilometers, can be formed by a baroclinic instability mechanism.

For this purpose a three-level baroclinic model is used as a minimum model since observations show that the characteristic vertical thermal structure consists of a lower layer with a lapse rate approaching the adiabatic stratification and an upper layer with a normal lapse rate. The basic state consists then of such a thermal stratification and two thermal winds, one in a lower and one in a higher layer.

The results show that the basic state described above may be baroclinically unstable with a minimum e-folding time at a wavelength of some hundreds of kilometers depending on how close the lapse rate in the lower layer is to the adiabatic lapse rate.

The structure of the baroclinic waves is investigated.

1. INTRODUCTION

Polar lows have been studied extensively over the past two decades or so. It is generally recognized that these low pressure systems occur on a smaller scale than the waves in the free atmosphere, which according to baroclinic stability theory have a wavelength of 3000–6000 km.

POLAR AND ARCTIC LOWS
Paul F. Twitchell, Erik A. Rasmussen,
and Kenneth L. Davidson (Eds.)

85

In the description of polar lows from satellite and other data (Rasmussen, 1985, 1988) it is pointed out that at least one type of polar lows is generated in connection with the passage of an arctic airmass over the polar ice edge into a region covered with somewhat warmer ocean water. The lower part of the airmass is apparently modified rather rapidly by the ocean. Through transfer of sensible heat from the ocean to the atmosphere a lower layer next to the water surface attains a stratification that is close to the adiabatic lapse rate. It is in this airmass that the rather shallow polar low is formed in a matter of some hours.

During the further development even smaller intense vortices may form in the already existing polar low. The smaller vortices are presumably created by enhanced air-sea interaction through transfer of sensible heat and evaporation, but in this paper we shall be concerned only with the formation of the initial low pressure system.

The question is how it is formed. On the one hand we may apparently exclude the standard form of baroclinic instability because the resulting scale of the most unstable baroclinic wave would be too large. On the other hand, a baroclinic zone exists certainly along the edge of the polar ice with low temperatures over the ice and somewhat higher temperatures over the adjacent ocean, giving a thermal wind increasing with height along the ice edge. Since the lower layer is rapidly modified to attain a nearly adiabatic stratification it is possible that the total atmosphere consisting of a statically stable layer on top of the rather shallow almost adiabatic layer may have stability properties that can explain the formation of rather short baroclinic waves. If so, we would have a possible mechanism for creation of the first phase of the polar low.

The mechanism just described has in fact been explored by Blumen (1979), who has used a generalisation of the original model by Eady (1952) to contain two layers. He finds that such a model gives a maximum baroclinic instability of much lower wavelengths when the lower layer has a stratification close to an adiabatic lapse rate. Blumen (1979), however, does not investigate the structure of these shorter baroclinic waves. It is therefore worthwhile to expand the investigation to consider these aspects.

The role of baroclinic instability in the formation of polar lows originated with Harrold and Browning (1969) who considered that the low developed in a shallow frontal zone, and that its small size is due to the small depth in agreement with baroclinic instability theory by Eady (1949) and Green (1960). Mansfield (1974) investigated this further, but these investigations do not seem to model the polar low in sufficient detail since they are in fact rather deep. Duncan (1977) used a quasi-geostrophic model to find normal mode solutions for unstable disturbances with small static stability in the lower layer and a low-level baroclinic zone. He found also shallow disturbances.

Sardie and Warner (1983) employed a three-level, linear model that included effects of surface friction and stable and convective latent heat release. The model is thus an attempt to resolve the controversy regarding the relative role of baroclinic instability and the conditional instability of a second kind (CISK) mechanism proposed by Ramussen (1979), but it is difficult to determine this relative role in a model including both effects. In a later study (1985) Sardie and Warner used a fully developed mesoscale model to simulate a Pacific and an Atlantic case of polar low development. They found that the purely baroclinic (dry) effects accounted for much of the development in the Pacific case but that the CISK mechanism was needed to obtain the full development, but also that the development in the model was very sensitive to the shape of the vertical heating profile. The Pacific case is thus in this investigation an example of a polar low development in a deep baroclinic zone. The less successful simulation of the Atlantic case, which was supposed to be an example of the more shallow baroclinic zone, leaves the question unanswered. In this regard it is useful to consider Reed's (1986) comments that in the Atlantic case there was at most a brief association with the shallow baroclinic zone at the ice boundary, and further that the low was in a filling stage for the major part of the investigation.

As pointed out by Rasmussen (1985) and further elaborated by Emanuel (1986) the development of a polar low may better be described as a two-stage air-sea interaction process where baroclinic instability (or some other type of instability) dominates in the first stage, while the second stage is strongly influenced by enhanced air-sea interaction leading to a rapid spin up creating the inner core of the smaller lows forming in the now existing baroclinic wave. While the second stage is treated in detail by Emanuel (1986), in this paper we shall treat the first stage. The model will be a linearized three-level quasi-nondivergent model where we shall neglect friction and heating to obtain the baroclinic instability in a pure form. The three-level model is chosen as the minimal vertical resolution, which permits two static stabilities and we shall thus be able to simulate the stability of the basic state where a shallow layer near the surface has become nearly adiabatic through rapid warming from below.

2. THE MODEL

The three-level quasi-nondivergent model has been used in several investigations (Cressman, 1961; Wiin-Nielsen, 1961). The only modifications that will be made here are that we shall let the levels be arbitrary and not necessarily equidistant in pressure and that the beta-effect will be neglected. We shall refer to the levels at which the streamfunction and the vorticity are carried by subscripts 1, 3, and 5. The vertical boundaries will be given by subscripts 0 and 6 while 2 and 4 will be the levels at which the vertical velocity and the temperature can be calculated (see Figure 1). Adopting the boundary conditions that $\omega = dp/dt$ vanishes at levels 0 and 6 we may write the vorticity equations in the form:

$$\frac{\partial \zeta_1}{\partial t} + \vec{V}_1 \cdot \nabla \zeta_1 = \frac{f_0}{p_2} \, \omega_2$$

$$\frac{\partial \zeta_3}{\partial t} + \vec{V}_3 \cdot \nabla \zeta_3 = f_0 \, \frac{\omega_4 - \omega_2}{p_4 - p_2} \qquad (2.1)$$

$$\frac{\partial \zeta_5}{\partial t} + \vec{V}_5 \cdot \nabla \zeta_5 = - f_0 \, \frac{\omega_4}{p_6 - p_4}$$

where the notations are:

$\zeta = \nabla^2 \psi$, the vorticity
$\psi =$ the streamfunction
$\vec{V} = \vec{K} \times \Delta \psi$, the horizontal nondivergent wind
f, the Coriolis parameter
f_0, a standard value of f
$\omega = dp/dt$, the vertical velocity
p, the pressure
t, the time

We introduce the two thermal fields with subscripts T and B by the relations

$$(\ )_T = (\ )_1 - (\ )_3, \ (\ )_B = (\ )_3 - (\ )_5 \qquad (2.2)$$

and we find using Eqs. (2.1) and (2.2):

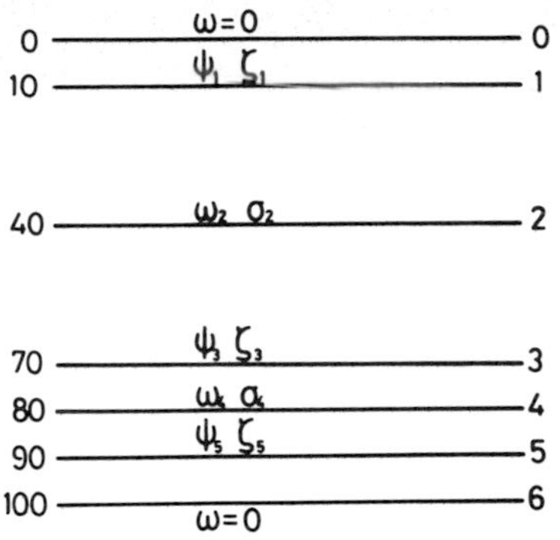

Figure 1. The distribution of the variables at the various levels in the model employed in the study.

$$\frac{\partial \zeta_T}{\partial t} + \vec{V}_3 \cdot \nabla \zeta_T + \vec{V}_T \cdot \nabla \zeta_3 + \vec{V}_T \cdot \nabla \zeta_T$$

$$= \frac{f_0 p_4}{p_2 (p_4 - p_2)} \omega_2 - \frac{f_0}{p_4 - p_2} \omega_4$$

(2.3)

$$\frac{\partial \zeta_B}{\partial t} + \vec{V}_3 \cdot \nabla \zeta_B + \vec{V}_B \cdot \nabla \zeta_3 + \vec{V}_B \cdot \nabla \zeta_B$$

$$= \frac{f_0 (p_6 - p_2)}{(p_6 - p_4)(p_4 - p_2)} \omega_4 - \frac{f_0}{p_4 - p_2} \omega_2$$

From the thermodynamic equation applied at levels 2 and 4 we find

$$\omega_2 = \frac{f_0}{(p_3 - p_1) \sigma_2} \left(\frac{\partial \psi_T}{\partial t} + \vec{V}_3 \cdot \nabla \psi_T \right)$$

(2.4)

$$\omega_4 = \frac{f_0}{(p_5 - p_3) \sigma_4} \left(\frac{\partial \zeta_B}{\partial t} + \vec{V}_3 \cdot \nabla \zeta_B \right)$$

Inserting Eq. (2.4) in Eq. (2.3) we find the equations:

$$\frac{\partial}{\partial t} \left(\zeta_T - r_2^2 \psi_T + q_4^2 \psi_B \right) + \vec{V}_3 \cdot \nabla \left(\zeta_T - r_2^2 \psi \right.$$

$$\left. + q_4^2 \psi_B \right) + \vec{V}_T \cdot \nabla (\zeta_3 + \zeta_T) = 0$$

(2.5)

$$\frac{\partial}{\partial t} \left(\zeta_B - r_4^2 \psi_B + q_4^2 \psi_B \right) + \vec{V}_3 \cdot \nabla \left(\zeta_B - r_4^2 \psi_B + q_4^2 \psi_T \right)$$

$$+ \vec{V}_B \cdot \nabla (\zeta_3 - \zeta_B) = 0$$

in which

$$q_2^2 = \frac{f_0^2}{\sigma_2} \frac{1}{(p_4 - p_2)(p_3 - p_1)} ; q_4^2 = \frac{f_0^2}{\sigma_4} \frac{1}{(p_4 - p_2)(p_3 - p_1)}$$

(2.6)

$$r_2^2 = \frac{p_4}{p_2} q_2^2 ; r_4^2 = \frac{(p_6 - p_2)}{(p_6 - p_4)} q_4^2$$

A further equation is obtained from the middle equation in (2.1). It is:

$$\frac{\partial}{\partial t} \left(\zeta_3 - q_4^2 \psi_B + q_2^2 \psi_T \right) + \vec{V}_3 \cdot \nabla \left(\zeta_3 - q_4^2 \psi_B + q_2^2 \psi_T \right)$$ (2.7)

We consider the system Eqs. (2.5) and (2.7). On this system we shall perform a perturbation analysis on a steady state defined by a zonal flow with the winds U_3, U_T, and U_B. The perturbations will be of the form

$$(\psi_3, \psi_T, \psi_B) = (\hat{\psi}_3, \hat{\psi}_T, \hat{\psi}_B) \exp[ik(x-ct)] \qquad (2.8)$$

Denoting $x=c-U_3$ $\lambda^2=q^2/k^2$ and $\mu^2=r^2/k^2$ we find by substitution of Eq. (2.8) in Eqs. (2.5) and (2.7):

$$\begin{bmatrix} \lambda_2^2 U_T - \lambda_4^2 U_B - x & \lambda_2^2 x & -\lambda_4^2 x \\ (1-\mu_2^2)U_T + \lambda_4^4 U_B & U_T - (1+\mu_2^2)x & \lambda_4^2 x \\ (1-\mu_4^2)U_T + \lambda_2^2 U_T & \lambda_2^2 x & -U_B - (1+\mu_4^2)x \end{bmatrix} \begin{bmatrix} \hat{\psi}_3 \\ \hat{\psi}_T \\ \hat{\psi}_B \end{bmatrix} = 0$$

$$(2.9)$$

If Eq. (2.9) is going to have nontrivial solutions it is a necessity that the determinant shall vanish. This condition leads to a cubic equation in x, which we shall solve. The equation is:

$$a_3 x^3 + a_2 x^2 + a_1 x + a_0 = 0 \qquad (2.10)$$

where

$$a_3 = (1 + \mu_2^2)(1 + \mu_4^2) - \lambda_2^2\lambda_4^2$$

$$a_2 = [(1 + 2\lambda_4^2)(1 + \mu_2^2) - 2\lambda_2^2\lambda_4^2] U_B$$

$$\quad - [(1 + 2\lambda_2^2)(1 + \mu_4^2) - 2\lambda_2^2\lambda_4^2] U_T \qquad (2.11)$$

$$a_1 = \lambda_4^2(1 + \mu_2^2 - \lambda_2^2) U_B^2$$

$$\quad + \lambda_2^2(1 + \mu_4^2 - \lambda_4^2) U_T^2$$

$$\quad - (1 + 2\lambda_2^2 + 2\lambda_4^2) U_B U_T$$

$$a_0 = U_B U_T (\lambda_2^2 U_T - \lambda_4^2 U_B)$$

Before we consider a number of cases it will be instructive to use Eq. (2.10) in the normal case where the spacing of the isobaric surfaces are equidistant in pressure. In that case we have:

$$p_i = \frac{i}{6} p_0 \tag{2.12}$$

and thus

$$q_2^2 = 9 \frac{f_0^2}{\sigma_2 p_0^2}; \quad q_4^2 = 9 \frac{f_0^2}{\sigma_4 p_0^2} \tag{2.13}$$

and

$$r_2^2 = 2q_2^2; \quad r_4^2 = 2q_4^2 \tag{2.14}$$

Eq. (2.11) becomes in this case:

$$a_3 = (1 + 2\lambda_2^2)(1 + 2\lambda_4^2) - \lambda_2^2\lambda_4^2$$

$$a_2 = (1 + 2\lambda_2^2 + 2\lambda_4^2 + 2\lambda_2^2\lambda_4^2)(U_B - U_T)$$

$$a_1 = \lambda_4^2(1 + \lambda_2^2)U_B^2 + \lambda_2^2(1 + \lambda_4^2)U_T^2 \tag{2.15}$$

$$\qquad - (1 + 2\lambda_2^2 + 2\lambda_4^2)U_BU_T$$

$$a_0 = U_BU_T(\lambda_2^2U_T - \lambda_4^2U_B)$$

For the considerations to follow it is furthermore convenient to have a specific stability stratification in mind. We shall in this regard assume that the layer from level 6 ($p=p_0$) to level 3 is characterized by a constant lapse rate γ_B, while the remaining layer from p_3 to $p=0$ has the lapse rate γ_T. The stability factor σ has for a constant lapse rate layer the form (Jacobs and Wiin-Nielsen, 1966)

$$\sigma = \frac{R^2T_0}{gp_0^2} (\gamma_d - \gamma)p_*^{-[2-(R\gamma/g)]}; \quad p_* = \frac{p}{p_0} \tag{2.16}$$

We may then calculate that

$$\sigma_4 = \frac{R^2T_0}{gp_0^2} (\gamma_d - \gamma_B)p_*^{-[2-(R\gamma_B/g)]} \tag{2.17}$$

and that

$$\sigma_2 = \frac{R^2 T_0}{g p_0^2} (\gamma_d - \gamma_T) p_{*3}^{(R/g)(\gamma_B - \gamma_T)} \, p_{*2}^{-[2-(R\gamma_T/g)]} \tag{2.18}$$

3. SOME SPECIAL CASES

Before the treatment of the general case we shall in this section consider a few special cases. For simplicity we shall select the model with an arrangement of the levels at equidistant pressure surfaces in which case Eqs. (2.12) to (2.15) apply.

As the first case we shall consider a situation in which $\gamma_B = \gamma_d$ indicates an adiabatic stratification of the lower layer. This means that q_4^2 will go to infinity and so will λ_4^2. We divide Eq. (2.10) by λ_4^2 and find for $\lambda_4^2 \to \infty$ that

$$a_3 \to 2 + 3\lambda_2^2;$$

$$a_2 \to 2(1 + \lambda_2^2)(U_B - U_T); \tag{3.1}$$

$$a_1 \to (1 + \lambda_2^2) U_B^2 + \lambda_2^2 U_T^2 - 2 U_B U_T$$

$$a_0 \to -U_T U_B^2$$

If we furthermore assume that $U_B = U_T = U_s$ we find for $z = x/U_s$ that z will satisfy the equation

$$z^3 + \frac{2\lambda_2^2 - 1}{3\lambda_2^2 + 2} z - \frac{1}{3\lambda_2^2 + 2} = 0 \tag{3.2}$$

This equation will, according to the classical theory, have complex roots if

$$\frac{1}{4} \frac{1}{(3\lambda_2^2 + 2)^2} + \frac{1}{27} \frac{(2\lambda_2^2 - 1)^3}{(3\lambda_2^2 + 2)^3} > 0 \tag{3.3}$$

By a simple evaluation it is seen that this inequality is satisfied for all values of $\lambda_2{}^2 > 0$, which is always the case. We have thus shown that all waves are unstable in this case.

This result says that a stable layer on top of an adiabatic layer leads to instability for all wave lengths in the absence of the beta effect.

A similar result is obtained if we reverse the property of the two layers in such a way that the adiabatic layer is on top of a stably stratified layer, or, to be exact, if $\lambda_2{}^2 \to \infty$ and $\lambda_4{}^2 > 0$, in which case we get an equation for the eigenvalues Z corresponding to Eq. (3.3) with $\lambda_2{}^2$ replaced by $\lambda_4{}^2$. It is thus equally true that an adiabatic layer on top of a stable stratified layer leads to instability for all wavelengths under the present assumptions.

These two results are naturally rather unrealistic. We would seldom or never experience adiabatic layers of the thickness assumed here. However these results indicate that we may possibly get similar and more realistic results in less extreme and more realistic cases.

A simple situation is one where $U_B = U_T = U_S$ and $\lambda_2{}^2 = \lambda_4{}^2 = \lambda_m{}^2$. In this case the three-level model should give results compatible with the two-level model without the beta effect. The equation corresponding to Eq. (3.2) is now

$$(1 + 4\lambda_m{}^2 + 3\lambda_m{}^4)Z^3 - (1 + 2\lambda_m{}^2 - 2\lambda_m{}^4)\, Z = 0 \tag{3.4}$$

with the roots

$$Z = 0; \; Z = {+ \atop -}\left(\frac{1 + 2\lambda_m{}^2 - 2\lambda_m{}^4}{1 + 4\lambda_m{}^2 + 3\lambda_m{}^4} \right)^{1/2} \tag{3.5}$$

The denominator in Eq. (3.5) is always positive for a stable stratification, while the numerator is negative for

$$\lambda_m{}^2 > \tfrac{1}{2}(1 + \sqrt{3})$$

Converted into wavelength, measured in the unit 10_6 and using a standard value $\sigma = \sigma_m$, we find the inquality above corresponds to instability for

$$1 > \frac{2\pi}{3}\left[\tfrac{1}{2}(1 + \sqrt{3})\right]^{\tfrac{1}{2}} \sqrt{\sigma_m} = 4.240 \text{ for } \sigma_m = 3$$

indicating the normal scale of baroclinic instability.

Many other special cases can be considered. Suppose for example that we assume that $U_T=0$. In that case we have a wind profile where the temperature gradient is in the lower half of the atmosphere, while the zonal wind is constant above 50 kPa. With $x=U_Bz$ we find in this case that Z satisfies the equation

$$[1 + 2(\lambda_2^2 + \lambda_4^2) + 3\lambda_2^2\lambda_4^2]Z^3 + [1 + 2(\lambda_2^2$$
$$+ \lambda_4^2) + 2\lambda_2^2\lambda_4^2]\, Z^2 + \lambda_4^2(1 + \lambda_2^2)\, Z = 0 \tag{3.7}$$

One root in Eq. (3.7) is $Z=0$ while the other two roots are obtained from the quadratic equation obtained from Eq. (3.7). Complex roots are formed if the discriminant of the quadratic equation is negative. The condition leads to the inquality

$$\lambda_4^4 > \frac{1 + 2\lambda_2^2}{4(1 + \lambda_2^2)} \tag{3.8}$$

Using $\gamma_B=\gamma_T=6.5\times10^{-3} \text{ km}^{-1}$ we may calculate σ_2 and σ_4 from Eq. (2.17) and Eq. (2.18) q_2^2 and q_4^2 from Eq. (2.13) and use $\lambda^2=q^2/k^2$ to calculate the terms in Eq. (3.8). It is then found that Eq. (3.8) is satisfied for $1\geq2.012$ or about 2012 km.

Otherwise, if we take $U_B=0$, we get with $x=U_TZ$ the equation:

$$[1 + 2(\lambda_2^2 + \lambda_4^2) + 3\lambda_2^2\lambda_4^2]Z^3 - [1 + 2(\lambda_2^2 + \lambda_4^2)$$
$$+ 2\lambda_2^2\lambda_4^2]Z^2 + \lambda_2^2(1 + \lambda_4^2)Z = 0 \tag{3.9}$$

which has the root $Z=0$. The additional two roots are complex if

$$\lambda_2^4 > \frac{1 + 2\lambda_4^2}{4(1 + \lambda_4^2)} \tag{3.10}$$

Using the same values as before we find that Eq. (3.10) is satisfied for $1 > 4.166$ or 4166 km.

As expected we get instability both when the wind shear is in the upper or in the lower half of the atmosphere, but the instability occurs at a smaller wavelength when the wind shear is found in the lower half of the atmosphere.

After these special cases we consider the normal case for the three-level model where the levels are equidistant in pressure. As we have seen in one of the special cases, we shall have baroclinic instability for all wavelengths if either σ_2 or σ_4 are zero. This case corresponds to what is sometimes called an ultra-violet catastrophe because the e-folding time will go to zero as the wavelength approaches zero. This situation is naturally to be avoided. Nevertheless, according to Rasmussen (1988), the polar low tends to develop in a situation where the lower part of the atmosphere has an almost adiabatic stratification, probably created by air-sea interaction as the cold air mass leaves the ice-covered ocean and is heated from below through transfer of sensible heat from the somewhat warmer ocean surface to the atmosphere. We simulate this situation by assuming that γ_B is reasonably close to $\gamma_d = 10^{-2}$ km^{-1}, while γ_T is assumed to be normal, say $\gamma_T = 6.5 \times 10^{-3}$ km^{-1}.

The problem of stability is solved by finding the roots of Eq. (2.10), where the coefficients in this equation are given by Eq. (2.15). The method of solution is straightforward by using the classical formulas for the roots of the cubic equation. At this point we shall restrict ourselves to showing the curve $T_e = T_e(1)$ where the e-folding time is defined by

$$T_e = (kC_i)^{-1} \tag{3.11}$$

In each of the following four cases we have used $U_B = U_T = 20$ m s^{-1} and $\gamma_T = 6.5 \times 10^{-3}$ km^{-1}. Otherwise the values are

(1): $\gamma_B - 9.75 \times 10^{-3}$ km^{-1}
(2): $\gamma_B = 9.9 \times 10^{-3}$ km^{-1}
(3): $\gamma_B = 9.99 \times 10^{-3}$ km^{-1}
(4): $\gamma_B = 1 \times 10^{-2}$ km^{-1}

The results are shown in Figure 2, which gives the e-folding time in days. The curve for case (4) shows $T_e = 0$ for $l = 0$. In the other three cases we find

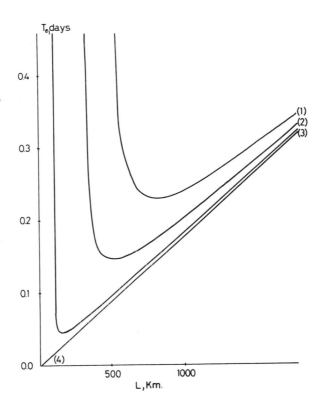

Figure 2. Curves showing the e-folding time as a function of wavelength in four cases. In all cases we have $U_B = U_T = 20$ m s^{-1} and $\gamma_T = 6.5 \times 10^{-3}$ km^{-1} while γ_B has the following values: (1) 9.75×10^{-3} km^{-1}, (2) 9.9×10^{-3} km^{-1}, (3) 9.99×10^{-3} km^{-1}, and (4) 10.0×10^{-3} km^{-1}. The levels are in the standard location for a three-level model.

in each case a minimum in T_e and a vertical asymptotic value at a finite value of l. The minimum corresponds to the wavelength of maximum instability. These wavelengths are about 800 km for (1), 500 km for (2), and 150 km for (3). We may thus conclude that the three-level model with γ_B close to γ_d has a short-wave baroclinic instability with maximum instability for a wavelength of a few hundred kilometers and e-folding times of a rather small fraction of a day (0.05 to 0.25 days).

These cases, shown in Figure 2, are not totally convincing although they look promising for our purposes. The reasons are that we have assumed that the whole lower half of the atmosphere has an almost adiabatic stratification, and that the wind profile is linear. Both of these assumptions may be unrealistic in comparison with actual situations. In the next section we shall relax these rather severe assumptions by adapting the three-level model with arbitrary levels to more realistic cases.

4. THE GENERAL CASE

In this section we shall investigate the stability and structure of the waves based on the more general formulation given in Section 2 particularly Eqs. (2.9)–(2.11). To simulate the basic state typical for the formation of the initial low pressure system in which the smaller systems develop later, we shall select the pressure levels in the model with $p_1 = 10$, $p_2 = 40$, $p_3 = 70$, $p_4 = 80$, and $p_5 = 90$ kPa. This means that the static stabilities are used at 80 and 40 kPa, where the first is representing the layer 70 to 100 kPa and the second the remaining part of the atmosphere. To simulate the real situation even closer one should select the lower layers nearer to the Earth's surface. An example would be the following selection 30, 60, 90, 93⅓, and 96⅔ kPa in which case the lower layer would have a static stability representing only a layer of 10 kPa. It is felt that such a selection would make the vertical truncation error too extreme in the upper part.

We start by considering changes in the lapse rate γ_B keeping $\gamma_T = 6.5 \times 10^{-3}$ km^{-1} in all cases. For this purpose we use $U_B = 8$ m s^{-1} and $U_T = 24$ m s^{-1}, which, considering the selection of levels (see Figure 1), corresponds to a linear wind profile. Three cases were computed $\gamma_B = 9.99 \times 10^{-3}$, $\gamma_B = 9.9 \times 10^{-3}$, and $\gamma_B = 9 \times 10^{-3}$ km^{-1}. The computed e-folding times measured in days are shown in Figure 3. The main difference is found for small values of the wavelength where the minimum value of T_e is found at larger values of the wavelength the smaller γ_B is. For the largest value of γ_B the minimum value of T_e is found at $L = 100$ km, while T_e is a minimum at $L = 500$ km for the middle value of γ_B. The smallest value of γ_B shows a minimum of T_e at $L = 1000$ km. We may thus conclude that γ_B should be reasonably close to the adiabatic lapse rate to get a maximum instability for wavelengths smaller than 1000 km. This means that a sufficient condition for the formation of the initial low is that an almost adiabatic layer is present close to the ground. Figure 3 shows also that two regions of instability exist. A short wave region, highly sensitive to the values of γ_B, and a longer wave regime corresponding to the well-known region of baroclinic instability.

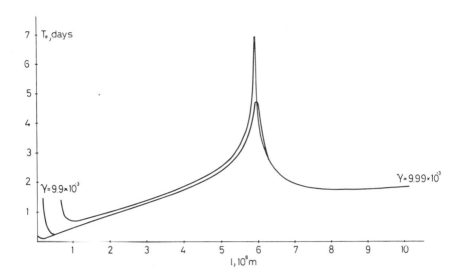

Figure 3. The e-folding time as a function of wavelength in these cases, all having $U_B=8$ m s^{-1}, $U_T=24$ m s^{-1}, $\gamma_T=6.5\times10^{-3}$ km^{-1} and $p_i=10$, $p_2=40$, $p_3=70$, $p_4=80$, $p_5=90$ kPa. γ_B has the values: 9.99×10^{-3} km^{-1}, 9.9×10^{-3} km^{-1} and 9×10^{-3} km^{-1}.

Variations of the total wind shear, which in Figure 3 is 40 m s^{-1}, from 100 to 0 kPa have the expected effect that a decrease in the wind shear increases the e-folding time. However, in this model there does not exist a minimum wind shear below which all waves are stable. As an example we may select $U_B=0.5$ m s^{-1}, $U_T=1.5$ m s^{-1}, $\gamma_B=9.99\times10^{-3}$ km^{-1}, and $\gamma_T=6.5\times10^{-3}$ km^{-1}. The result is shown in Figure 4 indicating the T_e in general is very large, but the minimum not seen in Figure 4 occurring at a small value of L is still as small as 1 day.

The next series of calculations deals with the so-called reversed wind shear. Four calculations were done as follows:

(1) $U_B = 6$ m s^{-1}, $U_T = 18$ m s^{-1}
(2) $U_B = 6$ m s^{-1}, $U_T = -18$ m s^{-1}
(3) $U_B = -6$ m s^{-1}, $U_T = 18$ m s^{-1}
(4) $U_B = -6$ m s^{-1}, $U_T = -18$ m s^{-1}

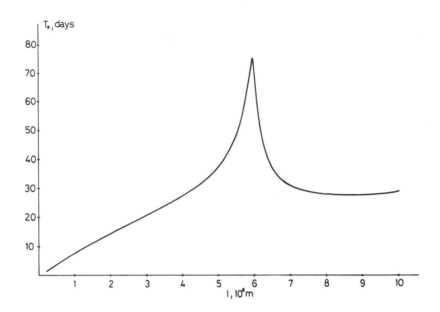

Figure 4. The e-folding time as a function of wavelength. Parameters:
$U_B=0.5$ *m* s^{-1}, $U_T=1.5$ *m* s^{-1}, $\gamma_T=6.5\times10^{-3}$ *km*$^{-1}$, $\gamma_B=9.9\times10^{-3}$ *km*$^{-1}$.

Figure 5 shows the resulting e-folding times. As expected calculation (1) is similar to the curves in Figure 3. It turns out also that $T_e(1) = T_e(4)$, which means that a basic current decreasing with height is just as unstable as one increasing with the same rate. The reversed wind shear introduced in the upper layer in (2) simulates a situation often found in reality (see Rasmussen, 1988, Figure 3). The sharp temperature gradient along the ice edge creates a lower baroclinic zone with a positive wind shear, while the basic current decreases aloft. For the sake of completeness, calculation (3) was also made. Calculations (2) and (3) turn out to be identical with respect to the e-folding time T_e. For small wavelengths we find that all four calculations are almost identical giving a minimum at $L=300$ km with $T_e=0.3$ days. However, calculations (2) and (3) with the reversed wind shears have a clear separation between the two regions of instability containing a band of stable states between $L\approx3200$ km and $L\approx4600$ km.

The fact that the e-folding times in cases (1) and (4) are the same, and that the two situations thus are equally unstable, does not mean that the waves are identical. One may study the difference by computing the relative structure of

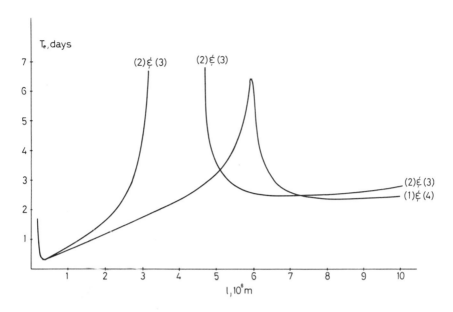

Figure 5. The e-folding time as a function of wavelength in four cases all having $\gamma_T=6.5\times10^{-3}$ km^{-1} and $\gamma_B=9.9\times10^{-3}$ km^{-1}. The thermal winds are as follows: (1) $U_B=6$ m s^{-1}, $U_T=18$ m s^{-1}; (2) $U_B=6$ m s^{-1}, $U_T=-18$ m s^{-1}; (3) $U_B=-6$ m s^{-1}, $U_T=18$ m s^{-1}; and (4) $U_B=-6$ m s^{-1}, $U_T=-18$ m s^{-1}.

the two baroclinic waves. This is done by returning to Eq. (2.9), knowing the eigenvalues, and proceeding to calculate the waves at, for instance, levels 1 and 5 assuming that the wave at level 3 is known. The results of such a calculation can for each case be expressed as relative amplitudes r_1-R_1/R_3, $r_5=R_5/R_3$ and relative phases θ_1 and θ_5 (assuming that $\theta_3=0$). The phase angles are expressed as the position of the ridge of the wave.

For case (1) in which we have a positive shear in both layers the curves $\theta_5(1)$ and $\theta_1(1)$ are given in Figure 6 as functions of the wavelength with $\theta_3(1)=0$. We may note immediately that for $l>5.6$ we find the usual structure of a long baroclinic wave that slopes westward with height because $\theta_5(1)<0$ and $\theta_1(1)<0$. For the short waves in which we are especially interested in this study we find both phase angles to be positive. This means that these short baroclinic waves slope westward in the lower layer and eastward in the upper layer (assuming here that U_B and U_T blow from west, which is

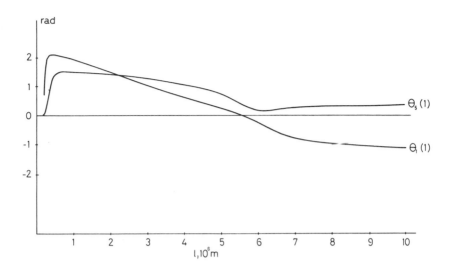

Figure 6. The phase differences as a function of wavelength for the case:
$U_B=6 \ m \ s^{-1}$, $\qquad U_T=18 \ m \ s^{-1}$, $\qquad \gamma_B=9.9\times10^{-3} \ km^{-1}$ $\qquad$ *and*
$\gamma_T=6.5\times10^{-3} \ km^{-1}$.

not a requirement since the Rossby parameter beta is zero). We may thus say that the short baroclinic waves transport sensible heat northward in the lower layer and southward in the upper layer. This is true because the heat transport in the lower layer may be written:

$$HT_8 = \frac{1}{2} \ \frac{C_p}{R} \ \frac{f_0}{g} \ \frac{p_5 + p_3}{p_5 - p_3} \ kr_5 \ \sin\theta_5$$

while the transport in the upper layer is

$$HT_T = \frac{1}{2} \ \frac{C_p}{R} \ \frac{f_0}{g} \ \frac{p_3 + p_1}{p_3 - p_1} \ kr_1 \ \sin\theta_1$$

The relative amplitudes r_1 and r_5 are shown as functions of the wavelength in Figure 7. We note that for the short waves r_5 is much larger than r_1 indicating that the maximum amplitude is in the lower level for the short waves. With our choice of levels we find also that

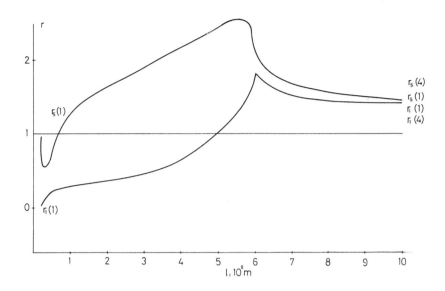

Figure 7. The relative amplitudes as a function of wavelength for the cases
$U_B=6 \ m \ s^{-1}$, $\quad U_T=18 \ m \ s^{-1}$, $\quad and \quad U_B=-6 \ m \ s^{-1}$. $\quad In \quad both \quad cases$
$\gamma_B=9.9\times10^{-3} \ km^{-1}$, $\gamma_T=6.5\times10^{-3} \ km^{-1}$.

$$\frac{p_3 + p_5}{2(p_5 - p_3)} = 4; \frac{p_3 + p_1}{2(p_3 - p_1)} = \frac{3}{4}$$

Based on these numbers and the curves in Figure 6 and Figure 7 one may easily calculate that the total heat transport in the layer from p_1 to p_5 is positive (northward).

For case (2) ($U_B=6$ m s^{-1}, $U_T=-18$ m s^{-1}) we shall limit ourselves to the short waves with wavelengths below the interval of neutral waves, i.e., $L<3300$ km. Figure 8 shows the phase differences as a function of wavelength. In this case obviously the wave at low level (5) is in front of the wave at level 3, which in turn is ahead of the wave at level 1. The wave slopes therefore westward through the whole depth of the atmosphere, and it is thus transporting sensible heat northward in the same way as the longer Rossby wave. Contrary to these waves it has its largest amplitude at the lower levels as seen from Figure 9 showing that r_5 is larger than r_1, for all wavelengths. Thus, seemingly, the situation with the negative or reversed wind shear on top of the positive wind shear close to the Earth's surface creates a deep baroclinic wave.

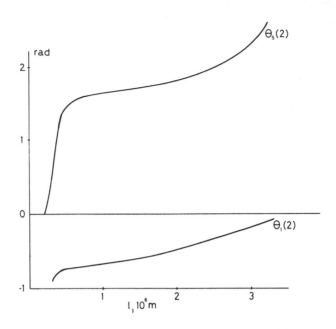

Figure 8. *The phase differences as a function of wavelength for the case:*
$U_B=6\ m\ s^{-1}$, $U_T=-18\ m\ s^{-1}$, $\gamma_B=9.9\times10^{-3}\ km^{-1}$, $\gamma_T=6.5\times10^{-3}\ km^{-1}$.

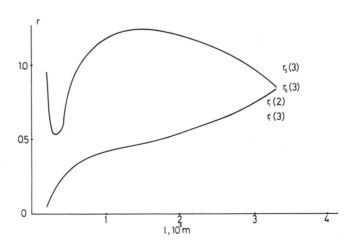

Figure 9. *The relative amplitudes as a function of wavelength for both cases:*
$U_B=6\ m\ s^{-1}$, $U_T=-18\ m\ s^{-1}$ and $U_B=-6\ m\ s^{-1}$, $U_T=18\ m\ s^{-1}$. *In both
cases:* $\gamma_B=9.9\times10^{-3}\ km^{-1}$, $\gamma_T=6.5\times10^{-3}\ km^{-1}$.

Case (3) where $U_B=6$ m s^{-1} and $U_T=+18$ m s^{-1} does not simulate the usual ice edge simulation. The amplitude ratios are identical in cases (2) and (3) as indicated in Figure 9. The relative phase angles in Figure 10 show an eastward slope with height, but we do not expect that this case will occur in connection with the polar low.

5. CONCLUSIONS

The investigation deals with the precursor for the formation of a polar low. According to synoptic and other studies the formation may be considered as a two-stage process where the first stage is the creation of a polar depression of a few hundred kilometers in dimension. Within this rather shallow low the real polar lows of an even smaller dimension are formed during the second stage.

In this paper we consider the possibility that the first stage is created by a special kind of baroclinic instability where the basic state consists of a lower, rather thin layer with an almost adiabatic lapse rate. On top of this layer is another layer with a more standard lapse rate.

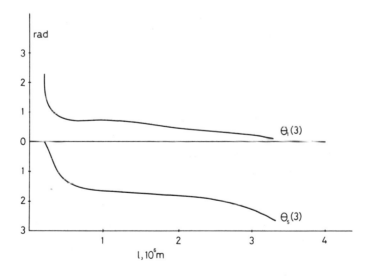

Figure 10. The phase differences as a function of wavelength for the case $U_B=-6$ m s^{-1}, $U_T=-18$ m s^{-1}, $\gamma_B=9.9\times10^{-3}$ km^{-1}, $\gamma_T=6.5\times10^{-3}$ km^{-1}.

The basic state described above seems to occur close to the ice edge at high latitudes where the lower layer has been modified to an almost adiabatic lapse rate by air-sea interaction. At the same time, there is a low-level baroclinic zone with low temperatures over the ice and higher temperatures over the ice and higher temperatures over the ocean creating a low-level thermal wind.

The model used to describe the situation is a three-level, quasi-geostrophic model on an f-plane where the reference levels are arranged to depict the relatively thin low layer and a much thicker higher layer. On the one hand the vertical resolution is just sufficient to have two different static stabilities. On the other, it is well known that these models require the static stability to be a function of pressure only. It is thus impossible to model the different stratifications that must exist over the ice and over the oceans. More complicated models are required for this purpose.

The results show that the basic state thus defined is baroclinically unstable with a minimum e-folding time for wavelengths of 100–200 km if the lapse rate in the lower layer is very close to the adiabatic lapse rate. Slightly smaller lapse rates create minimum e-folding times at 500 km. While the correction dimension can be obtained by the baroclinic instability mechanism, it is also true that the scale of the disturbances is highly sensitive to the assumed lapse rate.

The results show that the instability on the small scale is different from the normal Rossby scale instability, which occurs at large wavelengths. If the wind shear in the two layers are of opposite sign there are two regions of instability with a region of stability between them.

By calculating the relative structure of the baroclinic waves it is seen that the short waves slope eastward, while the longer Rossby waves slope westward for a positive wind shear in both layers. The amplitude is high at the low level and small at the high level.

A more realistic condition for the basic state is one where the wind increases in the lower part of the atmosphere and then decreases. In that case the baroclinic system will slope to the west with height.

We may thus conclude that if the right basic state is created by physical processes, i.e., a lower nearly adiabatic layer with a positive wind shear, then the baroclinic instability process can account for the formation of a wave with maximum amplitude at the low level forming the apparently necessary precursor for the later development of the even smaller polar lows, which may develop due to the CISK mechanism.

ACKNOWLEDGMENTS

The author has benefited from several discussions with Dr. E. Rasmussen who first mentioned the interesting problems connected with polar lows. Miss Kirsten Cornett has provided expert typing and has also drawn the figures.

REFERENCES

Blumen, W., 1979: On short-wave baroclinic instability. *J. Atmos. Sci.*, *36*, 1925–1933.

Cressman, G.P., 1961: A diagnostic study of a mid-tropospheric development. *Mon. Wea. Rev.*, *89*, 74–82.

Duncan, C.N., 1977: A numerical investigation of polar lows. *Quart. J. Roy. Meteor. Soc.*, *103*, 255–268.

Eady, E.T., 1949: Long waves and cyclone waves. *Tellus*, *1(3)*, 33–52.

Emanuel, K.A., 1986: A two stage air-sea interaction theory for polar lows. *International Conference on Polar Lows*, Oslo, 187–200.

Green, J.S.A., 1960: A problem in baroclinic stability. *Quart. J. Roy. Meteor. Soc.*, *86*, 237 251.

Harrold, T.W., and K.A. Browning, 1969: The polar low as a baroclinic disturbance. *Quart. J. Roy. Meteor. Soc.*, *95*, 710–723.

Jacobs, S.J., and A. Wiin-Nielsen, 1966: On the stability of a barotropic basic flow in a stratified atmosphere. *J. Atmos. Sci.*, *23*, 682–687.

Mansfield, D.A., 1974: Polar lows: The development of baroclinic disturbances in cold air outbreaks. *Quart. J. Roy. Meteor. Soc.*, *100*, 541–554.

Rasmussen, E., 1979: The polar low as an extratropical CISK disturbance. *Quart. J. Roy. Meteor. Soc.*, *105*, 531–549.

Rasmussen, E., 1985: A case study of a polar low development over the Barents Sea. *Tellus*, *37A*, 407–418.

Rasmussen, E., 1988: On polar lows, arctic instability lows and arctic cyclones, an observational study (submitted for publication).

Reed, R.J., 1986: Baroclinic instability as a mechanism for polar low development. *International Conference on Polar Lows*, Oslo, 141–150.

Sardie, J.M., and T.T. Warner, 1983: On the mechanism for the development of polar lows. *J. Atmos. Sci.*, *40*, 869–881.

Sardie, J.M., and T.T. Warner, 1985: On the development mechanisms of polar lows. *Tellus*, *37A*, 460–477.

Wiin-Nielsen, A., 1961: Diagnosis of divergence in a three-parameter numerical prediction model. *Mon. Wea. Rev.*, *89*, 67–73.

GRADIENT WIND ADJUSTMENT, CISK AND THE GROWTH OF POLAR LOWS BY DIABATIC HEATING

Aarnout van Delden[1]
Free University of Amsterdam
The Netherlands

ABSTRACT

The results of computations of the radial circulation and associated surface pressure tendencies, needed to retain gradient wind balance in a model of an axisymmetric cyclone, due to the action of diabatic heating and boundary layer pumping, are presented. These computations show that diabatic heating will not induce further deepening (i.e., intensification) of the cyclone when this cyclone is weak and has a cold core. In contrast, a relatively intense warm core balanced cyclone will deepen appreciably, depending on the degree of baroclinicity and on where exactly the heat sources are located.

To set the stage, a short review is given of the theory of geostrophic and gradient wind adjustment and conditional instability of the second kind (CISK).

1. INTRODUCTION

In this paper I intend to shed some light on the problem of geostrophic adjustment and gradient wind adjustment of a strongly rotating fluid system and its relation to the growth of polar lows. I will start with the most simple version of this problem, namely geostrophic adjustment in a one layer, slab-symmetric model. This will illustrate in which way the time and space scales of diabatic heating may determine the partition of energy between the so-called gravity inertia wave modes and the balanced cyclone mode. Subsequently, in Section 3, I will

---

[1]Present address: Institute of Meteorology and Oceanography, University of Utrecht, Utrecht, The Netherlands.

POLAR AND ARCTIC LOWS
Paul F. Twitchell, Erik A. Rasmussen,
and Kenneth L. Davidson (Eds.)

show semi-qualitatively in which way cyclone intensification is related to gradient wind adjustment of a multilayer axisymmetric system. Subsequently, I will argue that CISK (conditional instability of the second kind) is equivalent to perfect gradient wind adjustment of a balanced vortex to self-induced disturbances of this state of balance. In Sections 4 and 5, I will discuss some results of computations that show that CISK cannot serve as an explanation of the genesis and initial growth of a polar low, but may contribute strongly to the intensification in a later stage. The paper is concluded with a further discussion of the physics and limitations of CISK and its relation to other theories.

2. TIME AND SPACE SCALES OF FORCING AND THE ENERGY PARTITION IN GEOSTROPHIC ADJUSTMENT

Consider a shallow slab-symmetric layer of fluid (depth h and density ρ_1) below an infinitely deep motionless layer with density $\rho_2 < \rho_1$. The whole system is rotating with an angular velocity equal to Ω (see Figure 1). The equations of motion for the lower layer are (see e.g., Gill, 1982)

$$\frac{\partial u}{\partial t} + u\frac{\partial u}{\partial x} = -g'\frac{\partial h}{\partial x} + fv \tag{2.1}$$

$$\frac{\partial v}{\partial t} + u\frac{\partial v}{\partial x} = -fu \tag{2.2}$$

$$\frac{\partial h}{\partial t} = -\frac{\partial hu}{\partial x} + Q \tag{2.3}$$

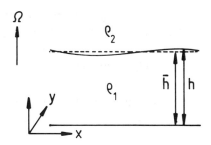

Figure 1. Geometry of the shallow slab-symmetric rotating layer of fluid (see text).

where u and v are, respectively, the velocities in the x- and y-direction, $f = 2\Omega$, Q represents the volume of mass extracted per unit time and area, and $g' = g(\rho_1 - \rho_2)/\rho_1$, where g is the acceleration due to gravity. Due to the slab symmetry, all y-derivatives are assumed zero. It is well known (see e.g., Gill, 1982, Section 7.3) that the linearized equations with $Q = 0$ support gravity inertia waves with phase speeds, c, given by,

$$c^2 = \frac{f^2}{\alpha^2} + g'\bar{h} \tag{2.4}$$

where α is the wavenumber and $\bar{h}$ is the mean depth of the lower layer. Evidently, the effect of rotation on the speed of these dispersive waves is more important than the effect of stratification if

$$L > \frac{2\pi\sqrt{g'\bar{h}}}{f} \equiv 2\pi\bar{h} \tag{2.5}$$

where $L = 2\pi/\alpha$ is the wavelength. R is called the Rossby radius of deformation. This important length scale also appears when we consider the frequency, ω, of gravity inertia waves, given by

$$\omega^2 = \alpha^2 c^2 = f^2 + \frac{4\pi^2 g'\bar{h}}{L^2} \tag{2.6}$$

It is easily deduced that the system possesses two fundamental limiting frequencies. If $L >> 2\pi R$, the waves are nearly inertial, with frequency,

$$\omega_f^2 = f^2 \tag{2.7}$$

If $L << 2\pi R$, then the waves become pure gravity waves with frequency,

$$\omega_b^2 = \frac{4\pi^2 g'\bar{h}}{L^2} \tag{2.8}$$

Note that the inertial frequency, ω_f, is independent of the horizontal scale of the disturbance, while the buoyancy frequency, ω_b, is not. This is the reason why the reaction of a rotating fluid system to large-scale disturbances is primarily inertial (in the form of inertial waves), which means that the reaction is noted most strongly in u and v (i.e., in the vorticity), whereas the reaction to small-scale disturbances is primarily buoyant (in the form of gravity waves), which means that it is noted most strongly in u and h (i.e., in the divergence). Somewhere in between these two extremes, where $\omega_b \approx \omega_f$, (or $L \approx 2\pi R$), we encounter a region in the spectrum of motions, which is of crucial importance for the dynamics of depressions driven by diabatic heat sources. Here, the reaction is manifest both in the vorticity field and in the divergence field. This is exactly what is needed for the intensification of a polar low or tropical cyclone, since these systems may be characterized as deep depressions (strong divergence; low pressure) with an *intense cyclonic circulation* (high vorticity).

The importance of the length scale, R, and the inertial time scale, $1/\omega_f$, can be illustrated with the results of some numerical integrations of Eqs. (2.1)–(2.3) with $v = u = 0$, $g' = 1$ m s^{-2} and $h = 2000$ m initially. We will vary the rotation rate, f (i.e., R), and the time scale of the mass extraction, while keeping the total volume of mass extracted during the integration time constant. The mass extraction term is specified as follows:

$$Q = \begin{cases} Q_0(x)/\tau & \text{if } t \leq \tau \\ \\ 0 & \text{if } t > \tau \end{cases} \tag{2.9}$$

where Q_0 is the total volume of mass extracted per unit area, and t is the time in which this is accomplished. Q_0 is plotted as the function of x in Figure 2. The horizontal scale of the forcing, L_Q, is approximately equal to 150 km.

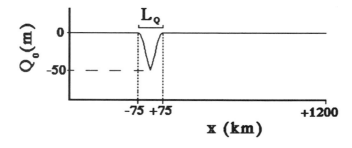

Figure 2. Total volume per unit area of mass extracted as a function of x.

Equations (2.1)–(2.3) are integrated on a grid of 240 points in the x-direction. The grid distance is 10 km and the timestep is 2 min. Details of the numerical procedure can be found in Pielke (1984). The Gadd (1978) scheme is implemented for the advection terms.

In Figure 3 the height, h, is plotted as a function of x every 8 min for three cases. In case (a) the rotation rate is representative for a motionless atmosphere at midlatitudes. The length scale of the forcing is very short compared to R. Also, the time scale of the forcing is very short compared to $1/\omega_f$. As expected, nearly all the potential energy imparted to the mass field in the first 8 min goes into gravity waves that propagate away at a speed approximately equal to 45 m s^{-1}.

In case (b), the rotation rate (f) is increased tenfold, while all other parameters are identical. Therefore, L_Q is greater than R. We see that, although there are dispersive gravity inertia waves propagating away from the forcing region, some potential energy is "caught" in the geostrophic mode. In case (c), which differs from case (b) only in the prescribed forcing time, τ, which is now 400 min ($>1/\omega_f$), the gravity inertia waves have such small amplitudes that they cannot be resolved by the graphics, although the same amount of potential energy as in case (b) is left in the geostrophic mode.

These numerical experiments serve to show that both the length scale of the forcing compared to the Rossby radius of deformation, and the time scale of the forcing compared to $1/\omega_f$, determine the type of reaction of a rotating fluid system. For a more detailed discussion of this topic, see e.g., Schubert et al. (1980) and references therein.

These conclusions are of great importance for the theory of polar lows, since diabatic heat sources (such as surface sensible heat fluxes and latent heat release) act as disturbances in the mass field. These disturbances will contribute significantly to the geostrophic mode and, thus to cyclone intensification, only if their length scale is comparable to or larger than the local Rossby radius of deformation. The time scale of the forcing relative to the local inertial time scale determines the amplitudes of the gravity inertia oscillations. We use the word "local" here because, as Ooyama (1982) and others have pointed out, not only the Earth's rotation has to be taken into account, but also the rotation of the flow itself. As the polar low intensifies, both the local Rossby radius and the local inertial frequency in the core will decrease, and disturbances with smaller and smaller length and time scales will be able to contribute to further deepening of the polar low without generating large amplitude gravity inertia waves.

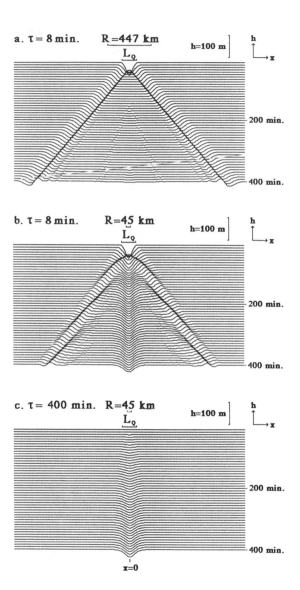

Figure 3. The reaction of the rotating shallow layer of fluid to forcing (Q), with a horizontal scale L_Q and a time scale, τ. The level of the free surface is shown every 8 min. (a) $f=0.0001$, $\tau=8$ min $(L_Q/R \approx 0.3;\ \tau f \approx 0.05)$; (b) $f=0.001$, $\tau=8$ min $(L_Q/R \approx 3;\ \tau f \approx 0.5)$; (c) $f=0.001$, $\tau=400$ min $(L_Q/R \approx 3;\ \tau f=2.4)$. The values of the other parameters are, $L_Q=150$ km, $g'=1$ m s^{-2} and $\bar{h}=2000$ m.

3. BALANCED MODELS OF CYCLONE INTENSIFICATION AND GRADIENT WIND ADJUSTMENT

In most analytical studies of cyclone intensification through diabatic heating (see e.g., Charney and Eliassen, 1964; Ooyama, 1969; Emanuel 1986a), it is assumed that geostrophic or gradient wind balance is *always* satisfied. This implies gravity inertia waves are filtered out of the model problem. We have seen that the validity of this assumption depends on whether $L_Q/R \geq 1$ and $\tau f \geq 1$.

Even if we assume that the system must always be in balance, however, it is not a trivial matter to understand why there should be a surface pressure decrease in a rotating sytem as a result of diabatic heating. After all, a polar low is not a single layer system. A deficit of mass (divergence) in one layer of the atmosphere could very well be compensated exactly by a surplus of mass (convergence) in a layer above or below, as happens in the case of ordinary shallow cumulus convection, where surface pressure changes are negligible.

Let us, to obtain more insight into this problem, assume that the troposphere in an axisymmetric polar low consists of two layers (see Figure 4). Suppose that the air in these two layers is rotating with the *same* tangential speed, v, about an axis $r = 0$, where r is the radial coordinate. Suppose also that the resulting centrifugal and Coriolis forces are in balance with a pressure gradient force, i.e.,

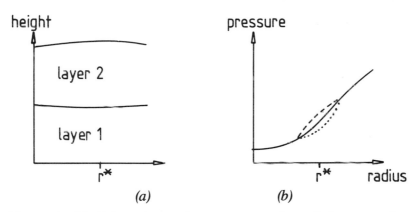

Figure 4. (a) Schematic two-layer representation of the troposphere. (b) Hypothetical pressure distribution in a balanced vortex. The dashed line indicates how the pressure distribution is altered in the upper layer after latent heat is released in an area centered at $r=r^*$. The dotted line indicates how the pressure distribution is altered in the lower layer due to outflow (from $r=r^*$) in the upper layer.

$$\frac{1}{\rho} \frac{\partial p}{\partial r} = \frac{v^2}{r} + fv \tag{3.1}$$

where p is the pressure and ρ the density. Suppose now that latent heat is released in an area centered at $r=r^*$. A reason for this location of the heating rather than at $r=0$ *will be given at the end of this section.* Due to expansion of air, isobaric surfaces are elevated. The horizontal pressure gradient in layer 2 (above the location of the heating) is altered in such a way that it increases for $r<r^*$ and decreases for $r>r^*$. Gradient wind balance is disturbed. To restore gradient wind balance, mass must be redistributed radially, *subject to the constraint of conservation of angular momentum.* The angular momentum per unit mass, M, is defined as

$$M = vr + \frac{1}{2} fr^2 \tag{3.2}$$

which implies that,

$$v = \frac{M}{r} - \frac{1}{2} fr \tag{3.3}$$

This, in turn, implies that the Coriolis force per unit mass (F_{cor}) plus the centrifugal force per unit mass (F_{cen}) acting on the radially moving air must obey the following relation:

$$F_{cen} + F_{cor} = \left(\frac{v^2}{r} + fv \right) = \frac{M^2}{r^3} - \frac{1}{4} f^2 r \tag{3.4}$$

The right-hand side of Eq. (3.4) is plotted in Figure 5 as a function of r for a representative value of M. *A fluid parcel with fixed mass moving radially and conserving angular momentum, M, will experience a net outward force given by the function plotted in Figure 5.* Evidently, ($F_{cen} + F_{cor}$) increases more than linearly with decreasing r.

Let us consider what happens in the region $r>r^*$. If the pressure gradient force in the upper layer is decreased by say $\Delta(\partial p/\partial r)$ due to latent heat release (see Figure 4b), air must flow outwards over a distance r_{out} to restore gradient

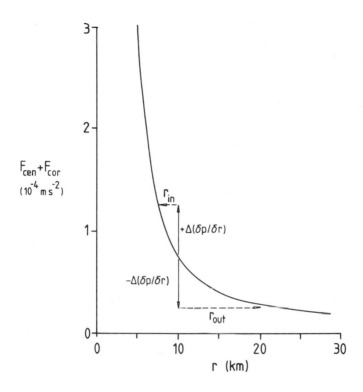

Figure 5. The radial dependence of the centrifugal force (F_{cen}) plus the Coriolis force (F_{cor}) under the constraint that the angular momentum (M) (Eq. (3.2) with $f=10^{-4}$) is constant, for $M=0.5.10^4$ m^2 s^{-1}. If the pressure gradient force is altered by, say $\Delta(\partial p/\partial r)$, air must flow radially outward or inward, depending on the sign of $\Delta(\partial p/\partial r)$, to restore gradient wind balance.

wind balance. Of course, this radial motion will automatically be attended with further changes in the pressure distribution, implying that the air may not have to flow exactly over a distance r_{out}. But let us, for simplicity, temporarily ignore this feedback effect.

The pressure change in each layer is proportional to the total divergence above. The divergence is given by $(\partial u/\partial r + u/r)$, where u is the radial velocity. Therefore, the outflow in the upper layer will lead to a pressure decrease in the lower layer. The pressure gradient in the lower layer will thus increase for $r > r^*$ (see Figure 4b). Air must then flow inward over a distance r_{in} (see Figure 5) to restore gradient wind balance. If the disturbance in the pressure gradient in the lower layer (due to the outflow in the upper layer) is approximately equal in absolute value to the initial disturbance in the pressure gradient in the upper layer (due to latent heat release), which seems to be a reasonable assumption, then $r_{out} > r_{in}$, i.e., the outflow is greater than the inflow, implying that the surface pressure will decrease.

Note that the same argument applied to $r < r^*$ yields a reverse circulation, i.e., inflow in the upper layer and outflow in the lower layer (eye formation). But, here also, the outflow must always be greater than the inflow in order to satisfy the two constraints, angular momentum conservation and "conservation" of gradient wind balance.

Although this semi-qualitative analysis ignores the variations in the depth of the inflow and outflow layer as well as all the important feedback processes that take place during perfect gradient wind adjustment, it does illustrate some essential features of this process. According to the above simplified analysis, *exact* gradient wind adjustment in an axisymmetric vortex, is very probably attended with a *surface pressure decrease.* Whether the surface pressure decrease exceeds, in absolute value, the surface pressure *increase* resulting from Ekman boundary layer pumping (one of the most important processes contributing to the filling and decay of the cyclone) is a question that has occupied many researchers in the past decades.

Among the early investigators were Charney and Eliassen (1964), Ooyama (1964), Ogura (1964), and Kuo (1965). They showed that, if frictional convergence in the center of a vortex could enhance the diabatic heating in the core, this would disturb gradient wind balance in such a way that the vortex was bound to grow, in spite of the countereffect of spin down due to boundary layer pumping. This mechanism is now called conditional instability of the second kind (CISK). It was also invoked by Rasmussen (1979) to explain the growth of polar lows. The theory of CISK has actually never been generally accepted as a theory for the growth of tropical cyclones (see e.g., Riehl, 1981; Emanuel, 1986a, 1987), and has led to much debate and confusion among those who do accept the basic philosophy (see e.g., Ooyama, 1982).

Probably, the confusion has one of its roots in the fact that CISK, as presented by Charney and Eliassen (1964), is a linear instability, i.e., given a suitable environment, a (gradient wind-) balanced infinitesimal symmetric disturbance will amplify spontaneously. However, as Charney and Eliassen recognized, CISK is not a theory for the genesis or formation of a hurricane depression in the same way as static thermal instability is a theory for the genesis of convection currents. In their introduction, Charney and Eliassen (1964) admit that there must be a pre-existing balanced depression of at least small amplitude before CISK can start accounting for further intensification. But, in order to apply perturbation analysis, they are forced to assume that the depression has in-finitesimal amplitude. Therefore their linear theory should not be taken too literally. It should be viewed as indicating the importance of processes in a small-amplitude system that could also be important in a finite-amplitude system.

More recently Schubert and Hack (1982), Shapiro and Willoughby (1982), and Hack and Schubert (1986) have shown that intense vortices in gradient wind balance react much more strongly to a specified heat source than weak balanced vortices. This could imply that CISK would become much more effective in a finite-amplitude system than in an infinitesimal system.

Thus, it appears worthwhile to extend the linear analysis of Charney and Eliassen (1964) and Ooyama (1964, 1969) to the finite amplitude case, without considering the problem as an initial value problem as is done in numerical simula-tions of balanced cyclones (e.g., Ooyama, 1969; Sundquist, 1970). A short review of the results of such an analysis (for details, see van Delden, 1989) with special emphasis on their application to polar lows, are described in the remaining part of this paper. This analysis is in fact very similar to those performed by Shapiro and Willoughby (1982) and Schubert and Hack (1982), except that boundary layer pumping is now included. Also, diabatic heating is assumed proportional to boundary layer convergence, as assumed in the theory of CISK, whereas in the former studies the diabatic heating rate *and* distribution were specified exactly.

I will present computations of the radial circulation and associated surface pressure tendencies, needed to retain gradient wind balance in an axisymmetric vortex, due to the action of diabatic heating. A slightly modified version of Ooyama's (1969) axisymmetric three-layer balanced cyclone model is used. In this model it is assumed that the frictional force in the boundary layer is deter-mined by the stress with the Earth's surface and not by eddy diffusion.

Although Charney and Eliassen (1964) and especially Ogura (1964) realized that the radial dependence of boundary layer convergence in a vortex was a con-troversial matter, they and many subsequent authors, studying CISK, chose the well-known formula due to Charney and Eliassen (1949), in which the vertical velocity above the Ekman boundary layer is proportional to the vorticity of the tangential flow at the top of the Ekman boundary layer. The reason for this was that this formula was easiest to handle in a linear perturbation problem. However, it is becoming increasingly clear that the exact radial dependence of boundary layer convergence in a vortex depends strongly on whether the dominant frictional force is eddy diffusion or the direct stress with the Earth's surface. Ogura (1964), Ooyama (1969), and Eliassen (1971) showed that frictional convergence in the core of a typical vortex attains a maximum value near the radius of maximum wind and tends to zero in the center of the vortex, if *surface stress* (which is proportional to the square of the wind velocity in the boundary layer) is the dominant frictional force. The detailed numerical (unbalanced) model calculations due to Yamasaki (1977) confirmed that surface stress was the most important frictional force in the boundary layer, as far as its effect on convergence is concerned. Yamasaki (1977) found that convergence in his model cyclone was indeed most notable at some distance from the center, leading him to conclude that frictional convergence is responsible for eye and, most important, eye-wall formation.

4. OOYAMA'S MODEL

The model consists of four axisymmetric superposed "shallow-water" layers (see Figure 6). The thickness of the lower layer, i.c., the boundary layer, is assumed constant. Frictional convergence in this layer will lead to a mass flux into the layer above. The thickness of the upper three layers may vary due to convergence or divergence, or mass fluxes between the layers. The upper free surface is assumed to be fixed, so that there is no pressure gradient and no tangential motion in the upper layer. Surface friction is the dominant frictional force in the boundary layer. The density of the boundary layer, ρ_0, is equal to the density of the layer above, while the density of layers 2 and 3 is a factor ϵ and α smaller, respectively, i.e., $\rho_0 = \rho_1 = \rho_2/\epsilon = \rho_3/\alpha \, (0 < \epsilon < 1, \alpha < \epsilon)$. Diabatic heating (Q) is represented as a mass flux from layer 1 to layer 2, and is directly proportional to boundary layer convergence, i.e.,

$$Q = \eta w^+, \; w^+ = w \text{ if } w > 0; \; w^+ = 0 \text{ if } w < 0 \qquad (4.1)$$

where w is the vertical velocity at the top of the boundary layer and η is a parameter that depends on the vertical distribution of equivalent potential temperature. Here η is treated as a free parameter. Due to diabatic heating, the upper layer grows at the cost of the lower layer. Thus, diabatic heating in the

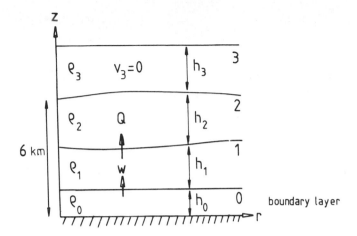

Figure 6: Structure of the four-layer version of Ooyama's (1969) model. The upper layer has been added to Ooyama's model to demonstrate that the external mode (interface between layers 2 and 3) is unimportant compared to the internal mode (interface between layers 1 and 2) (see van Delden, 1989).

model causes an upward bulging of surfaces of equal pressure. This effect is responsible for cyclone intensification (see Section 3). Sensible heat fluxes from the ocean also have this effect on the surfaces of equal pressure. Therefore, Eq. (6.1) is a rough parameterization of *the effect of sensible heat fluxes at the sea surface and latent heat release due to cumulus convection on the pressure distribution in a hydrostatically balanced column of air.* Cumulus convection itself is filtered out of the model due to the assumption of hydrostatic balance.

It is assumed that the tangential motion in the upper three layers is always in gradient wind balance with the radial pressure gradient. The tangential velocity distribution is prescribed by the following formula:

$$v_j = \frac{2v_j^* r}{r \cdot [1 + (r/r^*)^2]} \quad (j = 1,2); \; v_3 = 0 \tag{4.2}$$

where r^* is the radius of maximum tangential wind and v_j^* is the maximum tangential wind in layer j. The above simple profile allows for a relatively easy interpretation of the results, since the Rossby radius of internal deformation,

$$R_{\text{loc}} \equiv \frac{\sqrt{(1 - \epsilon)\, gh_1}}{\zeta_1 + f} \tag{4.3}$$

associated with the gradient wind adjustment of layers 1 and 2 is reasonably constant in the core of the vortex. In other words, the relative vorticity, ζ_j, is relatively constant for $r < r^*$ and decreases to zero rapidly with increasing r when $r > r^*$.

The radial pressure profile in each layer can then be computed by integrating the balance equation:

$$\frac{1}{\rho_j} \frac{\partial p_j}{\partial r} = fv_j + \frac{v_j^2}{r} \tag{4.4}$$

from $r=0$ outwards, assuming that $h_1(0) = h_2(0) = h_3(0) = \bar{h} = 2500$ m.

The equations expressing hydrostatic balance, gradient wind balance, conservation of mass and of angular momentum are only compatible if the radial motion satisfies certain diagnostic equations (see Ooyama, 1969; van Delden, 1989, for details). These equations can be solved by the computer for prescribed profiles of v_j (Eq. 6.2), yielding the radial motion, u_j, which is nonzero if there is diabatic heating. The surface pressure tendency, $\partial p_0 / \partial t$, is determined by

$$\frac{\partial p_0}{\partial t} = g\rho_0 \left(\frac{\partial (\psi_1 + \psi_2 + \psi_3)}{r\partial r} + w \right) \tag{4.5}$$

where

$$\Psi_j \equiv \frac{-\rho_j}{\rho_0}\, h_j u_j r \quad \text{and} \quad w = \frac{\partial}{r\partial r} \left\{ \frac{C_D |v_1| v_1 r}{\zeta_1 + f} \right\} \tag{4.6}$$

where C_D is the drag coefficient $(=0.0015)$ (see Ooyama, 1969, Eqs. 3.18–3.19).

5. SURFACE PRESSURE TENDENCIES ASSOCIATED WITH PERFECT AND CONTINUOUS GRADIENT WIND ADJUSTMENT

Figure 7 shows the computed surface pressure tendency and the vertical velocity at the top of the boundary layer as a function of radius in two cases of a barotropic vortex in which (a) $v_1^* = v_2^* = 22.5$ m s^{-1} and $r^* = 50$ km, and (b) $v_1^* = v_2^* = 15.5$ m s^{-1} and $r^* = 30$ km. The values of other parameters are $\bar{h} = 2500$ m, $h_0 = 1000$ m, $\epsilon = 0.9$, $\alpha = 0.6$, $f = 1.25 \times 10^{-4}$ s^{-1}, and $\eta = 2$. The choice $\eta = 2$ is representative for the tropics (see Ooyama, 1969). However, even though latent heating is much weaker at northern latitudes, it is still very likely that this value of η is also representative for polar low conditions, because here sea-surface sensible heat fluxes take over a large part of the role played by latent heat release in the tropics, as far as its effect on the hydrostatic pressure is concerned (see Section 6 for further discussion).

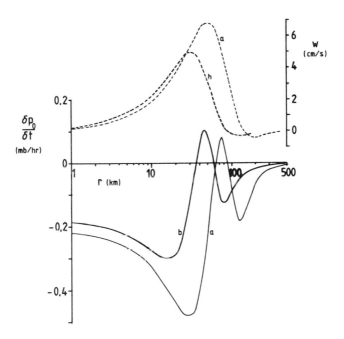

Figure 7: The vertical velocity, w (dashed line), at the top of the boundary layer and the surface pressure tendency (solid line) as a function of the radius in cases (a) and (b) (see text).

Cases (a) and (b) match as closely as possible two polar lows (vortex A and vortex C^*, respectively), discussed by Rasmussen (1988), although it should be stressed that it is very likely that these vortices were not barotropic ($v_1 \neq v_2$).

It can be seen that the maximum pressure decrease is not located in the vortex center, but near the region of maximum heating, where w reaches a maximum value in the order of 5 cm s^{-1}. A similar result was obtained by Shapiro and Willoughby (1982).

These are instantaneous pressure tendencies in a vortex with a velocity profile given by Eq. (6.2). Let us suppose that the profile remains the same (except that v^* varies) as the vortex intensifies (v^* increases). We can compute the central surface pressure tendency as a function of vortex intensity, v^*. This has been done for $r^* = 30$ km and $r^* = 50$ km. The result is plotted in Figure 8 (the solid lines). Evidently, the deepening rate increases with intensity until a certain optimum intensity is reached. Subsequently it decreases and becomes negative (a filling rate) above a certain maximum intensity. It appears that the filling sets

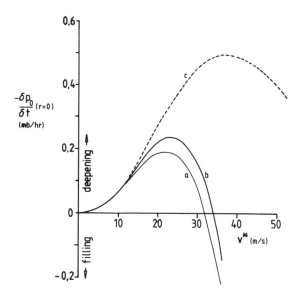

Figure 8. The central deepening rate as a function of vortex intensity ($v_1^* = v^*$) in a barotropic ($v_2^* = v_1^*$) balanced vortex with $r^* = 50$ km (a) and $r^* = 30$ km (b) (solid lines) and in a baroclinic balanced vortex with $v_2^* = (0.8)v_1^*$ and $r^* = 50$ km (dashed line). The values of other parameters are given in the text.

in when the local core Rossby radius, $R_{loc}(r=0)$, has decreased to a value that is much smaller than the radius of maximum heating, r_H. The value of $r_H/[R_{loc}(r=0)]$ when the deepening rate reaches a maximum is 2.14 in case (a) and 2.5 in case (b). When $\partial p_0/\partial t(r=0)$ becomes positive, the value of $r_H/[R_{loc}(r=0)]$ has increased to 3.4 in case (a) and to 3.9 in case (b). Since the balanced response of a rotating system to forcing is limited to a distance comparable to the Rossby radius (see Section 2), the heating cannot induce negative pressure tendencies in the center of the cyclone when it exceeds a certain intensity. The consequences of this result are discussed in more detail in van Delden (1989). The effect of vortex baroclinicity is illustrated by the dashed line in Figure 8. Here $v_2 = (0.8)v_1$, implying that the cyclone has a warm core (because of thermal wind balance), while all other parameters are identical to case (a). The deepening rate is greatly enhanced due to the decrease of the inertial resistance to the upper level outflow.

The dependence of the central deepening rate on the baroclinicity of the vortex is shown in Figure 9 from another point of view. Clearly, a warm core

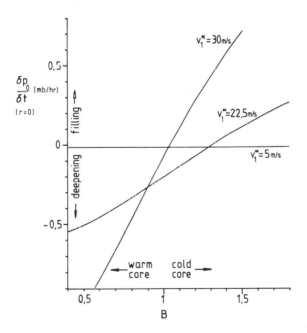

Figure 9. The central deepening rate as a function of vortex baroclinicity, $B(v_2=Bv_1)$, for three values of v_1^*. The values of the other parameters are exactly as in case (a) in Figure 8.

is accompanied by enhanced deepening rates. A weak vortex is hardly influenced by diabatic heating. A cold core vortex will most probably *fill* under the influence of diabatic heating. *All this implies that a polar low, which may grow initially due to baroclinic instability or lee-cyclogenesis, will somehow have to acquire a warm core before diabatic heating can contribute to further intensification.*

The maximum deepening rate experienced by the polar low/vortex A, investigated by Rasmussen (1988) was 8 mb in 6 hr (1.33 mb hr^{-1}). A maximum tangential velocity of 22.5 m s^{-1} was measured at the standard measuring height of about 10 m at the end of this period of strong deepening. The corresponding wind velocity above the boundary layer may very well be about 30 m s^{-1}. According to the model calculations (see Figure 9) the vortex would have to have a relatively warm core to make this deepening rate possible. It is difficult to assess from the coarse measurements whether this is the case, but it would be interesting to pay attention to this aspect in future observational studies.

6. DISCUSSION AND CONCLUSIONS

In this paper I have tried to communicate the idea that the growth of polar lows is related to the process of gradient wind adjustment. I have shown that the potential energy introduced into the atmosphere by diabatic heating is partitioned between gravity inertia waves and the so-called (gradient wind) balanced mode. This last mode of motion is naturally associated with the depression. Obviously the depression will grow if most of the energy is trapped directly in the balance mode. This appears to be the case if the heating has a length scale comparable to, or larger than the local internal Rossby radius of deformation. The amplitudes of gravity waves are small only when the time scale of the forcing is comparable to, or larger than, the local inertial time scale (see Figure 3).

Gradient wind balance is a basic assumption in the theory of CISK as described originally by Charney and Eliassen (1964), Ogura (1964), Ooyama (1964, 1969), and Kuo (1965). This assumption filters out gravity inertia waves. Therefore, the question of partitioning of energy between balanced and non-balanced modes is ignored in this theory. *In essence, CISK is equivalent to perfect gradient wind adjustment of a balanced vortex to self-induced disturbances of this state of balance.*

Although CISK was presented mathematically by Charney and Eliassen (1964) as a linear instability, it is obvious that it should not be interpreted as such, i.e., as a theory for the genesis of polar lows or any depression through diabatic heating. The finite amplitude analysis described in Sections 4 and 5 shows that the deepening rate due to CISK is very small when the vortex is weak. Diabatic heating may even induce a filling of the vortex if its core is relatively cold, i.e., if the tangential velocity increases with height. These results underline the fact that CISK cannot cause the genesis of polar lows; it can only *contribute* to the amplification in a later stage of the development of polar lows. It appears that this is supported by recent observational studies by Rasmussen (1985, 1988).

A few comments should be made on the other balance assumption, namely hydrostatic balance, and its relation to the parameterization of diabatic heating. Due to this assumption, convection, which is a result of *hydrostatic instability,* is not possible in the model. Nevertheless, the effect of latent heat release in cumulus clouds *on the hydrostatic pressure distribution* has to be included. This effect induces a bulging up of surfaces of equal pressure above the location of the heating, which sets off gravity inertia waves and gradient wind adjustment. *The sensible heat flux from the ocean, which is usually very large in polar low conditions, affects the hydrostatic pressure distribution over the total depth of the atmosphere in the same way as latent heat release.* The joint effects of latent heat release and sensible heat fluxes on the hydrostatic pressure distribution are parameterized in Ooyama's model simply by a term, Q (Eq. 4.1), which changes the stable density distribution (conserving mass) in such a way that the column of air expands upwards.

In the case of polar lows, Q will probably be determined by the sensible heat fluxes. Therefore, the assumption that Q is proportional only to boundary layer convergence is a bit questionable in these conditions. It seems just as appropriate to assume that Q is proportional to sensible heat flux, H, which can be related to the flow by the classical aerodynamic flux formula, $H = S|V|$, where V is the wind velocity at a certain height and S is a parameter, which, among other, is proportional to the difference between sea surface temperature and air temperature. This type of parameterization will maximize H near the radius of maximum wind, while H will be equal to zero at $r = 0$ (Økland and Schyberg, 1987). Therefore, the effects of this parameterization of diabatic heating on the vortex will be similar to the effects of Ooyama's parameterization. Thus, surface layer processes appear to be the basic cause of eye formation (see also Yamasaki, 1977, 1983).

The model calculations described in Sections 4 and 5 have shown that diabatic heating can only contribute significantly to the intensification of the vortex if it is located at a radius that is smaller than the internal Rossby radius in the core. Since the local internal Rossby radius in intense cyclones may be of the order of a few tens of kilometers, this implies that the important changes in such cyclones occur on a scale that approaches the scale of a typical cumulonimbus cloud. If the radius of maximum wind can be considered as a measure of the horizontal scale, many polar lows and hurricanes are indeed only a few times larger in the horizontal than in the vertical (see e.g., Rasmussen, 1988; Jørgensen, 1984). It will therefore be worthwhile to investigate nonhydrostatic effects, i.e., to allow for convective motions explicitly, as has been done by Yamasaki (1983).

Although Ooyama's (1969) model is relatively crude it appears to capture many essential dynamical characteristics of the growth and structure of a cyclone, such as a polar low, through diabatic heat sources. But, because it does not account for gravity waves and neglects the *compressibility of air,* it clearly has its limitations. The latter effect will be quite important in conditions where there are great horizontal, as well as vertical pressure gradients. Recently Emanuel (1986a, 1986b) has again stressed the important role of *thermodynamics* in tropical cyclones and polar lows. He proposes a new theory, called air-sea interaction instability (ASII), as an alternative to CISK. Emanuel points out that CISK requires conditional instability. By claiming that conditional instability is rare in the atmosphere, especially in the tropics, he concludes that CISK is a "false hypothesis" (Emanuel, 1987). I don't believe this is true. Ooyama (1969) imposes hydrostatic stability, thus excluding conditional instability explicitly. He then introduces a term that *only accounts for the vertical expansion of a hydrostatically balanced column of air due to a hypothetical heating. This may be any form of diabatic heating.* Diabatic heating in Ooyama's model disturbs the balance between horizontal pressure gradient force and the inertial forces. In Sections 3 and 5 we have seen that the *stable* atmosphere reacts to this effect by adjusting to a *new* state of gradient wind balance, which, in many cases, but not all, is such that the cyclone deepens. The exact expression for the diabatic heating (e.g., Eq. 4.1) is of course very debatable, but it is not necessary to interpret the heating, as specified by Eq. (4.1), as the release of convective available potential energy (CAPE), as Emanuel (1987) seems to do. The theory of CISK and the theory of ASII are different ways of looking at the same phenomenon and both probably contain elements of the truth.

ACKNOWLEDGMENTS

I would like to thank Erik Rasmussen for helpful comments on the subject of this paper.

REFERENCES

Charney, J.G., and A. Eliassen, 1949: A numerical method for predicting the perturbations of the middle-latitude westerlies. *Tellus, 1*, 38–54.

Charney, J.G., and A. Eliassen, 1964: On the growth of the hurricane depression. *J. Atmos. Sci., 21*, 68–75.

Eliassen, A., 1971: On the Ekman layer in a circular vortex. *J. Meteor. Soc. Japan, 49*, 784–789.

Emanuel, K., 1986a: An air-sea interaction theory for tropical cyclones, Part 1: Steady state maintenance. *J. Atmos. Sci., 43*, 585–604.

Emanuel, K., 1986b: A two stage air-sea interaction theory for polar lows. *Proceedings of the International Conference on Polar Lows*, Oslo, Norway, May 1986, 187–200.

Emanuel, K., 1987: Large-scale and mesoscale circulations in convectively adjusted atmospheres. *Proceedings of the Workshop on Diabatic Forcing*. ECMWF, Reading, UK, 323–348.

Gadd, A.J., 1978: A numerical advection scheme with small phase speed errors. *Quart. J. Roy. Meteor. Soc., 104*, 583–594.

Gill, A.E., 1982: *Atmosphere-Ocean Dynamics*. Academic Press, New York. 662 pp.

Hack, J.J., and W.H. Schubert, 1986: Nonlinear response of atmospheric vortices to heating by organized cumulus convection. *J. Atmos. Sci., 43*, 1559–1573.

Jørgensen, D.P., 1984: Mesoscale and convective scale characteristics of mature hurricanes, Part 1: General observations by research aircraft. *J. Atmos. Sci., 41*, 1268–1285.

Kuo, H.L., 1965: On the formation and intensification of tropical cyclones through latent heat release by cumulus convection. *J. Atmos. Sci., 22*, 40–63.

Ogura, Y., 1964: Frictionally controlled, thermally driven circulation in a circular vortex with application to tropical cyclones. *J. Atmos. Sci., 21*, 610–621.

Økland, H., and H. Schyberg. 1987: On the contrasting influence of organized moist convection and surface heat-flux on a barotropic vortex. *Tellus, 39A*, 385–389.

Ooyama, K., 1964: A dynamical model for the study of tropical cyclone development *J. Geofys. Intern., 4*, 187–198.

Ooyama, K., 1969: Numerical simulation of the life-cycle of tropical cyclones. *J. Atmos. Sci., 26*, 1–43.

Ooyama, K., 1982: Conceptual evolution of the theory and modelling of the tropical cyclone. *J. Meteor. Soc. Japan, 60*, 369–379.

Pielke, R.A., 1984: *Mesoscale Meteorological Modeling*. Academic Press, Orlando. 612 pp.

Rasmussen, E., 1979: The polar low as an extratropical CISK-disturbance. *Quart. J. Roy. Meteor. Soc.*, *106*, 313–326.

Rasmussen, E., 1985: A case study of a polar low development over the Barents Sea. *Tellus, 37A*, 407–418.

Rasmussen, E., 1988: On polar lows, arctic instability lows and arctic cyclones, An observational study. Manuscript, Geophysical Institute, University of Copenhagen, Denmark.

Riehl, H., 1981: Some aspects of the advance in the knowledge of hurricanes. *PAGEOPH, 119*, 612–627.

Schubert, W.H., J.J. Hack, P.L. Silvas Dias, and S.R. Fulton, 1980: Geostrophic adjustment in an axisymmetric vortex. *J. Atmos. Sci.*, *37*, 1464–1484.

Schubert, W.H., and J.J. Hack, 1982: Inertial stability and tropical cyclone development. *J. Atmos. Sci.*, *39*, 1687–1697.

Shapiro, L.J., and H.E. Willoughby, 1982: The response of balanced hurricanes to local sources of heat and momentum. *J. Atmos. Sci.*, *39*, 378–394.

Sundquist, H., 1970: Numerical simulation of the development of tropical cyclones with a ten-level model, Part 1. *Tellus, 22*, 359–390.

Van Delden, A., 1989: On the deepening and filling of balanced cyclones by diabatic heating. *Meteorology and Atmospheric Physics, 40* (accepted for publication).

Yamasaki, M., 1977: The role of surface friction in tropical cyclones. *J. Meteor. Soc. Japan, 55*, 559–571.

Yamasaki, M., 1983: A further study of the tropical cyclone without parameterizing the effects of cumulus convection. *Papers in Meteorology and Geophysics, 34*, 221–260.

BAROCLINIC INSTABILITY AND CISK AS THE DRIVING MECHANISMS FOR POLAR LOWS AND COMMA CLOUDS

George Craig and Han-Ru Cho
University of Toronto
Toronto, Canada

ABSTRACT

The relationship between those polar air stream disturbances dominated by baroclinic processes (comma clouds) and those where release of latent heat appears to play a more important role (polar lows) is explored using a linear model of baroclinic instability with a simple parameterization for cumulus heating. The importance of heating in the model atmosphere is determined by factors such as static stability and boundary layer moisture content. Depending on the amount of heating, the predicted system may be a purely baroclinic wave, a conditional instability of the second kind (CISK) disturbance, or anywhere in a continuous spectrum of intermediate types.

A comparison of the model results with observations of three polar lows and three comma clouds indicated that two of the polar lows corresponded to parameter values in the CISK regime, and two of the comma clouds corresponded to predominantly baroclinic systems. The remaining cases, consisting of a polar low and a comma cloud, appeared to be of a transitional nature, with both instability mechanisms playing significant roles.

1. INTRODUCTION

Most investigations into the physical processes responsible for the origin and intensification of polar air mass disturbances have centered around two mechanisms. One suggestion is that they form as a result of baroclinic instability (Mansfield, 1974; Duncan, 1977) in a manner similar to synoptic-scale

POLAR AND ARCTIC LOWS
Paul F. Twitchell, Erik A. Rasmussen, and Kenneth L. Davidson (Eds.)

131

midlatitude cyclones. The second mechanism relies on the release of latent heat in cumulus convection, and is generally referred to as CISK (Rasmussen, 1979).

Observed systems are also commonly divided into two classes (Locatelli et al., 1982; Businger, 1985). The members of the first group, sometimes called comma clouds, are similar in appearance to synoptic cyclones but smaller in size. In addition to the distinctive cloud pattern they sometimes show signs of a trough line or cold front (Reed, 1979; Reed and Blier, 1986a, 1986b). The second group, usually just called polar lows, tend to form further north and are more axisymmetric in shape. They often feature spiral bands of convective cloud and on occasion even a clear eye, reminiscent of tropical cyclones (Rasmussen, 1979, 1985). There is, however, no well-defined distinction between polar lows and comma clouds. Cumulus convection is a prominent feature in many comma clouds (Reed, 1979; Reed and Blier, 1986a) while baroclinicity is often present in environments where polar lows form (Rasmussen, 1985). In fact, there appears to be a spectrum of intermediate types of disturbances ranging between the two classes (Businger and Walter, 1988).

Theoretical modeling has yielded some relevant results on the interaction of baroclinic and convective processes. It has been shown that cumulus heating can act cooperatively in a baroclinic system, resulting in significantly increased growth rates (Mak, 1982; Sardie and Warner, 1983). Meanwhile Rasmussen (1979) found, using a model without baroclinicity, that CISK disturbances can grow under conditions where polar lows form. In the following study we will further explore the role of cumulus heating and CISK in a baroclinic environment using a simple linear model that incorporates both baroclinic and convective processes. The results will be compared with observations of a number of individual polar low and comma cloud cases.

2. MODEL RESULTS

We make use of the semi-geostrophic Eady model of baroclinic instability and include a simple parameterization of cumulus heating. The Eady model has continuous vertical structure but includes the assumptions of constant static stability, constant vertical wind shear, and a rigid lid at the upper boundary. The amount of cumulus convection is assumed to be controlled by low-level convergence of moisture. If it is then assumed that static stability, moisture content, and other properties of the ambient atmosphere do not change too rapidly, the heating rate can be set proportional to convergence below a fixed cloud base level. The heating is distributed vertically according to a given function, resulting in the following form for the heating rate $E = d\theta/dt$:

$$E = \left(\frac{\theta_o Q_\cdot}{gf} \right) \epsilon h(z)\ w(z_B) \tag{1}$$

where ϵ is a constant of proportionality, which will be referred to as the heating parameter, $h(z)$ is the normalized vertical heating profile depicted in Figure 1, and $w(z_B)$ is the vertical velocity at cloud base level z_B. The group of constants in square brackets serves to nondimensionalize ϵ. Q is the potential vorticity, which is a measure of stability in the semi-geostrophic system, θ_o is a fixed reference value of potential temperature, g is the gravitational acceleration, and f is the Coriolis parameter. Using this representation for the diabatic heating and writing in terms of geostrophic coordinates X, Y, Z, and T (Hoskins, 1975), the linearized governing equation is

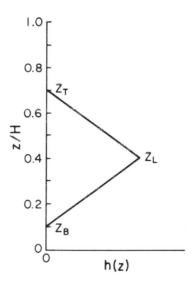

Figure 1. The prescribed vertical distribution of heating. Except where noted we will use the values $(z_B, z_L, z_T) = (0.1H, 0.4H, 0.7H)$. So that its integral from 0 to H is unity, $h(z)$ is normalized.

$$\left(\frac{\partial}{\partial T} + U\frac{\partial}{\partial X}\right)\left[\frac{1}{f^2}\left(\frac{\partial^2\phi'}{\partial x^2} + \frac{\partial^2\phi'}{\partial Y^2}\right) + \frac{f}{Q}\frac{\partial^2\phi'}{\partial Z^2}\right]$$

$$= -\frac{f}{Q}\epsilon h_Z(Z)\left[\left(\frac{\partial}{\partial T} + U\frac{\partial}{\partial X}\right)\frac{\partial\phi'}{\partial Z} - U_Z\frac{\partial\phi'}{\partial X}\right]_{Z=Z_B} \quad (2a)$$

$$\left(\frac{\partial}{\partial T} + U\frac{\partial}{\partial X}\right)\frac{\partial\phi'}{\partial Z} - U_Z\frac{\partial\phi'}{\partial X} = 0 \text{ at } Z=0, H \quad (2b)$$

where ϕ' is perturbation geopotential, U is the basic state wind, and U_Z is the basic state wind shear. This equation and its boundary conditions define an eigenvalue problem, which is solved according to the method described by Craig and Cho (1988).

The behavior of the solutions is summarized in Figure 2, which shows the evolution of the wavenumber of the fastest growing mode as the heating parameter ϵ is varied. The vertical heating profile was as shown in Figure 1 ($z_B=0.1H$, $z_L=0.4H$, $z_T=0.7H$). The plot can be divided into two regions:

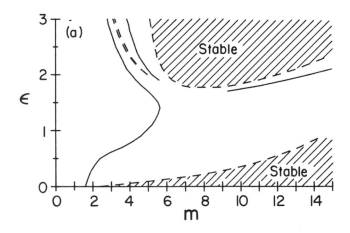

Figure 2. Summary of instability characteristics for the combined baroclinic-CISK model. Solid lines indicate wavenumber of fastest growing normal mode as a function of heating parameter ϵ. Dashed lines are boundaries of the unstable region.

for small values of ϵ there is a single maximum with wavenumber increasing (wavelength decreasing) as heating increases, while for larger values of ϵ the trend in wavenumber reverses and two secondary maxima appear. These local maxima correspond to waves with much smaller growth rates than the main branch and will not be discussed further.

The nature of the disturbance in these two regions is apparent in the vertical structure of the perturbation geopotential fields. Figure 3a (for $\epsilon = 0.3$)

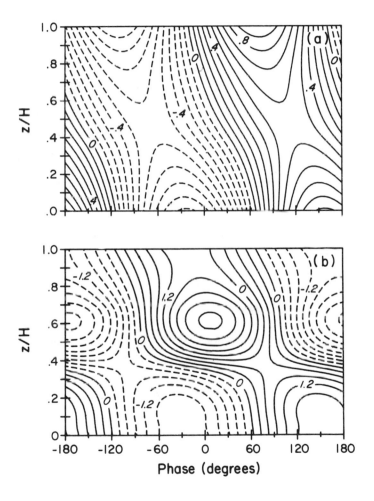

Figure 3. X-z cross sections of perturbation geopotential for the fastest growing wave with a) $\epsilon = 0.3$ and b) $\epsilon = 2.1$.

is typical of the structure of disturbances with values of the heating parameter in the lower region, and is dominated by a westward-tilting trough, indicative of baroclinic energy conversion. In fact, the structure is almost identical to that of a purely baroclinic wave in the Eady model (e.g., Gill, 1982; Figure 13.4). The disturbance shown in Figure 3a can thus be characterized as being caused predominantly by baroclinic instability. For values of ϵ in the upper region the geopotential field (e.g., Figure 3b, for $\epsilon = 2.1$) shows the typical features of a CISK mode: a low-level cyclone (negative geopotential anomaly) beneath an upper level anticyclone (postive geopotential anomaly). Unlike the more baroclinic type of waves, the amplitude is greatest in the heating region in the lower and mid-troposphere, which tends to confirm the dominant role of CISK for this disturbance.

We see then that both CISK and baroclinic instability can occur depending on the amount of heating. The occurrence of CISK only for sufficiently high heating rates corresponds to the behavior of CISK models without baroclinicity where instability is possible only if the heating exceeds a certain threshold value (Craig and Cho, 1988). In the present case, while a threshold level could be defined in the vicinity of $\epsilon = 1.5$ (for the heating profile in Figure 1), there is a smooth transition between the two types of behavior as ϵ is varied, giving rise to the possibility of a range of intermediate disturbances where the two mechanisms cooperate. Even well into the CISK region (Figure 3b) the structure of the disturbance shows traces of the westward-tilting trough of the baroclinic wave.

3. COMPARISON WITH CASE STUDIES

In order to explore how the model results correspond to observed systems, estimates of the nondimensional wavenumber m and heating parameter ϵ were calculated for six systems, which have been previously documented in the literature. Three of these systems were designated by the authors of the original studies as comma clouds (Reed and Blier, 1986a, 1986b; the land case of Wakimoto and Durkee, 1987), and three as polar lows (Rasmussen, 1985; Rabbe, 1987; Shapiro et al., 1987). The following formula derived by Craig and Cho (1988) was used to evaluate ϵ,

$$\epsilon = \frac{gL_c}{\theta_o c_p} \left(\frac{1}{H} \right) \left(\frac{f}{Q} \right) q_m (1 + r + \eta)\ (1 - b) \tag{3}$$

The interpretation of this relationship is that the heating induced by a given boundary layer convergence rate is determined by the moisture content of the subcloud layer q_m, with corrections for surface fluxes of moisture, water vapour going to moisten the air column rather than being condensed out, and moisture entrained in the cloud layer, denoted by r, b, and η, respectively. L_c is the latent heat of condensation for water and c_p the specific heat capacity of air at constant pressure. ϵ is also influenced by the static stability, as measured by $(Q/f)^{1/2}$, and by the depth of the troposphere, H. The values of these parameters for the six case studies are listed in Table 1. A detailed discussion of the data and methods used to obtain these estimates and of the accuracy of the results is given by Craig and Cho (1988). The uncertainty in the final values of ϵ is typically on the order of $\pm 50\%$.

The estimates of the heating parameter and observed wavenumber for the six cases are plotted in Figure 4 along with curves showing model predictions for a variety of vertical heating profiles. Two of the polar lows (Rasmussen, 1985; and Rabbe, 1987) can be fairly unambiguously identified as the results of CISK, while two of the comma clouds (Reed and Blier, 1986a, 1986b) appear to correspond to short wavelength baroclinic systems. The remaining two cases, one polar low and one comma cloud, cannot be clearly assigned to either region and are more likely of a transitional nature. It would seem that the classic polar low and comma cloud can be identified as resulting primarily from CISK and

TABLE 1. ESTIMATES OF PARAMETERS USED TO CALCULATE ϵ ACCORDING TO EQ. (3) FOR SIX CASE STUDIES OF POLAR LOWS AND COMMA CLOUDS*

	$(Q/f)^{1/2}$ $(\times 10^{-3} s^{-1})$	H (km)	g_m (gkg^{-1})	r	η	b	ϵ	λ (km)
Rasmussen (1985)	5.5	7.0	2.0	1.4	0.7	0.16	2.0	500
Rabbe (1987)	5.5	8.0	4.5	0.7	0.6	0.07	2.7	700
Shapiro et al. (1987)	7.5	7.0	2.3	1.2	0.8	0.16	1.1	500
Wakimoto and Durkee (1987)	10.8	9.0	11.0	0.0	1.1	0.28	1.1	1200
Reed and Blier (1986b)	11.3	9.0	5.0	1.3	0.8	0.15	0.9	1400
Reed and Blier (1986a)	10.5	8.0	4.0	1.1	0.8	0.15	0.7	2000

* The resulting value of ϵ and an estimate of the wavelength of the observed systems are given in the final two columns.

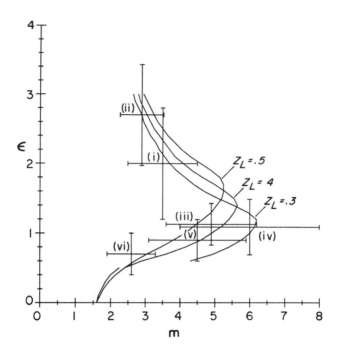

Figure 4. Observed values of wavenumber and heating parameter for the polar lows studied by (1) Rasmussen (1985), (2) Rabbe (1987), and (3) Shapiro et al. (1987), and the comma clouds studied by (4) Wakimoto and Durkee (1987), (5) Reed and Blier (1986b), and (6) Reed and Blier (1986a). Error bars give an estimate of the uncertainty in values derived from observations, as discussed by Craig and Cho (1988). Also shown are curves of the predicted wavenumber of the fastest growing mode as a function of ϵ for several levels of maximum heating z_L.

baroclinic instability, respectively. However a continuous spectrum of intermediate systems can and probably does occur.

In this simple model it is the nondimensional heating parameter that determines the dynamical nature of the resulting system. It is interesting to consider what physical quantities contributed to making ϵ higher for the CISK disturbances discussed above. As can be seen from the values in Table 1, the dominant factor is the static stability of the environment where the system formed: $(Q/f)^{1/2}$. Perhaps surprisingly, the boundary layer moisture content does not play a major role in determining whether CISK will occur. In fact, q_m is higher for the comma cloud systems that formed further to the south where the air and sea surface temperatures were warmer. Surface fluxes of moisture were important in all cases except that of Wakimoto and Durkee (1987), which formed over land, but there was no systematic difference between the polar low and comma cloud systems.

It should also be noted that the trend toward increasing nondimensional wavelength with increasing heating in the CISK region (Figure 4) is not a reflection of longer dimensional wavelengths for the observed CISK disturbances (Table 1). The wavelength is nondimensionalized using the Rossby radius of deformation, $L = (Q/f)^{1/2}H/f$, which is much smaller for the low stability environments where polar lows form.

4. SUMMARY AND CONCLUDING REMARKS

Within the context of the linear model used here, the interaction of baroclinic and convective processes may be understood as follows. If the heating rate is not too great a baroclinic wave will grow in the usual manner. Since cumulus heating is controlled by low-level convergence, it occurs where convergence and rising motion already exist and merely intensifies the existing circulation. The effects of convective heating are thus essentially the same as a reduction in static stability, which results in a faster growing disturbance with a shorter wavelength.

CISK occurs when the circulation induced by the heating is sufficient to supply the same or a greater amount of heating. It is thus only possible when the heating parameter, which is a measure of the efficiency of the heating in resupplying itself, is above a certain threshold. The contribution of baroclinicity can be seen from the fact that the transition to CISK takes place at a lower value of ϵ than the threshold of instability in the corresponding pure CISK model (Craig and Cho, 1988), due to the contribution of convergence forced by baroclinic instability. Since the convective and baroclinic processes cooperate at all stages, the onset of CISK is through a smooth transition and there is no precise point where one instability mechanism ceases to dominate and the other starts.

The six case studies of polar air mass disturbances considered in this study showed the full range of behavior discussed above. The systems called polar lows tended to be CISK-dominated while the comma clouds tended to be primarily baroclinic, however there were systems of a transitional nature given either name in the literature. In the rapidly changing air masses where polar lows often form it is even conceivable that a system could change character over time. For example, an initially baroclinic system could become CISK-dominated as surface fluxes of heat and moisture decrease the stability of the atmosphere and enhance cumulus convection.

REFERENCES

Businger, S., 1985: The synoptic climatolgy of polar low outbreaks. *Tellus, 37A,* 419–432.

Businger, S., and B. Walter, 1988: Comma cloud development and associated rapid cyclogenesis over the Gulf of Alaska: A case study using aircraft and operational data. *Mon. Wea. Rev., 116,* 1103–1123.

Craig, G., and H.-R. Cho, 1988: Cumulus convection and CISK in the extratropical atmosphere, Part I: Polar lows and comma clouds. *J. Atmos. Sci., 45,* 2622–2640.

Duncan, C. N., 1977: An investigation of polar lows. *Quart. J. Roy. Meteor. Soc., 103,* 255–268.

Gill, A.E., 1982: *Atmosphere-Ocean Dynamics,* Academic Press, London, 662 pp.

Hoskins, B.J., 1975: The geostrophic momentum approximation and the semigeostrophic equations. *J. Atmos. Sci., 32,* 233–242.

Locatelli, J.D., P.V. Hobbs, and J.A. Werth, 1982: Mesoscale structures of vortices in polar airstreams. *Mon. Wea. Rev., 110,* 1417–1433.

Mak, M., 1982: On moist quasi-geostrophic baroclinic instability. *J. Atmos. Sci., B39,* 2028–2037.

Mansfield, D.A., 1974: Polar lows: The development of baroclinic disturbances in cold air outbreaks. *Quart. J. Roy. Meteor. Soc., 100,* 541–554.

Rabbe, A., 1987: A polar low over the Norwegian Sea, 29 February–1 March 1984. *Tellus, 39A,* 326–333.

Rasmussen, E., 1979: The polar low as an extratropical CISK disturbance. *Quart. J. Roy. Meteor. Soc., 105,* 531–549.

Rasmussen, E., 1985: A case study of a polar low development over the Barents Sea. *Tellus, 37A,* 407–418.

Reed, R.J., 1979: Cyclogenesis in polar airstreams. *Mon. Wea. Rev., 107,* 38–52.

Reed, R.J., and W. Blier, 1986a: A case study of comma cloud develoment in the Eastern Pacific. *Mon. Wea. Rev., 114,* 1681–1695.

Reed R.J., and W. Blier, 1986b: A further case study of comma cloud development in the Eastern Pacific. *Mon. Wea. Rev., 114,* 1696–1708.

Sardie, J.M., and T.T. Warner, 1983: On the mechanism for the development of polar lows. *J. Atmos. Sci., 40,* 869–881.

Shapiro, M.A., L.S. Fedor, and T. Hampel, 1987: Research aircraft measurements of a polar low over the Norwegian Sea. *Tellus, 39A,* 272–306.

Wakimoto, R.M., and K.R. Durkee, 1987: A case of two mesoscale eddies: An oceanic versus a continental comma cloud. *Mon. Wea. Rev., 115,* 2202–2213.

ON THE DEVELOPMENT OF POLAR LOW WAVETRAINS

G.W. Kent Moore and W.R. Peltier
University of Toronto
Toronto, Ontario, Canada

ABSTRACT

Our previous work on the stability of deformation-induced frontal zones has led to the discovery of a new cyclone-scale mode of baroclinic instability. A crucial factor that led to this discovery was the use of primitive equations in the formulation of the stability problem. In fact, we have shown that this mode is filtered out by both the quasi-geostrophic and geostrophic momentum approximations to the primitive equations. In this paper, we apply our methodology to the problem of identifying the dynamic processes responsible for the development of polar low wavetrains. Observational evidence has shown that these wavetrains or families develop along shallow baroclinic zones that are situated north of the primary polar front. Usually three or four coherent disturbances make up a wavetrain. Each member of the wavetrain is typically observed to develop from a small amplitude perturbation into a fully developed cyclone with a characteristic wavelength of approximately 500 km. Such a development is indicative of the existence of a dynamic instability. Indeed we propose that the new cyclone-scale mode of baroclinic instability is responsible for the development of the polar low wavetrains. To demonstrate this, we will show that the stability characteristics of a typical baroclinic zone in which a wavetrain was observed to develop are very similar to those of the model frontal zones that we have previously studied. In addition, we will demonstrate that the wavelength and structure of the most unstable wave predicted by our theory are in good agreement with observations.

1. INTRODUCTION

The conventional interpretation of all midlatitude cyclones is that they are manifestations of the baroclinic instability mechanism first described by Charney

POLAR AND ARCTIC LOWS
Paul F. Twitchell, Erik A. Rasmussen,
and Kenneth L. Davidson (Eds.)

141

(1947) and Eady (1949). Charney (1975) writing in his preface to the "Selected Papers of J.A. Bjerknes" clearly recognized that this was not the case. He stated that there appeared to be some fundamental dynamic difference between the upper level long waves that quasi-geostrophic theory so successfully describes, and the shorter wavelength frontal cyclones upon which Bjerknes (1919) had focused his attention. However, until recently no clear theoretical explanation supporting the necessity of such distinction has been forthcoming.

Moore and Peltier (1987) described a detailed analysis of the stability of realistic atmospheric frontal structures against arbitrary three-dimensional perturbations. The frontal structures employed in this initial investigation were those generated by the action of a hyperbolic deformation field on a previously existing large-scale horizontal potential temperature gradient. The semi-geostrophic theory of Eliassen (1948) and Hoskins and Bretherton (1972) was employed to describe the process of frontogenesis. The frontal zones generated in this way, for which the assumption of uniform potential vorticity was also employed, were used as basic states for the purpose of the stability analysis. In this analysis the quasi-geostrophic approximation was *not* invoked and the complete nonseparable eigenvalue problem was solved *without* approximation.

The main result from this work was the demonstration of the existence of a short wave branch of unstable normal modes in the eigenspectrum. The fastest growing mode in this new branch was found to have a horizontal wavelength somewhat less than 1000 km, in close accord with the observed scale of frontal cyclones (Bjerknes and Solberg, 1922; Harrold and Browning, 1969; Reed, 1979). An analysis of the energy budget for the new cyclone scale mode demonstrated that it was also driven by the baroclinic instability mechanism.

The relationship between this new mode of baroclinic instability and the classical Charney-Eady mode has recently been the subject of a number of further investigations. The most important of these (Moore and Peltier, 1989a) has been the demonstration that the quasi-geostrophic approximation completely filters the new mode from the dynamic system while leaving the Charney-Eady mode only slightly affected. This can be understood on the basis of the fact that the short wavelength mode in the primitive equations system is boundary confined. It is therefore excluded from the instability spectrum of a constant potential vorticity basic state by the Charney-Stern theorem of quasi-geostrophic theory. When semi-geostrophic theory was employed in the stability analysis (Moore and Peltier, 1989b), it was found that it is similarly incapable of supporting the new short wavelength cyclone mode.

An important question that arises out of the discovery of this mode is that of the role it plays in the generation of observed cyclone scale disturbances, such as polar lows and comma clouds. Of particular interest are cases in which a coherent family or wavetrain of such disturbances develops (Harrold and Browning, 1969; Reed, 1979). A favoured location for the development of these wavetrains is the Norwegian Sea (Mansfield, 1974; Duncan, 1977). Recently, Reed and Duncan (1987) described one such case in which a family of four polar lows was observed to grow along a quasi-two-dimensional shallow baroclinic zone of low static stability. What makes their case so interesting is that each member in the wavetrain was observed to grow from a small perturbation into a fully developed polar low. They state that the time required for the disturbances to double in amplitude was in the range of 8 to 24 hr. More importantly, the polar lows in the wavetrain had a characteristic wavelength of approximately 500 m.

The growth of a family of waves from small amplitude perturbations to fully developed disturbances is indicative of the existence of a dynamic instability of the environment in which the waves appeared. In an attempt to identify this instability, Reed and Duncan (1987) performed a linear stability analysis of the underlying baroclinic zone. In this analysis, they made use of the quasi-geostrophic approximation. There are two apparent problems with their approach. First, the static stability in the environment in which the polar lows grow varies strongly in the horizontal. Second, the background Richardson number field is of order unity. Both of these factors imply that the quasi-geostrophic approximation to the primitive equations is invalid. If suitable modifications were made to the environment, i.e., the elimination of the observed horizontal variations in the static stability and along-front wind fields, then the results of their quasi-geostrophic analysis indicated that unstable perturbations with wavelengths on the order of 500 km could grow via the baroclinic instability mechanism. The doubling times for these quasi-geostrophic waves, however, were found to be long compared to the observed doubling times. This realization led Reed and Duncan (1987) to propose that some sort of convective instability was also needed to account for the rapid development of the disturbances.

We will show that when the primitive equations are employed in the stability analysis, the results are far more robust and in closer accord with the observations. Most importantly, they demonstrate that the stability characteristics of the baroclinic zones in which the polar low wavetrains develop are very similar to those of the frontal zones investigated by Moore and Peltier (1987). To accomplish this, we will consider the stability of two-dimensional nonseparable baroclinic zones to three-dimensional small amplitude perturbations. We begin by reviewing the theory of nonseparable baroclinic instability and its application to the phenomenon of frontal cyclogenesis.

2. THE THEORY OF NONSEPARABLE BAROCLINIC INSTABILITY AND FRONTAL CYCLOGENESIS

The problem that we are obliged to solve is that of the determination of the stability of a two-dimensional baroclinic zone, consisting of an along-front wind field $\bar{V}$ and corresponding potential temperature field $\bar{\theta}$, against arbitrary three-dimensional small amplitude perturbations that obey the full hydrostatic primitive equations.

There is an unfortunate inconsistency between the nomenclature used to describe the baroclinic zone and that used to describe the mechanisms by which unstable waves can grow on such a zone. The root of this inconsistency concerns the choice for the orientation of the coordinate system that is to be employed. Frontogenesis theory (Hoskins and Bretherton, 1972) assumes that the baroclinic zone varies in x and z but not y, while conventional baroclinic instability theory (Charney, 1947; Eady, 1949) assumes that the mean state is a function of y and z but not x! Prior to the work of Moore and Peltier, (1987), no one had considered the problem of determining the stability characteristics of realistic baroclinic zones. As a result, no one has been obliged to face this inconsistency. In this and our previous analyses we have chosen to retain the coordinate system that arises out of frontogenesis theory. As such, our x axis is in the cross-front direction and our y-axis is in the along-front direction. Provided that $\bar{V}$ and $\bar{\theta}$ are in thermal wind balance, viz:

$$f\frac{\partial \bar{V}}{\partial z} = \frac{g}{\theta_o} \frac{\partial \bar{\theta}}{\partial x} , \qquad (1)$$

then the mean state constitutes a steady two-dimensional solution to the hydrostatic primitive equations. It should be noted that in writing (1) we have employed the pseudoheight of Hoskins and Bretherton (1972) as our vertical coordinate. The stability analysis of such a mean state leads to the formulation of a nonseparable two-dimensional boundary value problem. The atmospheric dynamics group at Toronto has solved a number of such problems (Klaassen and Peltier, 1985; Moore and Peltier, 1987, 1989a, 1989b; Laprise and Peltier, 1989) by making use of ideas from the Floquet theory (Jordan and Smith, 1977) for differential equations with periodic coefficients. The assumption made is that the stability characteristics of a given state $(\bar{V}, \bar{\theta})$ are the same as those of a lattice of such mean states periodic in x. It then follows from this imposed periodicity, that the normal modes of the set of linear partial differential equations that describe the evolution of small amplitude perturbations have the following functional form:

$$F'(x,y,z,t) = Re\left[F^{\dagger}(x,z)\, e^{i(ax\, +\, by\,)}\, e^{st}\right]$$

$$F^{\dagger}(x,z) = F^{\dagger}(x + L,z)$$

(2)

where F' represents any of the five hydrodynamic fields that describe the perturbation, a is the cross-front Floquet number, b is the along-front wavenumber, s is the complex growth rate, and L is the underlying periodicity of the lattice.

Substitution of normal mode expansions (2) for each of the hydrodynamic fields into the full nonhydrostatic primitive equations linearized about a mean state $(\bar{V}, \bar{\theta})$ yields the following set of stability equations (Moore and Peltier, 1987):

$$(s + ib\,\bar{V})\, U^{\dagger} - fV^{\dagger} + (\partial_x + ia)\, \phi^{\dagger} = 0 \tag{3a}$$

$$(s + ib\bar{v})\, V^{\dagger} + (D^{\dagger} + f)\, \bar{V} + ib\phi^{\dagger} = 0 \tag{3b}$$

$$(s + ib\bar{V})\, W^{\dagger} - \frac{g}{\theta_o}\, \theta^{\dagger} + \partial_z\phi^{\dagger} = 0 \tag{3c}$$

$$(s + ib\bar{V})\, \theta^{\dagger} + D^{\dagger}\bar{\theta} = 0 \tag{3d}$$

$$(\partial_x + ia)\, U^{\dagger} + ibV^{\dagger} + \partial_zW^{\dagger} = 0. \tag{3e}$$

In this system the operator $D^{\dagger}$ is defined as:

$$D^{\dagger} = U^{\dagger}\partial_x + W^{\dagger}\, \partial_z. \tag{3f}$$

Subject to the hydrostatic approximation and making use of Galerkin expansions for each of the perturbation hydrodynamic fields $(U^{\dagger},$

$V^\dagger, W^\dagger, \theta^\dagger, \phi^\dagger$), the above system may be reduced to a matrix eigenvalue problem of the form (see Moore and Peltier, 1987 for details):

$$s\vec{\chi} = M\vec{\chi} \qquad (4)$$

where $\vec{\chi}$ is the vector of projections of the perturbation hydrodynamic fields onto the basis functions used in the Galerkin expansions and M is the complex stability matrix. In general M is a function of the Floquet number a, the along-front wavenumber b and the mean state ($\bar{V}, \bar{\theta}$). As described by Moore and Peltier (1987), we will restrict consideration to the case in which $a = 0$.

For a given mean state, (4) is solved to yield spectra of the growth rate ($\sigma = Re\ s$) and phase speed ($Cph = -Im\ s/b$) as a function of along-front wavenumber b. A mode with wavenumber b has a wavelength $\lambda = 2\pi/b$ and is said to be unstable if its growth rate is positive. If this is the case, then the mode will double in amplitude in a time $Td = \ell n(2)/\sigma$. One of the advantages of the Galerkin method is that it simultaneously finds all the normal modes for a given wavenumber b. As a result, it allows for the identification of the harmonics of the fundamental modes of instability. Although these harmonics are not in general physically realizable (having growth rates well below those of the fundamentals), nevertheless they are indicative of the symmetries contained within the underlying basic state. The power method employed by others (Duncan, 1977; Reed and Duncan, 1987) does not have this important capability.

An examination of the energy budgets of the unstable modes provides an understanding of the physical processes responsible for their growth. From (3), one can show that:

$$2\sigma\ KE' + ib\bar{V}\ KE' = RS + VHF + \partial_x\{U^\dagger\phi\} + \partial_z\{W^\dagger\phi^\dagger\} \qquad (5)$$

$$2\sigma\ PE' + ib\bar{V}\ PE' = HHF - VHF \qquad (6)$$

where:

$$KE'\ \text{(the eddy kinetic energy density)} = \frac{1}{2}[\{U^{\dagger 2}\} + \{V^{\dagger 2}\}] \qquad (7)$$

$$PE'\ \text{(the eddy potential energy density)} = \frac{1}{2}\{\theta^{\dagger 2}\} \qquad (8)$$

$$RS \text{ (the Reynolds stress term)} = -\frac{1}{2} [\partial_z \bar{V} \, Re\{ V^\dagger W^{\dagger *} \}$$

$$+ \partial_x \bar{V} \, Re\{ V^\dagger U^{\dagger *} \}] \tag{9}$$

$$VHF \text{ (the vertical heat flux term)} = \frac{1}{2} \, Re\{ W^\dagger \theta^{\dagger *} \} \tag{10}$$

and

$$HHF \text{ (the horizontal heat flux term)} = -\frac{1}{2} \partial_x \bar{\theta} \, Re\{ U^\dagger \theta^{\dagger *} \} \tag{11}$$

In the above, the along-front averaging operator $\{ \}$ is defined by:

$$\{ \psi \} = \frac{b}{2\pi} \int_0^{2\pi/b} \psi \, dy \tag{12}$$

3. THE GENESIS OF POLAR LOW WAVETRAINS

We are now in a position to apply this theory to determine the stability characteristics of the baroclinic zone upon which the polar low wavetrain described by Reed and Duncan (1987) was observed to develop. Figure 1a shows the cross sections of the along-front velocity and potential temperature fields for this zone. The cross sections were deduced from the operational analysis done by the European Center for Medium-Range Weather Forecasts (ECMWF). As described by Reed and Duncan (1987), this analysis was not able to resolve the individual polar lows. It therefore provides a representation of the synoptic scale environment in which the wavetrain developed. The Richardson number field

$$Ri = \frac{g}{\theta_o} \frac{\partial_z \bar{\theta}}{(\partial_z \bar{V})^2} \tag{13}$$

and the static stability field

$$S = \partial_z \bar{\theta} \tag{14}$$

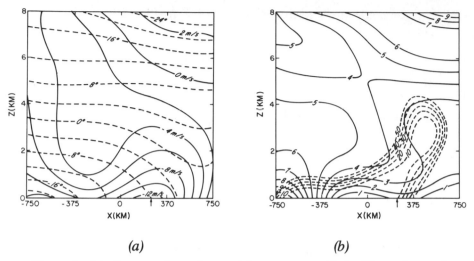

(a)   (b)

Figure 1: (a) Cross sections of potential temperature (dashed lines, °C) and along-front velocity (solid lines, m s⁻¹) representing the basic state in which the polar low wavetrain of Reed and Duncan (1987) was observed. (b) Cross sections of static stability (solid lines, °C km⁻¹) and Richardson number (dashed lines) derived from the fields shown in Figure 1a. The arrows at the bottom of each plot indicate the location in which the polar lows were observed to develop.

associated with this baroclinic zone are displayed in Figure 1b. The arrow indicates the storm track of the polar lows in both parts of Figure 1. Examination of this figure shows that the cyclones nucleated in a region in which both the static stability and Richardson number were small. Also evident is the large horizontal variation in both the along-front wind and the static stability. The latter important characteristic of the environment in which polar lows develop has been neglected in previous studies (Mansfield, 1974; Duncan, 1977; Reed and Duncan, 1987).

Displayed in Figure 2 are the spectra predicted by our theory for the growth rate and phase speed of the unstable waves that can develop in the baroclinic zone shown in Figure 1. It should be emphasized that no modifications to the baroclinic zone have been made. Inspection of these spectra shows that there are three distinct branches of unstable waves. The branch with a growth rate maximum at $b = 1.6$ (representing a wavelength of approximately 3000 km) represents the classical long wave Charney-Eady branch of baroclinic instability. That with a maximum at $b = 10$ (wavelength of approximately 500 km) corresponds to the cyclone scale branch. Note that also present is the first harmonic of the cyclone scale branch. As described in Moore and Peltier (1987), the waves

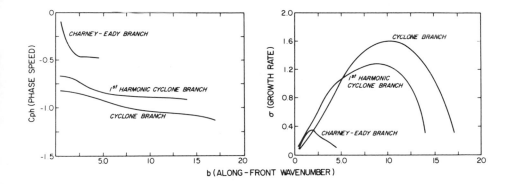

Figure 2: Nondimensional phase speed (C_{ph}) and growth rate (σ) vs. along-front wave number (b) spectra for the basic state shown in Figure 1. The phase speeds, growth rates, and wavenumbers have been scaled by 8 m s^{-1}, 10^{-5} s^{-1} and 1.25×10^{-6} m^{-1}, respectively.

in the Charney-Eady branch represent deep disturbances and as a result, they tend to have relatively low phase speeds (indicative of steering levels in the middle troposphere). In contrast, the waves in the cyclone scale branches are shallow and boundary confined and thus they tend to have higher phase speeds. Examination of Figure 2 shows that this is also the case for the present analysis. However, unlike the spectra described in Moore and Peltier (1987), the waves in the cyclone scale branch have much larger growth rates than those in the Charney-Eady branch. This result can be attributed to differences in the structure of the baroclinic zones analysed in Moore and Peltier (1987) as compared to the one considered here. Most importantly, the baroclinic zone under study here is relatively shallow, while those investigated by Moore and Peltier (1987) were quite deep.

From Figure 2, we see that there is a distinct maximum in the growth rate spectra. The wavelength of this most unstable normal mode, which is a member of the cyclone scale branch, is 500 km. The time required for it to double in amplitude is 9.6 hr and its phase speed is -8 m s^{-1}. The wavelength and doubling time are in good agreement with the observations made by Reed and Duncan (1987). The only serious discrepancy is in the phase speed, which is larger than the observed by a factor of approximately 2.

To identify the regions in which the normal modes develop and the mechanisms by which they develop, we present in Figures 3, 4, and 5 contour

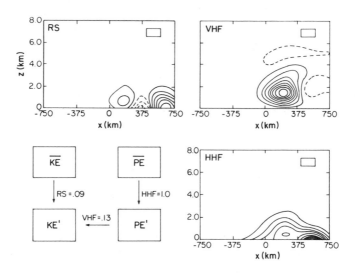

Figure 3: *The energy flux terms and energy budget for the most unstable wave in the Charney-Eady branch on the basic state in Figure 1. The box in the upper right corner of the flux term plots indicates the resolution of the Galerkin expansions.*

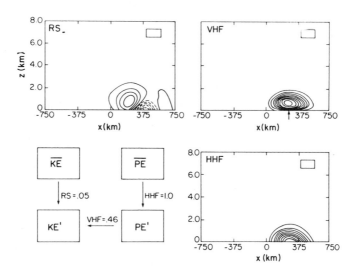

Figure 4: *As in Figure 3, but for the most unstable wave in the cyclone branch. The arrow under the VHF plot shows the location in which the polar lows were observed to develop.*

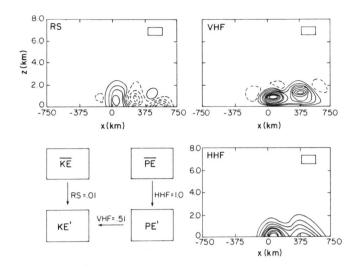

Figure 5: As in Figure 3, but for the most unstable wave in the 1st harmonic cyclone branch.

plots of the spatial distribution of the horizontal (*HHF*) and vertical (*VHF*) heat flux terms and the Reynolds stress term (*RS*) for the most unstable waves in each of the Charney-Eady, cyclone, and first harmonic cyclone branches. The sign convention used in Eqs. (5)–(11) has been adopted. Regions in which the terms are positive (negative) are contoured with solid (dashed) lines. The relative magnitudes of the various terms (normalized by *HHF*) are indicated in the corresponding energy box diagrams.

Inspection of these figures indicates that all three waves are growing by converting potential energy stored in the baroclinic zone into eddy kinetic energy. This conversion is accomplished by means of the baroclinic instability mechanism. As discussed above the Charney-Eady wave is a rather deep disturbance, while both cyclone waves are shallow and boundary confined. The waves in the fundamental branch have a singlet structure with only one maximum in the heat flux terms. By contrast, the waves in the first harmonic branch have a doublet structure with two maxima in the heat flux terms. Also illustrated in Figure 4 is the location in which the polar lows were observed to nucleate. This is exactly the region in which the cyclone wave has its maximum amplitude. It is important to note that in the quasi-geostrophic stability analyses of Reed and Duncan (1987), the region in which the unstable waves developed was very sensitive to the modifications made to the background baroclinic zone. As we have made no such modifications, our theory does not suffer from this unphysical sensitivity.

In summary, our results demonstrate that the stability characteristics of the baroclinic zone shown in Figure 1 are very similar to those of the frontal zones investigated by Moore and Peltier (1987). Of greatest importance is the fact that the baroclinic zone in Figure 1 is indeed unstable to waves in the cyclone scale branch of baroclinic instability. The structure and organization of the most unstable wave in this branch is very similar to that of the polar lows observed by Reed and Duncan (1987). This large measure of similarity leads us to propose that the cyclone scale branch of baroclinic instability discovered by Moore and Peltier (1987) is responsible for the initial development of coherent families of polar lows.

4. CONCLUSIONS

In this paper, we have demonstrated that the environment in which polar low wavetrains are observed to develop is unstable to the cyclone scale branch of baroclinic instability discovered by Moore and Peltier (1987). The predicated doubling time and wavelength of the most unstable wave in the branch are in good agreement with the observations made by Reed and Duncan (1987). Furthermore, the wave was observed to have its maximum amplitude in the region of the zone in which the static stability and Richardson number fields had their minimum. This is precisely the region in which the polar lows in the wavetrain were observed to nucleate. Our ability with a dry adiabatic theory to account for the rapid initial growth of such disturbances argues that their genesis need not involve a strong feedback with moist diabatic processes. Our prediction as to the phase speed of this wave is larger than was observed. However, it should be emphasized that our theory is valid only for the initial stages of the development of the cyclones. The observed phase speeds were those appropriate for finite amplitude disturbances. The finite amplitude behaviour of these disturbances are now being investigated with a nonlinear model that is initialized with the most unstable normal modes predicted by linear theory.

REFERENCES

Bjerknes, J., 1919: On the structure of moving cyclones. *Geofys. Publ.*, *1(2)*.
Bjerknes, J., and H. Solberg, 1922: Life cycle of cyclones and the polar front theory of atmospheric circulation. *Geofys. Publ.*, *1(2)*.

Charney, J.G., 1947: The dynamics of long waves in a baroclinic westerly current. *J. Meteor.*, *4*, 135–162.

Charney, J.G., 1975: Selected papers of J.A. Bjerknes, *Western Periodicals*.

Duncan, C.N., 1977: A numerical investigations of Polar Lows. *Quart. J. Roy. Meteor. Soc.*, *103*, 255–267.

Eady, E.T., 1949: Long waves and cyclone waves. *Tellus*, *1*, 33–52.

Eliassen, A., 1948: The quasi-static equations of motion. *Geofys. Publ.*, *17(3)*.

Harold, T.W., and K.A. Browning, 1969: The polar low as a baroclinic disturbance. *Quart. J. Roy. Meteor. Soc.*, *95*, 710–723.

Hoskins, B.J., and F. Bretherton, 1972: Atmospheric frontogenesis: Mathematical formulation and solution. *J. Atmos. Sci.*, *29*, 11–37.

Jordan, D.W., and P. Smith, 1977: *Nonlinear Ordinary Differential Equations*. Oxford University Press.

Klaassen, G., and W.R. Peltier, 1985: The onset of turbulence in finite amplitude Kelvin-Helmholtz billows. *J. Fluid Mech.*, *155*, 1–35.

Laprise, R., and W.R. Peltier, 1989: The linear stability of nonlinear mountain waves: Implications for the understanding of severe downslope windstorms. *J. Atmos. Sci.*, *46*, 556–577.

Mansfield, D.A., 1974: Polar Lows: The development of baroclinic disturbances in cold air outbreaks. *Quart. J. Roy. Meteor. Soc.*, *100*, 541–554.

Moore, G.W.K., and W.R. Peltier, 1987: Cyclogenesis in frontal zones. *J. Atmos. Sci.*, *44*, 384–409.

Moore, G.W.K., and W.R. Peltier, 1989a: Non-separable baroclinic instability, Part I: Quasi-geostrophic dynamics. *J. Atmos. Sci.*, *46*, 57–78.

Moore, G.W.K., and W.R. Peltier, 1989b: Frontal cyclogenesis and the geostrophic momentum approximation. *Geophys. Astrophys. Fluid Dyn.* (in press).

Reed, R.J., 1979: Cyclogenesis in polar air streams. *Mon. Wea. Rev.*, *107*, 38–52.

Reed, R.J., and C.N. Duncan, 1987: Baroclinic instability as a mechanism for the serial development of polar lows: A case study. *Tellus*, *39A*, 376–384.

POLAR LOWS AS ARCTIC HURRICANES[1][2]

Kerry A. Emanuel
Center for Meteorology and Physical Oceanography
Massachusetts Institute of Technology
Cambridge, Massachusetts, U.S.A.

Richard Rotunno
National Center for Atmospheric Research
Boulder, Colorado, U.S.A

ABSTRACT

Recent aircraft and dropsonde data show that polar lows, like hurricanes, occur within deep moist adiabatic atmospheres and possess warm cores. We propose that at least some polar lows are indeed arctic hurricanes. Using a recently developed theoretical model of the mature hurricane, we show that the observed difference between the moist entropy of the troposphere and that representing saturation at sea surface temperature can sustain moderately intense hurricanes. Unlike the environments of tropical hurricanes, much of this moist entropy difference -results from an air-sea temperature difference. Numerical experiments using an axisymmetric nonhydrostatic model confirm that intense hurricanes can develop in environments typical of those in which polar lows are observed to develop. We show that, like hurricanes, surface flux-driven polar lows cannot arise spontaneously but require an independent and presumably nonaxisymmetric dynamic mechanism for their initiation.

[1]This is a condensed version of an article published in *Tellus, 41A, January* 1989. Authors should cite original reference.

[2]The contribution of Kerry Emanuel was supported by National Science Foundation Grant ATM–8513871 and the Office of Naval Research under contract N00014–87–K–0291.

POLAR AND ARCTIC LOWS
Paul F. Twitchell, Erik A. Rasmussen,
and Kenneth L. Davidson (Eds.)

1. INTRODUCTION

Bergeron (1954) was perhaps the first to remark on the special properties of cyclogenesis over relatively warm water in winter. Noting that certain occluded cyclones were rejuvenated while passing over the North Sea and Baltic, he coined the term "extratropical hurricane" to describe these oceanic cyclones. Bergeron's early intuition is well served by recent satellite observations (e.g., see Rasmussen, 1985) that reveal small, intense vortices with clear eyes surrounded by deep cumulonimbi, and by research aircraft observations (Shapiro et al., 1987) that show that, like their tropical cousins, polar lows have warm, high θ_e cores.

Nevertheless, storms of this kind nearly always form in highly baroclinic environments and, unlike in the case of tropical cyclones, are preceded by powerful upper tropospheric precursors (Businger, 1985). Many polar lows appear to develop near, if not at, the location of the center of deep, cut-off cyclones (Rasmussen, 1985; Businger, 1985), while other small-scale vortices appear to form in polar air within strong zonal flow aloft (Reed, 1979). While it is agreed that all such cyclones form within convectively active polar air, the variety of baroclinic environments in which the storms originate has led to a proliferation of both terminology to describe them and theories for their cause. Some, such as Reed (1979) and Harrold and Browning (1969) have argued that baroclinic instability is the principle mechanism, with the reduced static stability of the low-level atmosphere responsible for both the small scales and rapid time evolution of the cyclones (Staley and Gall, 1977). Others, such as Rasmussen (1979, 1985), proposed that polar lows are the result of conditional instability of the second kind (CISK), perhaps acting in conjunction with baroclinic processes.

The importance of both baroclinic processes and surface fluxes has been highlighted in recent numerical simulations of polar lows. Sardie and Warner (1985) used the National Center for Atmospheric Research/Pennsylvania State University (NCAR/PSU) mesoscale model to simulate observed polar lows over both the Atlantic and Pacific Oceans. They concluded that (moist) baroclinic processes were operative in each case and dominated the development of the Pacific cyclone, while surface heat fluxes were essential in producing a disturbance of the observed amplitude in the Atlantic case. Grφnås et al. (1987) used the operational numerical weather prediction model of the Norwegian Meteorological Institute to simulate several observed cases of polar lows near Scandinavia. In those cases in which sensitivity studies were carried out, the absence of surface fluxes led to somewhat weaker cyclones, though usually some disturbance developed, presumably through (moist) baroclinic instability.

There is evidently more than one mechanism operating to produce the spectrum of phenomena called polar lows, although one mechanism may dominate the other(s) in a particular circumstance. One of these mechanisms is certainly baroclinic instability while the others involve CISK or air-sea interaction. The authors have recently argued against CISK as a mechanism for tropical cyclogenesis (Emanuel, 1986; Rotunno and Emanuel, 1987). In the first place, when viewed in the proper thermodynamic framework, the tropical atmosphere is very nearly neutral to deep moist convection (Betts, 1982), presumably being maintained in the convectively adjusted state by convection itself. The reservoir of available potential energy assumed by Charney and Eliassen (1964) and Ooyama (1964) in their original concept of CISK apparently does not exist in the tropical atmosphere. Secondly, tropical cyclones can be maintained in intense steady states without ambient conditional instability (Emanuel, 1986) and can intensify under the same conditions, provided a starting disturbance of sufficient amplitude exists (Rotunno and Emanuel, 1987). The finite-amplitude nature of the tropical cyclone viewed as an air-sea interaction instability is in good accord with the observation that tropical cyclones always arise from pre-existing disturbances such as easterly waves; this contrasts with the linear nature of CISK.

It seems likely that the convection observed in the environment of polar lows similarly serves to maintain a nearly moist adiabatic lapse rate, with no substantial, stored convective, available potential energy. Soundings near polar lows confirm this supposition (Shapiro et al., 1987). At the same time, the saturation moist entropy of the sea surface may be considerably higher than that of the atmosphere, as in the tropics, though in the case of the polar low environment much of this is due to an actual sea-air temperature difference. These observations lead us to ask whether hurricanes might be possible over the polar oceans in winter.

In the following section, we review the steady-state theory of Emanuel and Rotunno (1989), which is used to calculate the minimum sustainable central pressure of polar lows resulting purely from air-sea interaction. We use a polar low proximity sounding to estimate the central pressure of an observed low and compare this estimate with observations in Section 3. In Section 4 we present the results of numerical experiments using the nonhydrostatic axisymmetric model of Rotunno and Emanuel (1987). A discussion of these results and inferences regarding the role of other processes such as baroclinic instability are presented in Section 5.

2. THE MAXIMUM SUSTAINABLE PRESSURE DROP DUE TO SEA–AIR HEAT FLUX

Emanuel and Rotunno (1989) have developed an extension to the steady-state theory of Emanuel (1986), which accounts for sensible as well as latent heat flux from the sea surface. Broadly, the Bernoulli equation is integrated around a closed circuit, as illustrated in Figure 1. This shows that the total amount of mechanical energy available from the Carnot cycle is

$$\oint T ds = \epsilon T_s (s_c - s_a),\tag{1}$$

where s is the total moist entropy, which at low temperatures may be approximated by

$$s \equiv C_p \ln T + L_v w - R \ln p,\tag{2}$$

and ϵ is the thermodynamic efficiency, defined

$$\epsilon \equiv \frac{T_s \overline{T_o}}{T_s},\tag{3}$$

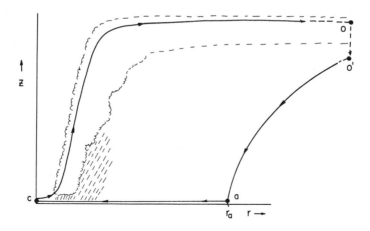

Figure 1. The integral path for the Carnot cycle. Points 0 and 0' are taken to lie at a very large radius.

where T_s is the sea surface temperature (at which entropy is transferred from sea to air) and $\overline{T_o}$ is the mean temperature at which entropy is lost by export or by radiation. In Eq. (1), the subscripts refer to "storm center" and "ambient," respectively, and in Eq. (2) C_p is the heat capacity at constant pressure, R is the gas constant, and L_v is the heat of vaporization.

In the steady state, all of the available mechanical energy is used to balance frictional dissipation. Assuming that most of this occurs in the subcloud layer near the surface, it is possible to equate Eq. (1) to the boundary layer dissipation, which in turn equals the total pressure drop along the surface.

All of the foregoing was shown by Emanuel and Rotunno (1989) to lead to a relation between the pressure drop of steady polar lows and the increase of temperature and humidity between the environment and the center of the polar low. This relation may be written:

$$ \ln x = - \frac{A}{x} - B , \tag{4} $$

where

$$ x \equiv \frac{p_c}{p_a} , $$

$$ A \equiv \frac{e_c}{p_a} \left\{ \frac{\epsilon}{1 - \epsilon} \frac{L_v}{\epsilon R_v T_c} \right\} , $$

$$ B \equiv \frac{T_c - T_a}{T_c} \left\{ \frac{\epsilon}{1 - \epsilon} \frac{C_{pd}}{R_d} \right\} - \frac{e_a}{p_a} \left\{ \frac{\epsilon}{1 - \epsilon} \frac{l_v}{R_v T_c} \right\} , $$

where p is the total pressure and e is the vapor pressure. The term A/x in Eq. (4) reflects the pressure dependence of the water vapor mixing ratio while the first term in B above represents part of the sensible heat input from the ocean. The sum of A/x and the last term in B is the latent heat input.

Equation (4) yields the central pressure of a cyclone maintained entirely by surface heat fluxes, given the central surface air temperature and humidity as well as those quantities evaluated in the ambient environment, and the entropy-weighted mean outflow temperature $\overline{T_o}$. While Eq. (4) is implicit for the ratio $p_o/p_a(\equiv x)$, it is easily solved iteratively. (Emanuel, 1988, showed that Eq. (4) has no solutions under certain conditions, but this will not be a problem in the present investigation.) The maximum sustainable pressure drop is obtained by setting T_c equal to the sea surface temperature and the central relative humidity $=1$.

3. COMPARISON WITH OBSERVATIONS OF A POLAR LOW

A beautiful example of a polar low that occurred in the Barents Sea north of Norway has been described by Rasmussen (1985). This case is particularly suited to comparison with theory since its environment was sampled by rawin-sonde observations within 200 km of the low center and since ship observations were made very close to the low center. The cyclone was typical of polar lows near Scandinavia, with a clear eye surrounded by deep convective clouds. Like many polar lows, this one formed near the center of a deep cut-off cyclone aloft.

The surface analysis at 0000 GMT on 14 December 1982 is reproduced in Figure 2. The cyclone apparently formed 24–36 hr preceding this time somewhere just southwest of Bear Island (the station reporting a temperature of $-8\,°C$ and dewpoint of $-9\,°C$). Satellite imagery presented by Rasmussen (1985) shows that the clear central eye had become less well defined by this time, but the central pressure was well known due to the proximity of a ship. The polar low formed in a region of generally low surface pressure associated with the cyclones at the eastern edge of the map. The ship near the polar low center is reporting a south wind of 20 m s^{-1}, a pressure of 980.2 mb and snow showers. Its temperature ($0\,°C$) and dewpoint ($-1\,°C$) are significantly higher than those of surrounding stations. If we assume that the wind field is in solid body rotation between the location of the ship and the storm center, then cyclostrophic balance gives a central pressure of 975–980 mb (independent of the distance between the storm center and the ship).

The Carnot cycle theorem reviewed in the previous section gives the pressure change between the storm center and the outermost closed isobar, where the air is considered to begin its inward movement toward the center. Inasmuch as the polar low in this case formed in a region of substantial ambient pressure gradient, an unambiguous estimate of the outer pressure is not possible. Based on the analysis shown in Figure 2 and on surface analyses over the preceding 24 hr, we take the outer pressure to be 993 mb.

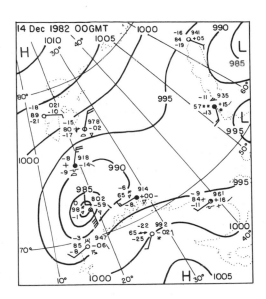

Figure 2. Surface analysis for 0000 GMT 14 December 1982 (from Rasmussen, 1985). Bear Island is near 75°N, 19°E reporting a temperature of −8°C.

The Bear Island sounding taken 12 hr before the analysis time of Figure 2 has been presented by Rasmussen (1985). At that time, satellite imagery (see Rasmussen, 1985) showed a clear central eye about 75 km in diameter located roughly halfway between Bear Island and the northern Norwegian coast. The Bear Island sounding was taken just inside the northern limit of the cloud shield associated with the polar low.

The Bear Island sounding exhibits a nearly moist adiabatic temperature lapse rate between 830 mb and the tropopause at 410 mb, perhaps reflecting moist adiabatic adjustment by deep cumulonimbi a short distance south over the open ocean. A nearly isothermal layer extends from the surface to 830 mb, reflecting arctic air that has traveled southward over ice. Bear Island is located at the southern extremity of the sea ice at this time.

Two obvious difficulties in using this or any other sounding as a reflection of the polar low environment are that the sounding is apt to be changing in time, even away from the cyclone, and the environment varies spatially as one moves around the perimeter of the disturbance. In this particular case, the polar low occurred near (but not necessarily at) the center of a deep cut-off low aloft so that the temperature does not vary much in the horizontal above about 800 mb. But, as is clear in Figure 2, a large north-south temperature gradient exists near the surface in association with the ice edge.

The structure of the Bear Island sounding does suggest, however, that most of the temporal and spatial variability of the atmospheric thermodynamic structure in the environment of the polar low will occur below 830 mb. While surface heating can lead to rapid modification of the sounding below this level, once an adiabatic lapse rate is established from 830 mb downward, further surface heating will lead to turbulent fluxes over a much deeper layer (essentially the whole troposphere) with a correspondingly much slower change in temperature and moisture. Inspection of Figure 2 and the Bear Island sounding reveal that the only surface air that is capable of rising buoyantly above 830 mb is air very close to the polar low center. We therefore feel confident in assuming that the Bear Island sounding accurately reflects the polar low environment *above* 830 mb.

The strategy outlined by Emanuel and Rotunno (1987) for dealing with the environment below 830 mb is simply to perform three separate calculations based on three assumptions about the environment at low levels. These are described as follows:

(1) Cold, Dry (CD). This assumes that the low-level environment is that of the Bear Island sounding. Since this is probably the coldest air at that distance from the low center, this represents an extreme estimate on the cold side.

(2) Warm, Moist (WM). Here we assume that the surface air has the temperature and dewpoint ($-8\,°C$, $-9\,°C$) of Bear Island at the analysis time of Figure 2, and that the sounding is adiabatic from the surface upward to the point at which the relevant moist adiabat intersects the 1200 GMT Bear Island sounding. This represents our "best estimate" of the low-level environment at 0000 GMT 14 December.

(3) Hot, Dry (HD). Here we take the lower atmosphere to be dry adiabatic from the 830-mb level down to the surface, with a lifting condensation level of 830 mb. This represents an extreme estimate on the warm side.

Each of these soundings yields the environmental estimates needed to calculate central pressure using Eq. (4). The central temperature and vapor pressure are taken from the 0000 GMT AMI ship observation near the low center (Figure 2), which recorded a temperature of $0\,°C$ and dewpoint of $-1\,°C$. (A continuous time series of temperature from ship AMI actually shows that the temperature increased to nearly 1 °C as the cyclone passed the ship (see Rasmussen, 1985), but the dewpoint fell to -2°C indicating almost no change in θ_e.)

Calculation of the thermodynamic efficiency, ϵ, requires knowledge of the mean outflow temperature $\overline{T_o}$. This is estimated, as in Emanuel (1986), by assuming that air flows out of the low at its level of neutral buoyancy with respect to the environment. To calculate $\overline{T_o}$, we begin by lifting *ambient* surface air to its level of neutral buoyancy (which may be at the surface) and finding the temperature at that level. Then the θ_e of the surface air is increased by a small increment (we choose 2 °C) and the new temperature at the new neutral buoyancy level is found. This procedure is repeated until we achieve the observed θ_e of the air at the low center. The resulting temperatures are then averaged to find $\overline{T_o}$ and thus ϵ, as defined by Eq. (3).

Clearly, the smallest estimate of ϵ will occur in sounding CD due to the small value of surface θ_e, while the largest will obtain from HD. Notwithstanding, the greatest difference between ambient and central temperature and vapor pressure will occur in sounding CD. These effects tend to cancel each other in the calculation of central pressure using Eq. (4).

Estimates of the central pressure made from the three soundings are listed in Table 1, assuming an ambient pressure of 993 mb. The estimates made using

TABLE 1. ESTIMATES OF CENTRAL PRESSURE FROM EQUATION (4) FOR THREE ENVIRONMENTAL SOUNDINGS*

Sounding	T_c(°C)	e_c(mb)	T_a(°C)	e_a(mb)	ϵ	p_c(mb)
CD	0	5.68	-12	2.45	0.0788	974
WM	0	5.68	-8	3.10	0.0912	978
HD	0	5.68	0	3.10	0.2123	979
WM	7	10.01	-8	3.10	0.181	922
						918**

T_c=central surface air temperature
T_a=ambient surface air temperature
e_c=central surface vapor pressure
e_a=ambient surface vapor pressure

*Assumed ambient pressure is 993 mb.
**Calculated using latent heat of sublimation. Values of p_c estimated above are the same to within 1 mb when heat of sublimation is used.

sounding WM and HD are very close to the observed central pressure of about 978 mb, although the estimate from sounding CD is not far off. It is evident that the difference in soundings at low levels has a noticeable, but not large, effect on the storm's intensity. The good agreement between the estimates of central pressure made using Eq. (4) and observations suggests, but does not entirely prove, that surface sensible and latent heat fluxes were instrumental in the development of this polar low. It could be argued that the high temperature and water vapor content near the low center were the result of horizontal advection, as in a purely baroclinic disturbance; but inspection of Figure 2 and previous surface analyses (Rasmussen, 1985) shows that no air with similar values of θ_e could be found in the environment (though a tongue of high θ_e air may have slipped between the surface observations).

The last entry in Table 1 shows the central pressure that would have resulted if the air near the polar low center had warmed up to and become saturated at the observed sea surface temperature of about 7 °C. This represents the minimum sustainable pressure permitted by the Carnot cycle mechanism. Its value of 922 mb is similar to minimum sustainable pressure estimates of tropical hurricanes. This shows that the thermodynamic potential for hurricanes is large in the polar low environment although, to be sure, the kinematic structure of the ambient environment is very different from that observed in hurricane genesis regions. Since polar lows in this region typically spend less than 2 days over relatively warm water it is not surprising that their central pressures do not approach the minimum value allowed by the Carnot cycle.

It is a straightforward matter to calculate the relative magnitudes of the sensible and latent heat input from the ocean for each of the ambient soundings listed in Table 1 and using the ship observations for thermodynamic data near the low center. The ratios of sensible to latent heat input for soundings CD and WM are 2:6 and 2:2, respectively, showing that the sensible heating is dominant. The ratio for sounding HD is 0:3 with all the sensible heating due to isothermal expansion. For the minimum sustainable pressure calculation, the sensible-to-latent heat ratio is 1:7.

The thermodynamic calculations presented here demonstrate that sea-air heat fluxes can indeed account for much of the energetics of polar lows. In the following section we explore the dynamics of polar lows resulting from air-sea interaction.

4. NUMERICAL SIMULATIONS

In designing numerical experiments suitable for examining the extent to which polar lows can develop through air-sea interaction, it is desirable to exclude other physical processes that might lead to cyclogenesis. For this purpose, we employ a tropical cyclone model whose axisymmetry precludes the operation of baroclinic or barotropic instability. We use the model described by Rotunno and Emanuel (1987) and by Emanuel and Rotunno (1989). Briefly, the model integrates the nonhydrostatic primitive equations expressed in a cylindrical coordinate system with uniform grid spacing in the radial and vertical directions. The model equations consist of conservation laws for momentum, mass, heat, water vapor, and liquid water. The microphysical scheme differs from that used in hurricane simulations in that all condensed water is assumed to be in the form of snow. As a result, we add the latent heat of sublimation and permit no evaporation. Condensate in excess of 1 g kg^{-1} is taken to fall at a constant terminal velocity of 1 m s^{-1}. The turbulence scheme employs an eddy viscosity that depends on the local rate of deformation and Richardson number. Surface fluxes are represented by bulk aerodynamic formulae with wind-dependent exchange coefficients. A sponge layer placed in the stratosphere well above the region containing the main circulation controls gravity wave reflections there while a wave radiation condition at the outer boundary minimizes the amplitude of reflected waves. Radiation is neglected. Moist convection is explicitly, albeit crudely, simulated by the model. Particular values of the parameters used in the simulations are listed in Table 2.

In estimating the initial thermodynamic environment we once again assume that the Bear Island sounding is representative of the polar low environment above 830 mb and perform three separate experiments corresponding to the soundings WM, HD, and CD. The differences between these soundings affect only the lowest thermodynamic grid points in the model.

To initialize the simulations we begin with a kinematic environment similar to that associated with hurricane formation; that is, with no substantial flow aloft. In Emanuel and Rotunno (1989) we examine the effects of an initial cold-core cyclone in the upper troposphere.

As a starting vortex we specify an azimuthal wind field of the form

TABLE 2. MODEL PARAMETERS PARAMETER

Parameter	Value	Description
l_0	200 m	Mixing length
l_H	2000 m	Horizonal mixing length
V	1 m s^{-1}	Terminal velocity of liquid water
		(=0 if $q_t < 1$ g kg^{-1})
τ_R	∞	No radiative damping
$c*$	30 m s^{-1}	Wave speed for open boundary condition
z_{sponge}	12.5 km	Bottom of "sponge" layer
α_{max}	0.013 s^{-1}	Maximum value of damping in "sponge" layer
z_{top}	15 km	Domain top
r_{outer}	1500 km	Domain outer radius
ΔZ	1 km	Vertical grid size
Δr	10 km	Horizontal grid size
Δt	20 s	Time step
f	0.000136 s^{-1}	Coriolis parameter at 70°N latitude

$$V = V_s F_s(r) \begin{cases} \left(1 - \dfrac{x}{z_t}\right) & \text{for } z \leq z_t \\[2ex] \dfrac{H}{z_t}\left(1 - \dfrac{z}{H}\right)\left(\dfrac{z_t - z}{H - z_t}\right) & \text{for } z > z_t , \end{cases} \tag{5}$$

where

$$F_s(r) \equiv \frac{2 \dfrac{r}{r_{max}}}{1 + \left(\dfrac{r}{r_{max}}\right)}$$

The temperature field is then determined by integrating the thermal wind equation outward from the center where the temperature is specified as discussed above. The initial circulation thus consists of a warm core vortex with maximum wind V_s occurring at the surface at a radius of r_{max} and decaying linearly with height to zero at $z = z_t$. The part of the flow above z_t is designed so that $\partial V/\partial z$ (and thus $\partial\theta/\partial x$) is continuous across z_t and such that V vanishes at $z = H$. In practice it results in a weak anticyclonic flow between z_t and H.

In a "control" experiment, hereafter denoted "A," we use sounding WM and set $V_s = 10$ m s^{-1}, $r_{max} = 50$ km, and $z_t = 6.5$ km. The time evolution of the maximum azimuthal wind is shown in Figure 3. After an initial period of adjustment the cyclone intensifies from about 8 m s^{-1} to hurricane force during the 24-hr period beginning 10 hr after initialization. Also shown in Figure 3 is the evolution of maximum wind for two experiments identical to A but using soundings HD and CD (denoted experiments B and C, respectively). As in the steady state theory, there is little difference in the final intensity of the storms.

The structure of the mature stage of the cyclone of experiment A is highlighted in Figure 4, which shows various fields averaged over the period between 80 and 90 hr after initialization. The fields bear a strong resemblance to those characteristic of hurricanes except that the circulation is shallower and the thermal structure shows a warm ring near the surface in addition to the warm core aloft. The former arises as a consequence of the strong sensible heat transfer from the sea surface, which is absent in true hurricanes. As a result of this warm ring stucture, the azimuthal velocity contours slope inward at low levels near the center. The condensate distribution reveals a ring of deep convection

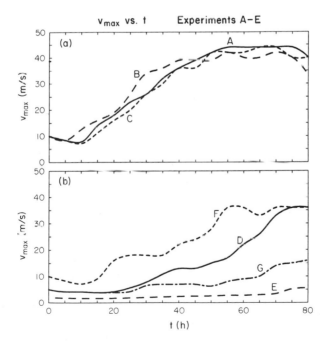

Figure 3. Maximum azimuthal velocity (v_{max}) as a function of time for experiments A–C (a) and D–G (b).

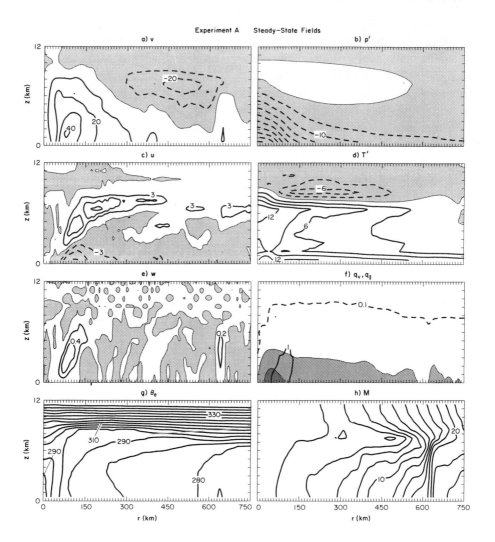

Figure 4. The 80–90 hr average fields for the nearly steady state reached by the control run (experiment A). All stippled regions indicate negative values of the fields. (a) Azimuthal velocity, contour interval 10 m s⁻¹; (b) pressure deviation from initial state, contour interval 5 mb; (c) radial velocity, contour interval 3 m s⁻¹; (d) temperature perturbation from initial state, contour interval 3°C; (e) vertical velocity, contour interval 0.2 m s⁻¹, (f) water vapor mixing ratio (q_v) and snow mixing ratio (q_ℓ). Light shading denotes $1 < q_v < 3g\ kg^{-1}$, dark shading indicates $q_v > 3g\ kg^{-1}$. Snow contours are 0.1 g kg⁻¹ (dashed) and 1 g kg⁻¹ (solid); (g) equivalent potential temperature, contour interval 5°C; (h) angular momentum in dimensionless units such that $M[r=r_{outer}]=100$, contour interval 2.

surrounding a clear central eye with considerable convective activity extending out to large radii. The θ_e field (Figure 4g) bears a strong resemblance to the θ_e cross sections through the polar lows that appear in the mesoscale numerical simulations described by Grønås et al. (1987) and Nordeng (1987).

As in our earlier simulations of hurricanes, the development of polar lows in our model is sensitive to the amplitude and geometric size of the starting vortex. We performed a number of experiments in which the size and amplitude of the initial vortex are varied; these are summarized in Table 3 and the evolution of the velocity maximum with time is shown in Figure 3b. In experiments D and E, the maximum velocity of the initial vortex is reduced to 5 and 2 m s^{-1}, respectively. The former intensified to an amplitude similar to that reached by A, whereas the latter did not amplify significantly. For this initial vortex size, the "critical amplitude" lies somewhere between 2 and 5 m s^{-1}. The sensitivity to initial amplitude is restricted to a determination of whether or not there is amplification; given an initial vortex of sufficient intensity, there will be amplification to a quasi-steady state, which is not itself sensitive to the magnitude of the starting vortex.

Experiments F and G begin with a vortex whose radius of maximum winds lies at 100 km. When the initial amplitude is 10 m s^{-1} (experiment F), the cyclone intensifies to almost the magnitude of the control experiment, whereas experiment G, beginning with half the amplitude of F, does not intensify to the

TABLE 3. INITIAL CONDITIONS USED IN THE NUMERICAL EXPERIMENTS

Experiment	r_{max} (km)	V_s (m s^{-1})	r_m (km)	V_m (m s^{-1})	Sounding
A	50	10		0	WM
B	,,	,,		,,	HD
C	,,	,,		,,	CD
D	,,	5		,,	WM
E	,,	2		,,	,,
F	100	10		,,	,,
G	,,	5		,,	,,
H	50	10	500	20	,,
I	100	,,	,,	,,	,,
J	50	5	,,	,,	,,

r_{max} = radius of maximum surface winds
V_s = maximum surface winds
r_m = radius of maximum winds at tropopause
V_m = maximum winds at tropopause

extent of F. Closer inspection of the velocity fields of experiment G reveals that the surface winds hardly increase at all; the increase of velocity shown in Figure 3 occurs in the upper part of the vortex. Apparently, the ring of convection develops at a larger radius and, as a consequence, the inward branch of the outflow aloft spins up a vortex inside the ring of convection at higher levels. The energy available from the Carnot cycle is used up in amplifying the circulation aloft, which does not have any feedback on the surface fluxes. This argument is quantified in Emanuel and Rotunno (1989).

The time scale of development of disturbances in the model is sensitive to the magnitude of the surface exchange coefficients. It may be supposed that the use of Deacon's formula, which was developed principally for the tropics, may seriously underestimate the fluxes in the highly unstable boundary layers typical of the polar low environment. We therefore made an attempt to incorporate more realistic formulations of surface fluxes that include the wind dependence of the surface roughness length and the bulk stability. A review of the literature on this subject unfortunately reveals a wide spectrum of results; there does not even appear to be agreement on whether the surface exchange coefficients for heat and moisture increase or decrease with wind speed. We did attempt to use the formulation of Louis (1979) to calculate the surface exchange coefficients of heat and momentum as a function of the bulk Richardson number and surface roughness length, the latter of which was estimated using Charnock's relation with a coefficient of 0.032. We found, rather remarkably, that the results were very similar to those given by Deacon's formula. At high wind speeds, the Richardson number is small and the wind dependence of the surface roughness yields results similar to Deacon's formula. At small wind speeds, the Richardson number is large but the surface roughness is small and again the dependence on wind speed is close to that given by Deacon. Moreover, all the coefficients were reasonably close to those measured by Large and Pond (1982) under unstable conditions.

We are discomfited by the foregoing. Measurements of surface fluxes under the highly disequilibrated sea states and thermodynamic states typically accompanying polar lows are scarce and we feel that the actual exchange coefficients may be far larger than current formulations indicate. To investigate the sensitivity of the model to the exchange coefficients we simply doubled them. The result is that the model storm intensifies about 40% faster than the control run, though it reaches the same intensity. We also reran experiment E and found no significant difference up to 40 hr; by 80 hr some higher winds developed as convection broke out over a large area of the domain, but no coherent vortex circulation occurred.

Observations of the type of polar low that we address in this study invariably show that it develops near or at the center of deep, cold core cut-off lows aloft (Businger, 1985). The particular polar low whose environment we specified for the numerical simulations discussed above was an example of one that developed nearly directly under an upper level cyclone (Rasmussen, 1985). Clearly, the cold air associated with such a system is favorable for the development of polar lows under our air-sea interaction hypothesis; but it is not clear what effects the circulations around a large-scale cut-off low might have on the development of polar lows. In order to examine such effects, Emanuel and Rotunno (1989) carried out several additional experiments in which we specified a cold-core upper cyclone in the initial condition. The upper cyclone is initialized with a maximum azimuthal velocity V_n occurring at a radius r_m.

We performed three experiments using different values of V_s and r_{max}; these are listed in Table 3 and the development of the *surface* velocity maxima with time is shown in Figure 5. Experiment H is identical to the control run, A, except for the presence of the cold core vortex aloft in the initial condition. The cyclone develops to almost the same amplitude as A. The fields, which averaged between 80 and 90 hr for experiment H, are displayed in Figure 6 and should be compared to those of A (Figure 4). The surface cyclone is much more concentrated in the presence of the cut-off low aloft and the latter has been destroyed between about 90 and 170 km from the center. In this case, the upper cyclone does not appear to have inhibited development of the polar low. Inspection of Figure 5 reveals, however, that the surface cyclone of experiment I, which is identical to F except for the cut-off low in the initial condition, does not develop

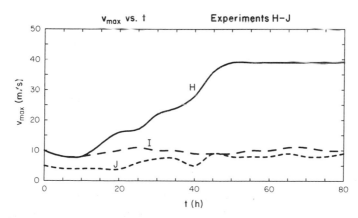

Figure 5. Development of maximum azimuthal velocity at lowest grid points for experiments H–J.

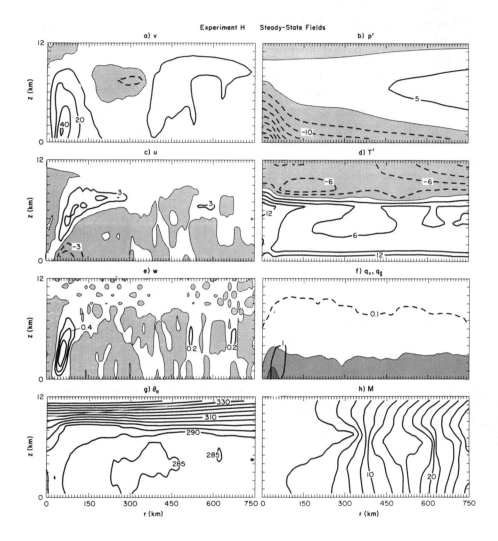

Figure 6. Same as Figure 4 but for experiment H.

at all, in contrast with F. Nor did experiment J develop, which can be compared to D. Close inspection of the velocity fields in these cases (not shown)
reveals that the lack of surface development appears to be for the same reasons
we gave in the case of experiment G; namely, that much of the energy available
from the Carnot cycle flows into the development of a cyclone aloft, inside the
convective ring, where it cannot feed back through the surface fluxes. This more

readily occurs in the presence of an initial cyclone aloft because the deformation radius is smaller for a given radius. For this reason, the "critical" initial radius of maximum surface winds is smaller, and the critical velocity at fixed r_{max} larger when an upper level vortex is present initially.

5. DISCUSSION

There is a certain thermodynamic similarity between polar air masses over relatively warm water and the tropical atmosphere in regions susceptible to hurricanes. In both cases, fluxes of sensible and/or latent heat from the underlying ocean surface can lead to substantial warming of a deep layer of the atmosphere. The particular environment of a polar low investigated by Rasmussen (1985) was shown to be capable of sustaining a pressure drop of up to 70 mb. Moreover, measurements of θ_e near the low center are consistent with the observed pressure drop.

Numerical simulations with an axisymmetric nonhydrostatic model show that air-sea interaction can produce cyclones whose structure is consistent with observations, but whose intensity is somewhat larger. The reasons for the different intensity are not clear, but several points come to mind. Sea surface temperatures near Norway (e.g., see Rasmussen, 1985) are relatively high only in a narrow tongue just west of the coast. The rapidly moving polar low may simply not remain over warm water long enough to achieve its potential intensity. The effect of the disturbances on the sea surface temperature is not known, but the strong winds and cyclonic circulation may induce enough mixing and upwelling to have a noticeable influence on the ocean temperature, as is observed with many tropical hurricanes. It is also true that only a few percent of all hurricanes intensify to the upper bound given by the Carnot theorem, but the reasons for this remain elusive.

We emphasize that, as in the case of tropical cyclones, the polar low environment appears to be stable to small amplitude axisymmetric perturbations; *disturbances of substantial amplitude are apparently necessary to initiate intensification by air-sea interaction*. To the extent that this is a valid finding, it points to the necessity of some presumably nonaxisymmetric dynamical process that operates in the early stages of cyclogenesis. Baroclinic instability is one obvious candidate since baroclinicity is almost always present to some degree in the pre-cyclone environment. It is conceivable that the baroclinic mechanism operates cooperatively with air-sea interaction; if not, the development of polar lows might be thought of as a two-stage process. Other disturbances, such as topographically generated cyclones, might also act as starting disturbances. The nature of the amplitude conditions for growth by the air-sea interaction mechanism is the subject of an on-going investigation by the authors.

Experiments with the numerical model show that the development of axisymmetric polar lows is sensitive to the initial kinematic as well as thermodynamic state of the atmosphere and to the formulation of the surface fluxes. In particular, the large inertial stability associated with cut-off lows aloft makes it necessary to start with a surface cyclone of higher amplitude and/or smaller radius of maximum winds; otherwise, the surface fluxes act to amplify the cyclone aloft rather than at the surface where it can feed back on the surface fluxes.

ACKNOWLEDGMENTS

We should like to thank Joel Sloman for typing the manuscript and Isabelle Kole for drafting several of the figures.

APPENDIX

AN EXPRESSION FOR MOIST ENTROPY VALID IN SATURATED AND UNSATURATED AIR

The following quantity can be shown to be conserved during reversible moist or dry adiabatic expansion:

$$s = (C_{pd} + QC\ell) \ln T + \frac{L_v w}{T} - R_d \ln p_d - wR_v \ln RH , \qquad \text{(A1)}$$

where C_{pd} and C_ℓ are the heat capacities of dry air and liquid water, respectively, Q is the total water content, L_v is the heat of vaporization (a function of temperature), w the vapor mixing ratio, R_d and R_v are the gas constants for dry air and water vapor, respectively, and RH is the relative humidity.

Differentiation of Eq. (A1) yields

$$T \, ds = (C_{pd} + QC_\ell \, dT + L_v dw - \frac{R_d T}{p_d} \, dp_d + w dL_v$$

$$- \frac{L_v w}{T} \, dT - w R_v T \, d \ln RH - R_v T \ln RH \, dw. \qquad (A2)$$

The last term of Eq. (A2) vanishes since reversible changes in w can only occur at $RH=1$. The temperature dependence of L_v is given by

$$dL_v = (C_{pv} - C_\ell) \, dT \, ,$$

where C_{pv} is the heat capacity of water vapor at constant pressure. Also, the Clausius—Clapeyron equation may be written:

$$L_v \frac{dT}{T} = R_v T \frac{de_s}{e_s} = R_v T \left(\frac{de}{e} - \frac{d \, RH}{RH} \right) ,$$

since $e_s = e/RH$. Using these two expressions in Eq. (A2) gives

$$T \, ds = 1(C_{pd} + w C_{pd} + \ell C_\ell) \, dT + L_v dw - \frac{R_d T}{p_d} \, dp_d$$

$$- w R_v T \frac{de}{e} . \qquad (A3)$$

Finally, we note that since $w = \alpha_d/\alpha_v$, where α_d and α_v are the specific volumes of dry air and water vapor, respectively, Eq. (A3) can be written

$$T \, ds = (C_{pd} + w C_{pv} + \ell C_\ell) \, dT + L_v dw - \alpha_d dp , \qquad (A4)$$

which is a direct statement of the first law of thermodynamics written as entropy changes per unit mass of dry air. This proves that Eq. (A1) is a uniformly valid expression for entropy of moist air.

In the relatively cold environments of polar lows we neglect the effect of water substance on heat capacity and on density. If one follows through the derivation of Eq. (A4), it can be seen that consistency demands that we neglect the temperature dependence of L_v/T in the second term on the right of Eq. (A1), the last term in Eq. (A1), and the difference between p and p_d. Thus a consistent low temperature approximation to Eq. (A1) is

$$s \simeq C_{pd} \ln T + \frac{L_{v0}}{T_0} w - R_d \ln p , \qquad \text{(A5)}$$

where L_{v0} and T_0 are constants.

REFERENCES

Bergeron, T., 1954: Reviews of tropical hurricanes. *Quart. J. Roy. Meteor. Soc.*, *80*, 131–164.

Betts, A.K., 1982: Saturation point analysis of moist convective overturning. *J. Atmos. Sci.*, *39*, 1484–1505.

Businger, S., 1985: The synoptic climatology of polar low outbreaks. *Tellus*, *37A*, 419–432.

Charney, J.G., and A. Eliassen, 1964: On the growth of the hurricane depression. *J. Atmos. Sci.*, *21*, 68–75.

Emanuel, K.A., 1986: An air-sea interaction theory for tropical cyclones, Part I: Steady-state maintenance. *J. Atmos. Sci.*, *43*, 585–604.

Emanuel, K.A., 1988: The maximum intensity of hurricanes. *J. Atmos. Sci.*, *45*, 1143–1155.

Emanuel, K.A., and R. Rotunno, 1989: Polar lows as arctic hurricanes. *Tellus*, *41A*, 1–17.

Grφnås, S., A. Foss, and M. Lystad, 1987: Numerical simulations of polar lows in the Norwegian Sea. *Tellus*, *39A*, 334–353.

Harrold, J.W., and K.A. Browning, 1969: The polar low as a baroclinic disturbance. *Quart. J. Roy. Meteor. Soc.*, *95*, 710–723.

Large, W.G., and S. Pond, 1982: Sensible and latent heat flux measurements over the ocean. *J. Phys. Oceanogr.*, *12*, 464–482.

Louis, J.-F., 1979: A parametric model of vertical eddy fluxes in the atmosphere. *Boundary Layer Meteor.*, *17*, 187–202.

Nordeng, T.E., 1987: The effect of vertical and slantwise convection on the simulation of polar lows. *Tellus*, *39A*, 354–375.

Ooyama, K., 1964: A dynamical model for the study of tropical cyclone development. *Geofis. Int.*, *4*, 187–198.

Rasmussen, E., 1979: The polar low as an extratropical CISK disturbance. *Quart. J. Roy. Meteor. Soc.*, *105*, 531–549.

Rasmussen, E., 1985: A case study of a polar low development over the Barents Sea. *Tellus, 37A,* 407–418.

Reed, R.J., 1979: Cyclogenesis in polar airstreams. *Mon. Wea. Rev., 107,* 38–52.

Rotunno, R., and K.A. Emanuel, 1987: An air-sea interaction theory for tropical cyclones, Part II: Evolutionary study using a nonhydrostatic axisymmetric numerical model. *J. Atmos. Sci., 44,* 542–561.

Sardie, J.M., and T.T. Warner, 1985: A numerical study of the development mechanism of polar lows. *Tellus, 37A,* 460–477.

Shapiro, M.A., L.S. Fedor, and T. Hampel, 1987: Research aircraft measurements of a polar low over the Norwegian Sea. *Tellus, 39A,* 272–306.

Staley, D.O., and R.L. Gall, 1977: On the wavelength of maximum baroclinic instability. *J. Atmos. Sci., 34,* 1679–1688.

ON THE GENESIS OF POLAR LOWS

Hans Økland
Institute of Geophysics, University of Oslo
Oslo, Norway

ABSTRACT

No doubt, some polar lows develop as perturbations on a baroclinic zone. This happens usually along the east coast of Greenland. Over the eastern part of the Norwegian Sea and over the Barents Sea the baroclinicity is weak, and the vortices that develop there must be caused by some other mechanism, not fully understood. The most obvious suggestion is that heating by convection may be the energy source, and the major part of the paper is devoted to a discussion of this idea.

Much evidence supports the assumption that comparatively deep convection takes place at the centre of the low, and the paper discusses the origin of the conditionally unstable layer. At greater distance from the centre the convection is shallow because of a low-level inversion. Here the surface air may be rapidly heated and moistened. The paper also discusses the consequences of an asymmetric shape of the low.

1. INTRODUCTION

The existence of mesoscale vortices embedded in cold air flows has been recognized by Scandinavian meteorologists for a long time. The small, and often very intense, lows seemingly develop over the ocean where the air mass is in a state of rapid transformation over the comparatively warm water. Baroclinicity, however, is not very pronounced, and so it was thought that the cyclogenesis was associated with the low static stability in the convective layer. For this reason, the phenomenon was sometimes referred to as instability lows (Rabbe, 1975).

POLAR AND ARCTIC LOWS
Paul F. Twitchell, Erik A. Rasmussen,
and Kenneth L. Davidson (Eds.)

179

British meteorologists, at the same time, had noted certain disturbances that cross the British Isles from northwest or north. They were called polar lows, presumably because they develop in the interior of the polar air mass and obviously are not connected to the main polar front. A case study by Harrold and Browning (1969) showed a structure resembling a baroclinic wave.

In a way the different names reflect the two different "explanations" put forward as the cause of these disturbances: Based on a case study, Mansfield (1974) showed that the evolution of the low was consistent with baroclinic wave theory, while two Scandinavians, Økland (1977) and Rasmussen (1979), attempted to show that heating by convection might be the cause. The difference in opinion can perhaps be explained in terms of geography. As we shall see, the preferred place for a baroclinic zone is the Greenland–Jan Mayen–Iceland area, while the baroclinicity is weak over the central and eastern part of the Norwegian Sea. (We are, of course, referring exclusively to weather situations in which polar lows occur.) Accordingly, most polar lows passing Great Britain probably are of the baroclinic type, while those observed in the Scandinavian area may have a different origin, not yet fully understood. However, we do not exclude the possibility that the two processes may work in combination.

In the absence of a sufficiently strong baroclinicity the heat flux from the water surface might be the energy source. If so, it is necessary that the heating has a spatial variation. This brings up the question of the horizontal scale. In linear instability theory (CISK) a simple feedback mechanism between the rate of heating and the intensity of the vortex is postulated. The dominating scale then emerges as the dimension of the vortex with the greatest growth rate. More likely, the scale is imposed by spatial variations in the external conditions. In any case, the intensification must be rapid enough to explain the explosive developments sometimes observed.

It will be shown later that in a characteristic case the turbulent fluxes of sensible and latent heat take about equal parts in the air mass transformation in general. Attention has been drawn mostly to the latent heat, because of the analogy there appears to be between polar lows and tropical cyclones (i.e., small scale, strong winds, and small baroclinicity). Generally speaking, a cold climate makes the influence of latent heat less likely since the water vapour mixing ratio is limited by the low saturation level. However, this may be compensated by low static stability, which can be seen by considering the production of available potential energy. Following Lorenz (1955), we may write

$$A = \frac{1}{2} \int_0^{p_0} \overline{T'^2} \; / \; [\overline{T}(\Gamma_d - \Gamma)] \; dp \qquad (1)$$

where the bar indicates average over an isobaric surface, while $\Gamma d = g/c_p$, $\Gamma = \partial T / \partial z$ and T' is the departure of temperature from its isobaric mean value. Localized heating may increase the numerator, and the effect on A is great if the denominator is small.

That release of latent heat may be an important energy source is demonstrated by Rabbe (1987) in his case study of a polar low that enters the coast of Norway. Heavy precipitation, concentrated in an area less than 100 km across, was observed where the polar low centre crossed the coastline. Note, however, that when the convection is vigorous, the lapse rate is kept close to the adiabats, the dry one below the condensation level, and the moist one above. The sensible heat and the net release of latent heat contribute jointly to the warming of an average air column through the whole depth of the convective layer.

In Section 3 we shall be mostly concerned with the influence of the heat fluxes from the ocean. However, we first discuss briefly the role of baroclinic instability in the light of the information derived from case studies and simulations.

2. THE POLAR LOW AS A BAROCLINIC
WAVE DISTURBANCE

Extremely cold air often flows southward over the ice-covered ocean east of Greenland. Since the glacier reaches the average height of the 700-mb surface, the flow below this level is blocked towards the west, while relatively warm air is present over the open waters farther to the east and southeast. The cold air stream is usually quite strong and often extends to the latitude of Iceland. The polar lows, which develop in this flow, appear to be baroclinic waves in their initial structure (Reed and Duncan, 1987). However, since the shear, defined as the vertical derivative of the geostrophic wind, is opposite to the surface wind, the disturbances may move "backward" and the wind and pressure fields have a somewhat unfamiliar appearance.

Simulations of polar lows of this kind have been made by Grønås et al. (1987). The simulations appeared to be quite successful in some respects, but did not quite reproduce the intensity. This was particularly true for the shallow lows with small horizontal scales, and points to the possibility that some specific mechanism was missing or poorly taken care of in the model. The guess of the authors was that the release of latent heat was the problem.

At this point we refer to a paper by Haugen (1986). He attempted to simulate the growth of a baroclinic wave, starting with idealized initial data resembling the flow structure east of Greenland, and discovered that during the cyclone formation an area with conditionally unstable stratification developed near the low centre in the initially stable air. Since he used an adiabatic integration scheme, the reason for this instabilization must be chiefly vertical stretching associated with the cyclogenesis. Thus, areas that favour deep convection may be created inside the developing baroclinic wave.

3. THE HEAT FLUX FROM THE SEA

In this section we discuss some general aspects of heating by convection from the sea surface. We are specifically interested in the depth of the convective layer, since there is much evidence in favour of the view that most polar lows are associated with convection of a rather deep nature, considering the latitude and climate. In fact, Wilhelmsen (1986) counted 38 cases of "gale-producing lows" in the period 1977–1982, and found a conditionally unstable lapse rate between the surface and the 500-mb level in all cases. Similar facts are revealed by many case studies (Rabbe, 1987; Økland, 1987). We shall pursue this view in the following.

Figure 1 shows the temperature and dewpoint soundings from (1) Bear Island and (2) Bodφ, 12 hr later, in an outbreak of cold air. A trajectory calculated from the geostrophic wind brings a point located near Bear Island at the time of the sounding to a point 150 km northwest of Bodφ during the following 12 hr (Økland, 1976). Calculations of the heat fluxes through a horizontal surface moving along the sea level trajectory give the total amount of heat as 15×10^3 kJ m^{-2} and 13×10^3 kJ m^{-2} for sensible and latent heat, respectively. The calculations are based on the so-called bulk formulas, using observed sea surface temperature and the transfer coefficients $C_H = C_E = 1.4 \times 1.0^{-3}$. The air temperature was calculated repeatedly along the trajectory, assuming that the heat was transferred to a vertical air column. These numbers may be compared to the difference in enthalpy between the two soundings in Figure 1, assuming that the change below 700 mb is caused entirely by the heat fluxes from the ocean. It turns out

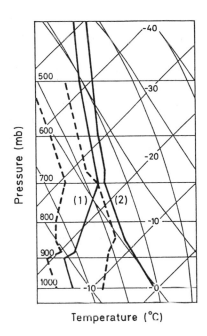

Figure 1. Soundings of temperature (full lines) and dew point (dashed lines) for (1) Bear Island (74.3⁰N 19.0⁰E), 0000 GMT 21 November 1975 and Bodφ *(67.2°N 14.2°E), 1200 GMT 21 November 1975.*

that an air column with unit cross section has increased its content of sensible heat by 25×10^3 kJ, and of latent heat by 5×10^3 kJ. The comparatively small value of latent heat must be due to condensation. The total amount is seen to be in excellent agreement with the flux computations.

The sounding from Bear Island (Figure 1) is quite typical for cold air outbreaks, and may be considered as characteristic for the air mass in its source area over the ice-covered Arctic Ocean. We note specifically the low-level inversion. At greater heights (above 800 mb, say) the stability varies from case to case. On Figure 1 it is absolutely stable, but may very well be conditionally neutral or unstable. We shall return to this point on several occasions in the following.

When the air enters the warm ocean, the low-level inversion is gradually eroded by a convective layer, which increases in thickness and temperature downstream. This process may be evaluated by means of formulas given by Økland (1983). Under certain simplifying assumptions he finds that the downstream change taking place in a moving air column is measured by a characteristic length scale, L, along the trajectory and a height scale, H, for the convective layer thickness. Both scales are proportional to the initial air-sea temperature difference and inversely proportional to the potential temperature lapse rate, $\partial\theta/\partial z$. Using data from sounding (1) in Figure 1, we find that $H=4200$ m. This, in fact, represents an asymptotic value, which is approached rapidly in the first part of the trajectory, but very slowly later on. Specifically, a thickness of 3100 m is reached after a trajectory length of about 700 km, in fair agreement with the observed values.

The above result demonstrates that for a comparatively stable initial sounding, the convective layer depth may be quite moderate even in the southern part of the Norwegian Sea. However, if the lapse rate is small at greater heights, the depth increases much more rapidly as soon as the lower, stable part has been eroded. In fact, if the atmosphere above some height is conditionally unstable, as the sounding marked (1) in Figure 2, the increase is explosive. The conclusion is that deep convection is reached only in those cases where a layer of weak stability is present on top of the low-level inversion. That this is not always the case is demonstrated by the sounding marked (2) in Figure 2. Also, case studies show that the depth of the conditionally neutral layer increases dramatically during the approach of a polar low (Økland, 1987). We intend that this is what causes the low, rather than being a consequence of its existence.

Before leaving this section we shall discuss an observation often referred to in papers about polar lows. On a map showing the relative topography, for instance between the 1000-mb and 700-mb surfaces, there usually appears to be a certain baroclinicity over the ocean in cold air outbreaks. This is caused by the downstream increase in temperature and depth of the convective layer, and it is highly questionable if this baroclinicity can support baroclinic wave development, taking into account that the real thickness of the baroclinic air mass equals the height of the convective layer top, and the very intense vertical mixing.

Frontogenesis along the ice edge, which is also often mentioned, must be a rare event, possible only when the wind is very weak or blowing slowly parallel to the ice border. The previously mentioned baroclinic zone east of Greenland is primarily of other origin, but may to some extent be strengthened by the proximity of the ice border and the ocean front between the East Greenland Current and the Gulf Stream water.

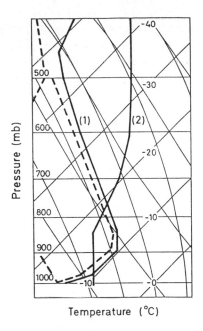

Figure 2. As Figure 1, but for (1) Bear Island, 0000 GMT 3 December 1980, and (2) Bear Island, 0000 GMT 15 November 1983.

4. VORTEX GENERATION BY ORGANIZED CONVECTION

In conditionally unstable air heating from the sea surface can sustain convection on the scale of the cumulus or cumulonimbus. However, the scale of the polar low is at least two orders of magnitude larger. On this scale the air mass in essence is stable for Benard type convection. There is as yet no clear indication of an eye-wall circulation such as tropical cyclones. The "eye" sometimes observed on satellite pictures of polar lows may have other causes. For instance, the fact that the fluxes from the sea are greatest where the wind speed has its maximum, may produce subsidence in the cyclone centre (Økland and Schyberg, 1987). In absence of a nondiluted cyclone-scale circulation driven by release of latent heat in the ascending part, a "mean" circulation may be forced by the average heating from many cumulonimbi (Eliassen, 1952). In this case no part of the circulation is buoyant. The scale of the perturbation is determined by the scale of the average heating rate, which may simply be introduced by the external conditions as an initial perturbation of finite amplitude.

Skepticism has been voiced about the validity of CISK as an "explanation" of tropical cyclones. It may very well be that a highly nonlinear structure like the hurricane cannot be satisfactorily understood on the basis of a linear theory. The same may be true for the mature polar low. However, since heating by cumulonimbus convection certainly takes place as a part of the air mass transformation, it is conceivable that some sort of coupling between the heating rate and the vortex may be present.

In the previous section we found that even if midtropospheric instability is present, no deep convection takes place before the low-level stable layer has been destroyed. As soon as this happens, the convection rapidly grows to much greater heights. Examples may be found on most good infrared satellite pictures of a polar low. High and cold clouds usually tower over the central part of the low, while warmer stratocumulus clouds cover large areas of the surrounding ocean (see for instance Shapiro et al., 1987).

Surprisingly few measurements have been made of the cloud top temperature in polar lows. However, Rabbe (1987) finds clouds as cold as $-51°C$, placing the tops above the 400-mb level. Of course, one cannot be certain that high clouds found over a polar low really is the top of the cumulonimbi (i.e., anvil cirrus). In fact, in some cases they have such large extension that they probably have other sources. Nevertheless, the clear comma structures, often observed, strongly indicate a connection to the polar low circulation.

Next, there is the question about the source of the low midtropospheric stability, sometimes (but not always) found in the arctic air mass. We have already mentioned that a developing baroclinic wave may create areas with conditionally unstable air. However, the source area for the arctic air mass is mostly characterized by little baroclinic activity, actually a condition for the production of the low-level inversion. Nevertheless, the middle troposphere may interact barotropically with synoptic scale disturbances at lower latitudes, so that vorticity and associated temperature gradient are advected to the polar area. In fact, Rasmussen (1985) demonstrates that at least one of the polar lows he studied was triggered by an upper tropospheric trough. Since such troughs are usually not noticeable at the surface, they must have a cold core and a relatively low static stability. It is our belief that this is the nucleus for the polar low development, rather than the cyclonic vorticity associated with the trough.

Above we have tried to show that deep convection seems to be a characteristic feature of polar lows, especially the small and intense ones. However, we have not yet explained why shallow convection should be less effective. In fact, any low-level source of vapour produces heating if the latent heat is released in clouds and, at the first sight, the shallow clouds do not seem to be less effective than the deep ones. On the contrary, the shallow convection might produce stronger heating since the mass of air to be heated is smaller. This circumstance is reflected in many studies of CISK in connection with polar lows (Bratseth, 1985).

In another place this author suggested that the release of precipitation could be a clue to the problem discussed above (Økland, 1987). Most of the known processes that produce precipitation are much more effective in deep clouds than in shallow ones. Therefore, as an average over the air volume containing cumulonimbi, the precipitation acts as a sink in the amount of liquid and frozen water, and presumably also reduces the mean relative humidity. A small relative humidity enhances the evaporation from the sea surface, thus increasing the amount of latent heat. Condensation products that stay in the air may evaporate in the sinking part of the convection, thus cooling the air. We conclude that the net heating by release of latent heat is substantial only in deep convection.

On the background of the previous discussion we propose the following structure of a polar low, which intensifies because of heating by convection. In the centre of a developing vortex the low-level stable layer is missing, either because it has been eroded by vigorous convection, or because the air has another origin. The last possibility may for instance take place if there is a strong northerly wind to the west of a stationary trough. If the air mass at greater heights has small static stability, the convection will be deep and form precipitation. This may happen over a fairly restricted area where the midlevel stability has a minimum, or over a local area of strong winds or high sea surface temperature. In both cases the initial development is triggered by an existing disturbance, for instance a trough in the lower and middle troposphere.

Outside the area with deep convection the low-level inversion usually persists in a large part of the peripheral sections of the low. Here the heating is confined to the shallow convective layer, and because the release of precipitation is small, the same will be true for the content of the water substance. However, the flux from the sea surface is comparatively large because of the strong wind in the low. The Ekman layer flux towards low pressure brings this air into the central part of the low where the heat and moisture feed the deep convection. A schematic picture of this idea is given in Figure 3.

The processes, which we have tried to describe, may be further intensified if we consider the asymmetry of most polar lows. This asymmetry, pictured in Figure 3, may be caused by the fact that the low is embedded in a northerly

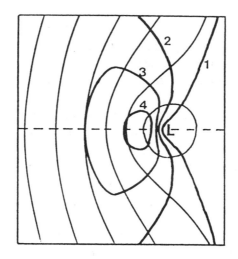

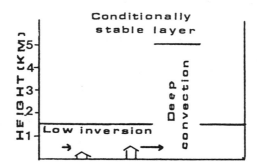

Figure 3. Sketch of a polar low caused by convection. Top plate shows isobars (thin) and isotachs (thick). The numbers on isotachs are relative. Bottom plate shows vertical cross section through the low center. See text for further explanation.

air flow, or located near the shear zone between strong northerly flow to the west and weak winds towards the east. The asymmetry causes an area of strong wind to the west of the low. In this wind maximum the heat flux from the sea is large. In fact, this is nicely demonstrated in the case study by Shapiro et al. (1987), although their case has a somewhat more complicated structure. The mass flux in the Ekman layer is known to be proportional to the wind speed. Since the heat flux from the sea and the Ekman layer inflow both depend on the wind speed, we are obviously dealing with a nonlinear effect and also a feed-back mechanism, since the wind speed depends on the intensity of the low.

That an asymmetric structure is more effective than a symmetric one may be seen by considering the following equation

$$\overline{F_S\, F_E} = \overline{F_S}\, \overline{F_E} + \overline{F_S'\, F_E'} \tag{2}$$

where F_S and F_E are the flux from the sea and the Ekman layer flux, respectively. Furthermore, the bar indicates mean along concentric circles around the low centre, and primed variables are departures from this mean. The last term, representing the effect of the asymmetry, is positive because of the correlation between the two fluxes.

5. CONCLUSIONS

In many studies of polar lows, theories for tropical cyclones have been adapted with the proper modification. However, it is not sufficient to show that the theories apply in arctic regions. In addition, it must be explained why similar lows do not develop in middle latitudes, for instance in maritime polar air masses. In fact, at the west coast of Norway there is maximum frequency of thunderstorms in the winter. This thunderstorms occur in strong winds blowing from the Atlantic, and indicate deep and vigorous convection. Troughs of relatively small dimensions are frequent. Yet, polar lows are never observed. We maintain that this is so because the low-level inversion is missing in this case.

Earlier in this paper we stated that ideas about polar low development should be shown to predict the rapid intensification often observed. This we have not been able to do regarding the propositions we have set forth in this paper. The mechanism we have described is not well suited to analytical methods. Presumably, a suitable numerical model is necessary.

REFERENCES

Bratseth, A.M., 1985: A note on CISK in polar air masses. *Tellus, 37A*, 403–406.
Eliassen, A., 1952: Slow thermally or frictionally controlled meridional circulation in a circular vortex. *Astronomica Norvegica 5*, 19–60.
Grϕnås, S., A. Foss, and M. Lystad, 1987: Numerical simulations of polar lows in the Norwegian Sea. *Tellus, 39A*, 334–353.
Harrold, T.W., and K.A. Browning, 1969: The polar low as a baroclinic disturbance. *Quart. J. Roy. Meteor. Soc., 95*, 710–723.

Haugen, J.E., 1986: Numerical simulations with an idealized model. *Proceedings, International Conference on Polar Lows,* The Norwegian Meteorological Institute, Oslo, Norway.

Lorenz, E.N., 1955: Available potential energy and the maintenance of the general circulation. *Tellus, 7,* 157–167.

Mansfield, D.A., 1974: Polar lows: the development of baroclinic disturbances in cold air outbreaks. *Quart. J. Roy. Meteor. Soc., 100,* 541–554.

Økland, H., 1976: *An Example of Air-Mass Transformation in the Arctic and Connected Disturbances of the Wind Field.* Rep. DM-20, Department of Meteorology, University of Stockholm, Stockholm, Sweden.

Økland, H., 1977: *On the Intensification of Small-Scale Cyclones Formed in Very Cold Air Masses Heated by the Ocean.* Institute Rep. Series No. 26, Institute of Geophysics, University of Oslo, Norway.

Økland, H., 1983: Modelling the height, temperature and relative humidity of a well-mixed planetary boundary layer over a water surface. *Boundary-Layer Meteor., 25,* 121–141.

Økland, H., 1987: Heating by organized convection as a source of polar low intensification. *Tellus, 39A,* 397–407.

Økland, H., and H. Schyberg, 1987: On the contrasting influence of organized moist convection and surface heat flux on a barotropic vortex. *Tellus, 39A,* 385–389.

Rabbe, Å., 1975: Arctic instability lows. *Meteorologiske Annaler, No. 6,* Meteorologisk Institutt, Oslo, Norway, 303–329.

Rabbe, Å., 1987: A polar low over Norwegian Sea 29 February–1 March 1984. *Tellus, 39A,* 326–333.

Rasmussen, E., 1979: The polar low as an extra tropical CISK disturbance. *Quart. J. Roy. Meteor. Soc., 105,* 531–549.

Rasmussen, E., 1985: A case study of a polar low development over the Barents Sea. *Tellus, 37A,* 407–418.

Reed, J.R., and C.N. Duncan, 1987: Baroclinic instability as a mechanism for the serial development of polar lows: A case study. *Tellus, 39A,* 376–384.

Shapiro, M.A., L.S. Fedor, and T. Hampel, 1987: Research aircraft measurements of a polar low over the Norwegian Sea. *Tellus, 39A,* 272–306.

Wilhelmsen, K., 1986: Climatological study of gale-producing polar lows near Norway. *Proceedings, International Conference on Polar Lows,* The Norwegian Meteorological Institute, Oslo, Norway, 31–39.

THEORETICAL INVESTIGATION OF SPIRAL FEATURES IN MESOLOW CIRCULATIONS

William H. Raymond
Cooperative Institute for Meteorological Satellite Studies
University of Wisconsin
Madison, Wisconsin, U.S.A.

ABSTRACT

Many cold air vortices are associated with a comma or spiral cloud con-figuration. These vortices develop within a day of a cold air outbreak and are commonly found in situations where cold air is over warm water. Their early development stage is thought to be enhanced by stretching of low-level absolute vorticity and by diabatic heating processes. To understand better how the spiral feature is related to the mesolow and its environment, an instability study is performed using the two-dimensional barotropic perturbation vorticity equation in cylindrical coordinates. A normal mode approach is utilized to predict the eigenvalues and eigenfunctions. Experiments are conducted using different azimuthal wave numbers. The environmental flow is three dimensional, axisym-metric, and contains convergence. For a perturbation azimuthal wave number of one the growing modes take on the familiar comma-spiral pattern.

1. INTRODUCTION

Comma cloud patterns are common to the polar vortices (Rasmussen, 1983). Those with deep convection contain more of a spiral pattern (Forbes and Lottes, 1985). (This suggests possibly that the comma is just an under-represented spiral.) Rasmussen (1981) has also reported on one case that had a hurricane-like struc-ture. These cold air vortices are found over warm water (Sardie and Warner, 1983) and have a typical spacing or wavelength of 1000–1500 km (Reed, 1979). They develop in an air mass that is moderately baroclinic and conditionally unstable below 850 mb, and the notch of the comma lies in the region of strong

POLAR AND ARCTIC LOWS
Paul F. Twitchell, Erik A. Rasmussen,
and Kenneth L. Davidson (Eds.)

191

horizontal shear on the poleward side of the jet stream (Reed, 1979). Also it has been suggested that stretching of ambient absolute vorticity is important in the early development of these vortices (Forbes and Lottes, 1985).

In addition to polar lows other phenomena also exhibit spiral patterns. In many cases these spirals are associated with disturbances having strong rotation. These include, in addition to hurricanes, the spiral generated on the water surface by waterspouts (Golden, 1974) and in tornadic mesolows (Anderson, 1985). Consequently, rotation may also be an important element in many circumstances.

There has been a large number of theoretical investigations to explain spiral structures in tropical disturbances, especially hurricanes. With respect to the latter, Wexler (1947), Haurwitz (1947), and Kuettner (1959) thought that the elongation of convective cells in the tangential direction was caused by vertical wind shears. MacDonald (1968) pointed to possible Rossby-like waves in the hurricane vortex. Tepper (1958) and Abdullah (1966) hypothesized that the spiral bands are gravity waves. Anthes (1972) related these bands to internal gravity waves. Kurihara and Tuleya (1974) also related the band structure to internal gravity waves while Mathur (1975) suggested that the bands are related to latent heat release. Kurihara (1976) analyzed growing perturbations created from the different physical factors, given an analytical three-dimensional axisymmetric mean flow field. Willoughby (1977, 1978) elaborated further on the role of outward and inward propagating inertia-buoyancy waves. Modeling the comma formation has also been successful, e.g., comma-shaped vorticity features have been reproduced in numerical forecast models of tropical storms (Tuleya and Kurihara, 1984).

We have carried out an instability study to understand better how the spiral feature is related to the mesolow and its environment. The two-dimensional barotropic perturbation vorticity equation in cylindrical coordinates is solved using a normal mode approach to predict the eigenvalues and eigenvectors. Because vortex stretching is observed in association with the early stages of the polar vortex (Forbes and Lottes, 1985) we focus on the importance of a mean flow containing convergence.

In this study we consider the influence of the converging environmental flow. It is easily established from an analysis of the vorticity equation that vertical stretching leads to growing modes and enhanced vorticity. In our experiment we focus on the pattern or horizontal shape of these growing modes. Details of the instability study are now presented.

2. THE EXPERIMENT

2.1 The Perturbation Equations

We want to find the eigenvectors describing the solution to the linearized depth averaged vorticity equation in the presence of a mean axisymmetric field containing convergence. To do this we first separate the total horizontal wind components u_T and v_T into a perturbation and a mean, i.e.,

$$u_T = u(r,\phi,t) + U(r,z), \tag{1a}$$

$$v_T = v(r,\phi,t) + V(r,z). \tag{1b}$$

Here the lower case denotes the perturbation, and the upper case is the mean, while r, ϕ, z, and t are the radial, azimuthal, vertical, and time coordinates, respectively. Note that the perturbation quantities are barotropic while the mean windfield is steady. The latter is axisymmetric but varies in the vertical and radial directions. It is the purpose of this study to examine the barotropic response under conditions when the mean field contains convergence.

The linearized depth averaged barotropic equation in cylindrical coordinates for the perturbation vorticity ζ is

$$[\partial/\partial t + U_a \partial/\partial r + (V_a/r)\, \partial/\partial \phi]\zeta$$

$$+ u[\partial/\partial r]\zeta_a + \zeta D_a + (\zeta_a + f)D = 0, \tag{2}$$

Here the subscript denotes a vertical average, hence ζ_a and D_a are the vertically averaged mean vorticity and divergence fields while D is the perturbation divergence. In this report we limit ourselves to the case when the mean field contains a radial inflow component only, i.e., there is no tangential environmental wind or vorticity component. Nevertheless we present the full set of equations.

The solution procedure is now outlined. Equation (2) is placed in non-dimensional form using r_o and U_{max} representing a characteristic radial measure and a velocity scale, respectively. Next we assume a streamfunction approach where

$$u = -(1/r)\,[\partial/\partial \phi]\psi \text{ and } v = [\partial/\partial_r]\psi. \tag{3}$$

Consequently D is set to zero in Eq. (2) and

$$\zeta = [\partial^2/\partial r^2 + (1/r)\, \partial/\partial r + (1/r^2)\, \partial^2/\partial\phi^2]\,\psi. \tag{4}$$

Utilizing a normal mode approach we let

$$\psi = \Psi(r)\, \exp[im(\phi - \alpha t)]. \tag{5}$$

Here m is an azimuthal wave number, and the eigenvalue α has both real and imaginary parts,

$$\alpha = \alpha_{real} + i\alpha_{img}. \tag{6}$$

The problem then is to determine the eigenvalues and eigenvectors describing the solution of the perturbation vorticity [Eq. (2)] for modes that grow in time. This requires that we solve a third order ordinary differential equation for the variable $\Psi(r)$. Note that in our investigation no horizontal wave number is imposed on our solution. However three boundary conditions are required. At $r=0$ we take $u=v=0$, while at r_{max} we require either that the perturbation vorticity $\zeta=0$ or the perturbation tangential velocity $v=0$. These conditions are then written in terms of the streamfunction utilizing Eqs. (3) and (5). In the numerical calculations the grid step size is 0.005 and the number of discrete grid points used in the radial direction is 1000. This number of grid points was selected after testing various combinations between 100 and 10,000, utilizing grid step sizes varying between 0.05 and 0.0005. The influence of the domain size was also examined by trying different values for r_{max}. The eigenvalues are determined in the usual numerical manner.

2.2 The Mean Equations

The steady axisymmetric equations, in dimensional form, describing the mean flow for the radial, tangential, and vertical velocities, i.e., $U(r,z)$, $V(r,z)$, and $W(r,z)$ respectively, are

$$U[\partial/\partial r]\, U + W[\partial/\partial z]\, U - V^2/r - fV$$

$$= -[\partial/\partial r]\,\Phi + K[\partial^2/\partial z^2]\, U, \tag{7a}$$

$$U[\partial/\partial r]\,V + W[\partial/\partial z]\,V + UV/r + fU = K[\partial^2/\partial z^2]\,V, \tag{7b}$$

$$[\partial/\partial r]\,U + U/r + [\partial/\partial z]\,W = 0 . \tag{7c}$$

Here K is an eddy diffusivity coefficient while Φ is the geopotential height.

To study the importance of convergence we must include the boundary layer. Consequently vertical diffusion terms are included in Eq. (7). Nevertheless, it is reasonable to assume over a finite vertical depth much larger than the boundary layer that the vertical average flow is nearly in gradient balance. Consequently in an average sense gradient balance, i.e.,

$$V^2/r + fV = [\partial/\partial r]\,\Phi, \tag{8}$$

is a realistic approximation.

In this study we are not concerned with the details of the mean flow other than wanting our choices for $U(r,z)$ and $V(r,z)$ to be realistic and requiring that steady-state solutions exist. If for the sake of argument we assume that Eq. (8) holds true then Eq. (7a) implies that

$$U[\partial/\partial r]\,U + W[\partial/\partial z]\,U = K[\partial^2/\partial z^2]\,U. \tag{9}$$

Then Eqs. (7b), (7c), (8), and (9) are four equations that describe four unknowns. Equations (9) and (7c) uniquely define U and W. The uncoupling of the radial and vertical wind components from the tangential velocity is analogous to the first term in the similarity expansion used in Raymond and Kuo (1982). From these two equations it can be shown that U is proportional to z provided U is also linear in r and diffusion is ignored. Because a steady solution exists of this special case there should be no difficulty with the existence of solutions to Eqs. (7a), (7b), and (7c) provided the hydrostatic assumption is invoked. Kurihara's (1976) analytical mean fields describe such a steady axisymmetric three-dimensional situation.

In this study we assume that over some finite depth in the convergence layer the depth-averaged expressions for U and V are

$$U_a(r) = -\beta r, \tag{10a}$$

$$V_a(r) = \begin{cases} \gamma r, \ r < r_o, \\ \\ \gamma r^{-\delta}, \ r > r_o. \end{cases} \tag{10b}$$

The above formulation is thought to hold over a considerable depth during the early development phase. In the current study $\gamma = 0$ and $\beta = 0.05$.

3. DISCUSSION OF RESULTS

3.1 An Analytic Solution

An analytic solution to the perturbation vorticity equation, Eq. (2) written in a normal mode formulation, can be determined for the special case when the mean flow contains constant convergence and solid rotation. The mathematical procedure requires the expansion of ζ into a real and imaginary part and the collection of the real and imaginary terms into two first-order equations. Then by reducing this system to one second-order differential equation we can easily obtain the analytical solution. Away from $r = 0$ the real part of the solution for the perturbation vorticity is

$$\zeta_{real} = r^{-A}\{c^1(1 + A/B) \cos[B \ ln(r)]$$

$$+ c_2(-1 + A/B) \sin[B \ ln(r)]\}, \tag{11}$$

where c_1 and c_2 are constants and

$$A = (D_a + m\alpha_{img})/(-\beta), \tag{12a}$$

$$B - m(a_{real} - \gamma)/(-\beta). \tag{12b}$$

For ζ to approach zero as r goes to infinity requires that $m\alpha_{img} < -D_a$. Here $m\alpha_{img}$ is the growth rate, which must be restrained when $D_a < 0$. Otherwise the solution on a finite area is not attainable. This restriction can be taken as an upper bound on the growth rate. In all of the numerical cases we have tested this requirement has always been satisfied.

3.2 Numerical Solutions

One eigenvector solution obtained from Eq. (2) for an azimuthal wave number $m = 1$ is shown in Figure 1. Note the comma pattern in the vorticity field. Also illustrated in Figure 1 are the wind vectors, which show a jet-like feature with the largest velocities in the vicinity of the comma. The flow pattern of the wind into the comma is realistic (see Tuleya and Kurihara, 1984).

Kurihara (1976) found that there are two types of spirals. Those with the spiral extending clockwise are called N-type while those with a counterclockwise extension are denoted as S-type. For each of these there are two modes. The G-mode is associated with $\alpha_{real} > 0$ for the N-type and with $\alpha_{real} < 0$ for the S-type. The H-mode refers to just the opposite configuration. Outward propagating gravity waves are associated with the G-mode while inward propagating bands

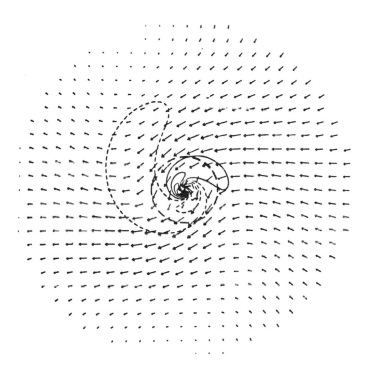

Figure 1. Vorticity (solid, positive and dashed, negative) and wind vector pattern for a growing barotropic mode. The azimuthal wave number is m=1. The circular heavy dashed line is the contour of the characteristic radius r_o.

are identified with the H-mode (Kurihara, 1976). Following the notation of Kurihara (1976) the comma configuration in Figure 1 is an N-type and has a H-mode because the real part of the eigenvalue is negative. The S-type G-mode counterpart is shown in Figure 2a. The other two combinations have also been obtained but are not shown.

For the eigenvector solution shown in Figures 1 and 2a the e-folding time, i.e., $1/\alpha_{img}$, is 22.5 hr while the period for one complete cycle is 79.8 hr. In relating our calculations to a dimensional time we assume scale factors $r_o = 10^5$ m and $U_{max} - 10$ m s^{-1}. The corresponding vertically averaged environmental divergence is $D_a = 1.0 \times 10^{-5}$ s^{-1}.

Shown in Figure 2b is the vorticity pattern for an azimuthal wave number $m = 2$ in Eq. (5). Note that the vorticity is concentrated near the center of the domain. This is in contrast to eigenvectors obtained using an environmental field that contains a tangential wind component (not shown). Solutions obtained using the nonzero γ in Eq. (10b) show patterns that are more reminiscent of hurricane-like structures. In these solutions the vorticity bands are located at a distance larger than r_o, which represents the radius of the maximum mean tangential wind. (The r_o radius is signified by the heavy dashed circle in the figures.) These results will be presented in detail elsewhere.

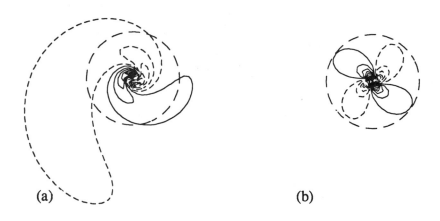

(a)   (b)

Figure 2. The vorticity pattern is shown for an azimuthal wave number m=1 (a) and for wave number m=2 (b).

In our investigation of the barotropic vorticity equation we have computed the first 100 eigenvalues and eigenvectors and in all cases the patterns are all similar. Other processes are also important, like latent heat release, and these will obviously influence the growth rates and the pattern. Nevertheless Kurihara (1976) found the different physical processes stimulated similar patterns.

We have chosen in this report to highlight the horizontal pattern in the barotropic perturbation vorticity fields for two different azimuthal wave numbers when there is convergence in the environmental flow. The vorticity pattern for the azimuthal wave number $m = 1$ appears very realistic. Our investigation concludes that comma vorticity configurations can be generated as a consequence of vortex stretching.

REFERENCES

Abdullah, A.J., 1966: The spiral bands of a hurricane: A possible dynamic explanation. *J. Atmos. Sci.*, *23*, 367–375.

Anderson, C.E., 1985: The Barneveld tornado: A new type of tornadic storm in the form of a spiral mesolow. Preprint, *Fourteenth Conference Severe Local Storms*, Indianapolis, Ind., American Meteorological Society, 289–292.

Anthes, R.A., 1972: Development of asymmetries in a three-dimensional numerical model of the tropical cyclone. *Mon. Wea. Rev.*, *100*, 461–476.

Forbes, G.S., and W.D. Lottes, 1985: Classification of mesoscale vortices in polar airstreams and the influence of the large-scale environment on their evolution. *Tellus*, *37A*, 132–155.

Golden, J.H., 1974: The life cycle of Florida Keys' waterspouts. *J. Appl. Meteor.*, *13*, 676–692.

Haurwitz, B., 1947: Internal waves in the atmosphere and convection patterns. *Ann. N. Y. Acad. Sci.*, *48*, 727–748.

Kuettner, J., 1959: The band structure of the atmosphere. *Tellus*, *11*, 267–294.

Kurihara, Y., 1976: On the development of spiral bands in a tropical cyclone. *J. Atmos. Sci.*, *33*, 940–958.

Kurihara, Y., and R. E. Tuleya, 1974: Structure of a tropical cyclone development in a three-dimensional numerical simulation model. *J. Atmos. Sci.*, *31*, 893–919.

MacDonald, N.J., 1968: The evidence for the existence of Rossby-like waves in the hurricane vortex. *Tellus*, *20*, 138–150.

Mathur, M.B., 1975: Development of banded structure in a numerically simulated hurricane. *J. Atmos. Sci.*, *32*, 512–522.

Rasmussen, E., 1981: An investigation of a polar low with a spiral cloud structure. *J. Atmos. Sci.*, *38*, 1785–1792.

Rasmussen, E., 1983: A review of mesoscale disturbances in cold air masses. In D.K. Lilly and T. Gal-Chen (Eds.), *Mesoscale Meteorology - Theories, Observations and Models,* Reidel, 247–283.

Raymond, W.H., and H.L. Kuo, 1982: Simulation of laboratory vortex flow by axisymmetric similarity solutions. *Tellus, 34,* 588–600.

Reed, R.J., 1979: Cyclogenesis in polar airstreams. *Mon. Wea. Rev., 107,* 38–52.

Sardie, J.M., and T.T. Warner, 1983: On the mechanism for the development of polar lows. *J. Atmos. Sci., 40,* 869–881.

Tepper, M., 1958: A theoretical model for hurricane radar bands. Preprint, *Seventh Weather Radar Conference,* Miami, American Meteorological Society, K56-65.

Tuleya, R.E., and Y. Kurihara, 1984: The formation of comma vortices in a tropical numerical simulation model. *Mon. Wea. Rev., 112,* 491–502.

Wexler, H., 1947: Structure of hurricanes as determined by radar. *Ann. N. Y. Acad. Sci., 48,* 821–844.

Willoughby, H.E., 1977: Inertia-buoyancy waves in hurricanes. *J. Atmos. Sci., 34,* 1028–1039.

Willoughby, H.E., 1978: A possible mechanism for the formation of hurricane rainbands. *J. Atmos. Sci., 35,* 838–848.

CHAPTER 3 – MODELING / SIMULATIONS

Introduction

The size of a polar low is small compared with the characteristic wavelength of baroclinic waves in the atmosphere. The present upper air stations supplemented with satellite and buoy observations are probably sufficient to provide a reasonably accurate analysis of the normal baroclinic waves, but the situation is quite different for polar lows.

In the research mode it is possible to study individual polar lows by obtaining observations from specialized platforms such as research aircraft, but in daily operational meteorology it is necessary to analyze the lows using standard observations from the ordinary network supplemented, if possible, with observations from satellites and buoys. Such observations will be used in operational data analysis and assimilation schemes as input to models. Large-scale, global analysis procedures use normally optimum interpolation schemes, but they have the undesirable property of filtering out systems smaller than a certain scale. They are therefore not directly applicable to mesoscale systems.

Present systems to analyze polar low situations use multivariate analyses made by a successive correction method with an effective initialization performed with a dynamic method. A very essential part of the whole procedure is the 6-hr forecasts made with a limited area model with a rather small grid size (about 50 km). The first paper in this chapter, by Midtbφ et al., describes the use of such an analysis scheme in a particular case.

As just mentioned, a mesoscale model is used as part of the continuing process of analysis and so on. Needless to say, the properties of the mesoscale model are of primary importance for the success of such an alternating procedure. This leads us directly to the description of the various physical processes incorporated through parameterization and the horizontal and vertical resolution necessary for an essentially correct description of the structure, formation, and development of polar lows.

The remaining three papers in the chapter (Nordeng et al.; Langland and Miller; and Pedersen) deal with various aspects of the physical processes and with resolution questions. It appears from these papers that the processes that are necessary for the understanding of polar lows are the air-sea interactions, a proper description of the atmospheric boundary layer, and a proper convection/condensation prescription. In addition, a rather high horizontal and vertical resolution is needed for a reasonably accurate description of a polar low.

While the various cases, which have been studied in detail, indicate that very signficant progress has been made in the general under-standing and numerical simulation of polar lows in the last decade or so, it is also apparent that some refinements in parameterizations are needed. It is, for example, stressed that new parameterization schemes are needed for the convection/condensation processes on the smaller mesoscale to account for and to forecast the hurricane-like disturbances that sometimes appear on the smaller scale.

Other studies indicate that polar lows can be described correctly only if the surface conditions are observed with sufficient accuracy. For example, errors in the sea surface temperature of a degree of two can have marked influence on the simulation. It seems also to be important in some cases that the ice edge is correctly located in the analysis.

Concerning the horizontal resolution it appears that grid sizes of 30 km or even less may be necessary, while operational models approach 20 vertical levels.

While the alternating process of analysis and short range forecasts do much to improve the initial state, if the forecast model is good, it is stressed repeatedly that a real problem is the verification of the forecasts due to the sparsity of observations and their sometimes poor quality. This problem can be alleviated only by better and more dense observational networks, which in turn will make the analysis procedures more accurate in general.

The general trend is certainly to use the primitive equations for simulations and forecasts, but the last paper in the chapter is an example of the use of quasi-geostrophic and semi-geostrophic equations for diagnostic studies of the vertical velocities in polar lows.

ANALYSES OF POLAR LOWS IN THE NORWEGIAN SEA WITH A MESOSCALE LIMITED AREA NUMERICAL WEATHER PREDICTION SYSTEM

Knut Helge Midtbø, Sigbjørn Grønås, Magne Lystad, and Thor Erik Nordeng
The Norwegian Meteorological Institute
Oslo, Norway

ABSTRACT

Problems are encountered when analyzed polar lows are discussed. The benefit of using data from a mesoscale numerical weather prediction model for this purpose is demonstrated on one polar low case.

1. INTRODUCTION

Shapiro et al. (1987) analysed a polar low based on detailed observations from a research aircraft. In this case extensive subjective analyses, projected to one time step, were made. Operationally, polar lows have to be analysed from conventional observations and satellite imagery. Analyses of this kind are presented in the literature, e.g., Harrold and Browning (1969), Rasmussen (1985), and from the Norwegian Polar Lows Project, Lystad (1986).

In numerical weather prediction (NWP) the objective analysis is an updating of a short range prediction by available observations. It is shown by Hollingsworth (1987), that in such a process, the model prediction contributes substantially to the analysis, even in data dense areas. Statistical interpolation (OI) used in most assimilation systems, has the effect of filtering out observed systems smaller than a certain scale. Both this well-known property of the OI method and the lack of observations contribute to problems when analysing mesoscale systems like polar lows.

POLAR AND ARCTIC LOWS
Paul F. Twitchell, Erik A. Rasmussen,
and Kenneth L. Davidson (Eds.)

At the Norwegian Meteorological Institute a limited area mesoscale NWP system is operational. The practical use of this system for forecasting polar lows is discussed by Midtbø and Lystad (1987). In this paper we look into the potential and limitations of this system when making an objective analysis of polar lows. One example of an analysis of a polar low over the Norwegian Sea is presented and evaluated. Emphasis is put on the three-dimensional description given by the NWP model.

2. THE MESOSCALE LIMITED AREA NWP SYSTEM

We are running two limited area models. The first is called LAM150 since the grid size is 150 km. The integration area is large, and prognoses from global models are used for lateral boundary conditions. LAM150 is run every sixth hour and up to 42 hr three times a day (0000, 1200, and 1800 GMT).

The second routine is a mesoscale NWP system called LAM50. It has been operational since 1985 and is run four times a day. Boundary values are taken from the LAM150 routine from the same cycle. Both in LAM150 and LAM50, 11 pressure levels are used for analysis and 10 sigma levels in the prediction model. The prediction model, initalization, and analysis procedures are the same for LAM150 and LAM50 and are developed at The Norwegian Meteorological Institute. The first model version was described by Grønås and Hellevik (1982). The explicit time integration method is according to Bratseth (1983), together with a pressure gradient averaging described by Brown and Campana (1978). The model has proper dispersion properties, and no explicit lateral diffusion is used. The model physics are described by Nordeng (1986) and are computed every 15 min in both routines.

Efficient initialization is performed with a dynamic method proposed by Bratseth (1982) and further tested by Bratseth (1987). Multivariate analyses (Grønås and Midtbø, 1986, 1987) are made by a successive correction method proposed by Bratseth (1986). The geostrophic coupling and the scale of the three-dimensional error statistics are reduced with the number of iterations. In this way details are introduced in the analysis where the data density is high. The data assimilation system used in the case presented here (January 1987) has geostrophic constraints in the multivariate coupling between mass and wind field. A more sophisticated balance is used in a later version of the system. Our experience with LAM50 is very promising (Grønås et al., 1987b), and LAM50 gives a better prediction than LAM150. The model is found valuable for polar low forecasting (Grønås et al., 1987a; Nordeng, 1987).

3. ANALYSIS OF A POLAR LOW 23–26 JANUARY 1987

This case is selected from the operational routine and is an example of a slowly developing polar low. The development took place over the open sea southwest of Spitsbergen near 75°N 5°E during a 3-day period from 23–26 January 1987. The general weather situation and the results from the data assimilation with the LAM50 model will be described below.

The weather maps from the first day of the 3-day period, 23 January 1987, are shown in Figure 1. An area of high surface pressure covers France and the British Isles. The polar front stretches from southern Greenland eastwards north of Iceland and toward southern Norway. Another high covers the Spitsbergen area and extends towards the North Pole. Two days later a deep low is situated at the coast of northern Norway. Strong cold air outflow takes place in the eastern part of the Norwegian Sea. Sea level pressure analysis from 0000 UTC 25 January 1987 is shown in Figure 2. The polar low is situated west of Spitsbergen and it is intensifying. The situation 1200 UTC and 1800 UTC 26 January are shown in the surface maps in Figures 3 and 4. The satellite image in Figure 7 is from 1129 UTC on the same day.

Analysis and 6-hr prognosis repeated four times a day give an objective description of the polar low development in this case. The data assimilation

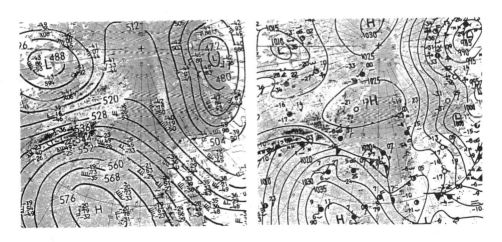

Figure 1. Analysis 0000 UTC 23 January 1987, 500 hPa (left) and MSLP (right), (from Europaischen Wetterbericht by Deutscher Wetterdienst). The polar low development took place southwest of Spitsbergen.

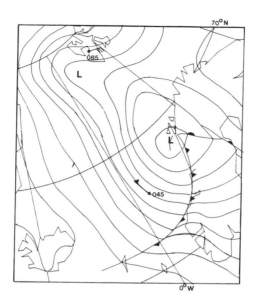

Figure 2. Subjective analysis of mean sea level pressure 0000 UTC 25 January 1987.

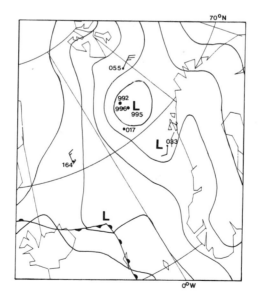

Figure 3. Subjective analysis of mean sea level pressure 1200 UTC 26 January 1987.

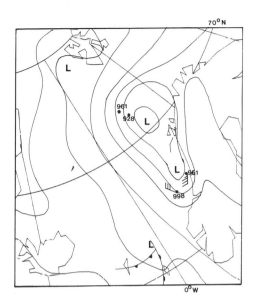

Figure 4. Subjective analysis of mean sea level pressure 1800 UTC 26 January 1987.

scheme uses data from radiosonde stations, synops, and drifting buoys. A few maps show results from the model runs. Most maps are selected from 26 January when the polar low reached its mature stage near 72°N 10°E.

The radiosoundings have the largest impact on the analyses, but the network is so coarse that only the large-scale weather systems are described sufficiently well by this type of observation. Observations from the surface are more dense both in space and time and should therefore have the potential to describe smaller scale weather systems. There are, however, problems to distinguish between incorrect observations and small-scale variations. Another problem is to extend the information from the surface to higher levels. The impact of the observations at one time is shown in Figure 5 by plotting the corrections (increments) added to the initial guess (prognosis) to obtain the analysis. This field of increments is typical for the case studied here, and it is seen that the influence from the observations are on a rather large horizontal scale. Since the resulting analysis describe smaller scales than seen in the increments, those details must have been created by the model.

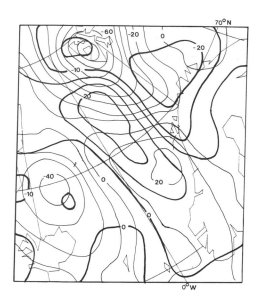

Figure 5. Height increments (m) added to the first guess (6-hr prognosis) to obtain the objective analysis 1200 UTC 26 January 1987. The thick and thin lines are increments in 1000 and 500 hPa, respectively.

The details in the model description is demonstrated in Figure 6 where a 6-hr prognosis valid at 1200 UTC 26 January is shown. Three parameter fields are plotted: mean sea level pressure (MSLP), temperature in 850 hPa and 3 hr of accumulated precipitation ending at 1200 UTC. The prognosis shows that the central pressure has dropped to below 995 hPa and that there is strong cold air advection to the northeast and west of the polar low centre. The warmest air is located near the centre. A band of precipitation stretches north to south and turns around the centre on the east and north side. There is an area free of precipitation in the centre of the low. As usual, there are too few observations to verify the prognostic fields. The subjective analysis (Figure 3) has a central pressure near 995 hPa and is very similar to the prognosis. The computed precipitation cannot be verified over sea, but the satellite image, Figure 7, gives an indication of the areal distribution. Several features in the precipitation band and in the cloud cover look similar, for example the open eye in the center of the low. There is also an indication of a secondary development just south of 70 °N, which is seen as a hook in the cloud cover. The prognosis shows an area of maximum precipitation and a turn of the precipitation band. The details of the prognosis seem therefore to be justified.

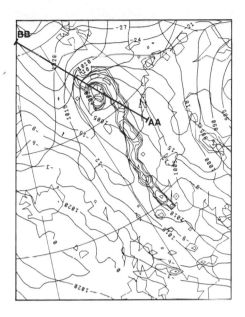

Figure 6. Six-hour prognosis valid 1200 UTC 26 January 1987. The map shows sea level pressure (hPa), temperature (°C) in 850 hPa and 3-hr accumulated precipitation (in mm). The line from AA to BB indicates the position of the cross section shown in Figure 8.

The three-dimensional structure of the polar low may be studied by displaying different parameters in a vertical cross section through the low. In Figure 6 we have placed a cross section from northwest (BB) to southeast (AA) through the center of the polar low. The result is seen in Figure 8 where three parameters are displayed: potential temperature (mainly horizontal isolines, except for the boundary layer and areas near the centre), vertical velocity (closed isolines), and vectors showing the displacement of an air parcel in the cross section during a 1-hr period. Following an air parcel from the north towards the centre we see that it is heated on its way to the centre of the low where it is forced to ascend. Strongest vertical motion is to the south of the centre where the maximum value is $1.5 \ 10^{-2}$ hPa s^{-1}. In the centre there is descending motion. The variation of the static stability may also be deduced from the figure.

Another way of utilizing the data from a NWP system is to compute trajectories of air parcels. Figure 9 shows trajectories in the lowest model layer,

Figure 7. Infrared satellite image from NOAA-9 1129 UTC 26 January 1987 (received by Tromsø Satellite Station, Tromsø, Norway). A part of Spitsbergen is seen to the left. The coast of northern Norway is seen to the right of the mid-point of the image.

all ending in the area where the polar low has its strongest development. The air originates from different areas over the ice and is differently heated from below before ending up in the same area.

We have compared the satellite image (Figure 7) and the precipitation pattern (Figure 6). Several other images and maps (not displayed here) have also been investigated. Based on our investigation, we conclude that the model description in this case is quite realistic. We must, however, make one important reservation. As in many other polar low cases there are too few traditional observations for a detailed verification of the model simulation. We will also mention that at 1800 UTC 26 January the secondary low seen in Figure 4 is described as a weak trough in the model analysis. This indicates that there might be some deficiencies in the model description of this secondary low 6 hr earlier.

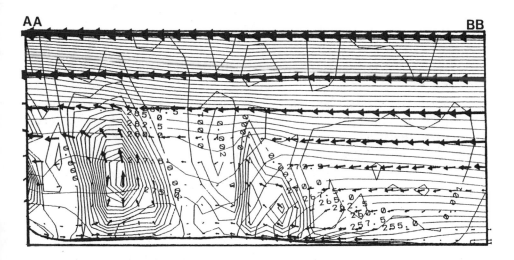

Figure 8. Cross section from south (AA) to north (BB) through the centre of the polar low 1200 UTC 26 January 1987 (see Figure 6). The cross section is computed from a 6-hr prognosis valid 1200 UTC. The top of the figure corresponds to 200 hPa while the bottom line is 1000 hPa. Isolines are drawn for potential temperature (K) and omega (hPa s^{-1}). Maximum vertical velocity is of the order of 10 cm s^{-1}. Arrows show displacements of air parcels during 1-hr period.

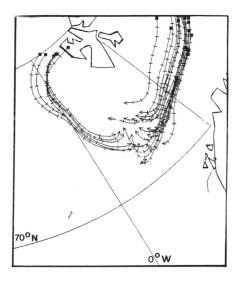

Figure 9. Trajectories in the lowest area layer in the LAM50 model starting 0000 UTC 25 January 1987 and ending 24-hr later. Starting points are indicated by squares and every second hour is indicated by a mark (|).

4. DISCUSSION

By using standard observations and satellite images one can follow the track and get a good idea about the intensity of a polar low near the surface. The interaction between the surface and the upper air (the three-dimensional structure) is subject to uncertainty and cannot be properly analysed from sparse observations by traditional subjective methods. It would be helpful to have a conceptual model to fit with the observations and in that way build up a three-dimensional structure. A problem, however, is to come up with a conceptual model that covers the whole range of different polar low developments.

The NWP data assimilation system has the advantage of interpolating the data in a dynamically consistent way. In addition the model creates data of equal value as the observations. The result must in many cases be viewed as the best possible description of the state of the atmosphere on the synoptic and larger scales.

We believe that numerical models will be increasingly important for polar low case studies. Both increased resolution and more sophisticated physical parametrization will hopefully lead to better simulations of polar lows in the future. However, even more detailed measurements than those given by Shapiro et al. (1987) will be necessary in order to verify all the details in the simulations.

REFERENCES

Bratseth, A.M., 1982: A simple and efficient approach to the initialization of weather prediction models. *Tellus, 34,* 352–357.

Bratseth, A.M., 1983: Some economical, explicit finite-difference schemes for the primitive equations. *Mon. Wea. Rev., 111,* 663–668.

Bratseth, A.M., 1986: Statistical interpolation by means of successive corrections. *Tellus, 38A,* 439–447.

Bratseth, A.M., 1987: *Efficient Dynamical Initialization of a Limited Area Model.* Rep. No. 65, Institute of Geophysics, University of Oslo, Norway.

Brown, J.A., and K.A. Campana, 1978: An economical time-differencing system for numerical weather prediction. *Mon. Wea. Rev., 106,* 1125–1136.

Grønås, S., and O.E. Hellevik, 1982: *A Limited Area Prediction Model at the Norwegian Meteorological Institute.* Tech. Rep. No. 61, The Norwegian Meteorological Institute, Oslo, Norway.

Grønås, S., and K.H. Midtbø, 1986: *Four Dimensional Data Assimilation at the Norwegian Meteorological Institute.* Tech. Rep. No. 66, The Norwegian Meteorological Institute, Oslo, Norway.

Grønas, S., and K.H. Midtbø, 1987: Operational multivariate analyses by successive corrections, *Short- and Medium-Range Numerical Weather Prediction*. Special Volume of the Journal of the Meteorological Society of Japan, Tokyo, Japan.

Grønås, S., A. Foss, and M. Lystad, 1987a: Simulation of polar lows in the Norwegian Sea. *Tellus, 39A,* 334–353.

Grønås, S., A. Foss, and K.H. Midtbø, 1987b: The Norwegian mesoscale NWP system. *Proceedings of the Symposium Mesoscale Analysis and Forecasting, Vancouver, Canada, 17–19 August 1987,* ESA SP-282, 481–486.

Harrold, T.W., and K.A. Browning, 1969: The polar low as a baroclinic disturbance. *Quart. J. Roy. Meteor. Soc., 95,* 710–723.

Hollingsworth, A., 1987: *Objective Analysis for Numerical Weather Prediction*. Special Volume of the Journal of the Meteorological Society of Japan, Tokyo, Japan.

Lystad, M., 1986: *Polar Lows in the Norwegian, Greenland and Barents Sea.* Final Report of the Polar Lows Project, The Norwegian Meteorological Institute, Oslo, Norway.

Midtbø, K.H., and M. Lystad, 1987: Forecasting polar lows using a numerical mesoscale model. *Proceedings of the Symposium on Mesoscale Analysis and Forecasting, Vancouver, Canada, 17–19 August 1987,* ESA SP-282, 487–490.

Nordeng, T.E., 1986: *Parameterization of Physical Processes in a Three-Dimensional Numerical Weather Prediction Model.* Tech. Rep. No. 65, The Norwegian Meteorological Institute, Oslo, Norway.

Nordeng, T.E., 1987: The effect of vertical and slantwise convection on the simulation of polar lows. *Tellus, 39A,* 354–375.

Rasmussen, E., 1985: A case study of a polar low development over the Barents Sea. *Tellus, 37A,* 407–418.

Shapiro, M., L.S. Fedor, and T. Hampel, 1987: Research aircraft measurements of a polar low over the Norwegian Sea. *Tellus, 39A,* 272–306.

ON THE ROLE OF RESOLUTION AND PHYSICAL PARAMETERIZATION FOR NUMERICAL SIMULATIONS OF POLAR LOWS

Thor Erik Nordeng, Anstein Foss, Sigbjørn Grønås, Magne Lystad, and Knut Helge Midtbø

The Norwegian Meteorological Institute
Oslo, Norway

ABSTRACT

The role of horizontal resolution and physical parameterization in numerical simulations of polar lows are illuminated by comparing results obtained from various experiments. The study suggests that new parameterization schemes for convection/condensation have to be constructed for the mesoscales in order to forecast the smallest scale hurricane-like disturbances.

1. NUMERICAL SIMULATIONS OF POLAR LOWS

Analyses of polar lows, e.g., Shapiro et al. (1987), show a strong vertical circulation connected to the inner core of the disturbance. In this process, fluxes of heat from the ocean are efficiently fed into the system and released, mainly through condensation at medium levels. Recently, Emanuel and Rotunno (1989) suggested that this process is self-amplifying and named it Air-Sea Interaction Instability. Theoretical analysis and numerical experiments show that the process can intensify and maintain polar lows after some other processes have initiated the development. A successful numerical simulation of this process depends crucially on a correct description of surface fluxes and condensation processes. Simulations of polar low developments, e.g., Grønås et al. (1986), Nordeng (1987), and Langland and Miller (1988) have so far failed to describe the intensity of the circulation. Due to the small horizontal scale of the disturbance, it is suggested that this comes from poor horizontal resolution in the models. To look into the effects of physical parameterization and resolution, we have rerun

POLAR AND ARCTIC LOWS
Paul F. Twitchell, Erik A. Rasmussen,
and Kenneth L. Davidson (Eds.)

217

some of the cases from Grønås et al. (1986) with a finer resolution of the
Norwegian mesoscale limited area model (25 km mesh, 16 levels). New versions
of the parameterization of the boundary layer and convection are included.

In Section 2 we will present results from these experiments together with
some other model results. Section 3 presents a successful simulation of an
explosively deepening small-scale polar low, while the results are discussed and
suggestions for further work are presented in Section 4.

2. THE EXPERIMENTS

2.1 Physical Parameterization

The physical parameterization used in Grønås et al. (1986) was rather
simple. Nevertheless, polar low disturbances were forecasted at approximately
the right time and place for all test cases. The following lists characterize the
old and new versions of the Norwegian mesoscale limited area models. It should
be stressed that both versions use the hydrostatic equations of motion.

(A) OLD:
- coarse vertical resolution ($\Delta p \sim 80$ mb)
- stability dependent surface fluxes (Louis, 1979)
- vertical exchange coefficients bounded ($K \le K_{max} (\Delta p)$) because
  of explicit time integration of diffusion equation
- roughness length constant over sea
- large-scale condensation, instantaneous fall-out of rain
- partial moist adiabatic adjustment

(B) NEW:
- higher vertical resolution near ground ($\Delta p \sim 10$ mb)
- stability dependent surface fluxes (Louis et al., 1981)
- exchange coefficients according to Blackadar (1979)
- diffusion equation solved implicitly (Iversen and Nordeng, 1987)
- Charnock's formulation of roughness length over sea
- large-scale condensation, cloud liquid water storage
- partial moist adiabatic adjustment
- Kuo-convection (Geleyn, 1985)
- evaporation of rain

- short wave and long wave radiation (Nordeng, 1986)
- radiation interacts with model generated clouds (Nordeng, 1986)
- surface temperature equation

We used the same initial and lateral boundary data as in Grønås et al. (1986), i.e., European Centre for Medium Range Weather Forecasts (ECMWF) large-scale analyses of geopotential height, wind, and relative humidity at pressure levels together with actual sea surface temperatures and ice-edge analyzed at The Norwegian Meteorological Institute (DNMI).

The cases that have been rerun are all from 1984: 26–27 February (case 1), 29 February–1 March (case 2), and 6–7 March (case 3). The first case was observed and analyzed by Shapiro et al. (1987).

Forecasts with the NEW model are visualized in Figures 1–3. For case 2 there is virtually no difference from experiments with the OLD model. Case 3 is modeled better with the NEW model than the OLD model (Figure 4), while case 1 was modeled somewhat better with the OLD model. Verifying surface analyses taken from Shapiro et al. (1987), Rabbe (1986), and Hoem and Hoppestad (1986) are shown in Figures 5–7.

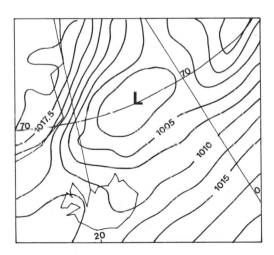

Figure 1. Thirty-nine-hour forecast of sea level pressure at intervals of 2.5 mb from 0000 GMT 26 February with the 50-km resolution model, NEW physics, for a section of the integration area.

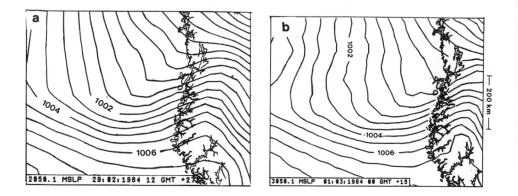

Figure 2. (a) Twenty-seven-hour forecast of sea level pressure at intervals of
1 mb from 1200 GMT 29 February with the 50-km resolution model, NEW
physics, for a section of the integration area. (b) Fifteen-hour forecast of sea
level pressure at intervals of 1 mb from 0000 GMT 1 March with the 25-km
resolution model, NEW physics, for a section of the integration area.

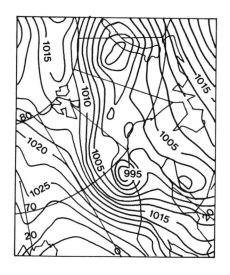

Figure 3. Thirty-six-hour forecast of sea level pressure at intervals of 2.5 mb
from 1200 GMT 5 March with the 50-km resolution model, NEW physics.

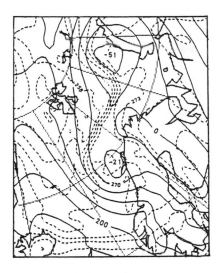

Figure 4. Thirty-six-hour forecast of geopotential height of 1000 mb at intervals of 20 m and potential temperature from the lowest model level at intervals of 3 K from 1200 GMT 5 March with the 50-km resolution model, OLD physics (from Grønås et al., 1986).

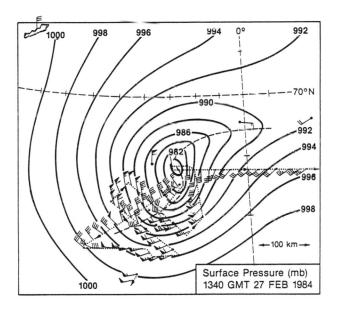

Figure 5. Surface pressure analysis (mb) at 1340 GMT 27 February 1984 (from Shapiro et al., 1987).

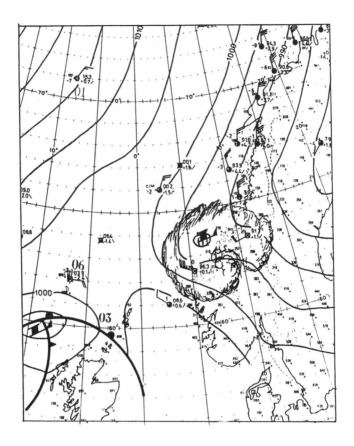

Figure 6. Surface pressure analysis (mb) at 1500 GMT 1 March 1984 (from Rabbe, 1986).

2.2 Resolution

Since the model forecasted case 3 significantly better than case 1 and case 2, we have looked into the developing mechanism for this low.

At the start of the integration there is a large-scale low in the Barents Sea with a wedge of warm air west of Spitzbergen. While the main low moves eastward, the polar low develops in the westernmost area. Figure 8 shows relative vorticity at 700 mb, mean temperature between 500 and 1000 mb, and some contours of vertical velocity after 12 hr of integration. We find a nice correspondence between vorticity advection with the thermal wind and vertical velocity in the area of the polar low development. The initial development of this low may therefore be explained from classical quasi-geostrophic theory. There

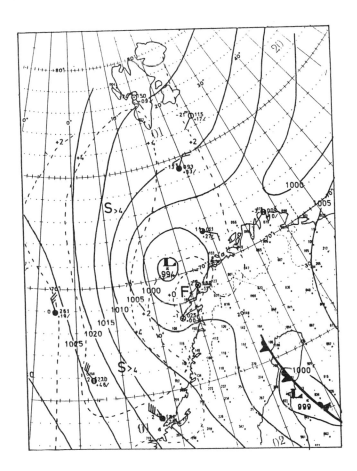

Figure 7. Surface pressure analysis (mb) at 0000 GMT 7 March 1984 (from Hoem and Hoppestad, 1986).

is a transition downward in scale, and at least the initial stage of the polar low development is well resolved by a 50-km resolution model. The simulation is fairly successful also with the OLD model, but the NEW physics gives a significantly improved simulation.

Cases 1 and 2 are smaller scale cyclones than case 3. A natural hypothesis is therefore that we lack resolution to resolve them. These cases were therefore rerun with a horizontal resolution of 25 km and 16 levels. We found some improvements compared to the 50-km runs, stronger vorticity, winds, etc., but the difference between observed and modeled surface pressure is much larger

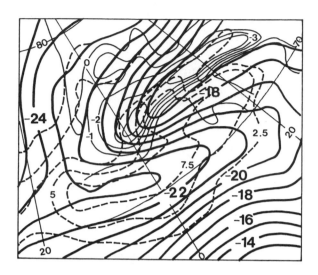

Figure 8. Twelve-hour forecast of relative vorticity of 700 mb at intervals of 2.5 10^{-5} s^{-1} (broken lines), mean temperature in the layer from 500 to 850 mb at intervals of 1 K (heavy lines) and vertical velocity at intervals of 1 10^{-3} mb s^{-1} (thin lines) from 1200 GMT 5 March 1984.

than the improvements. Figure 2b corresponds to Figure 2a, but with a 25-km resolution.

2.3 Sea/Atmosphere Interaction

To see the impacts of sea/atmosphere interaction, we made some sensitivity studies of sea surface temperature (SST) and surface roughness. The NEW model uses Charnock's relation for surface roughness for momentum as well as latent and sensible heat fluxes. Recent observations (e.g., Large and Pond, 1981; Geernaert et al., 1987) and numerical simulations (Nordeng et al., 1988) indicate that the roughness length for heat should be independent of wind speed while the Charnock relation is valid for momentum for well developed (saturated) waves only. For case 1 we used the 50-km resolution model (NEW) with $z_0 = 0.032 - u_*^z/g$, $z_0 = 0.016 - u_*^2/g$ and $z_0 = 1.10^{-4}$ m. Not surprisingly, we got the strongest development by using a constant roughness length (less friction). The improvements were small (~ 2.5 mb), but could explain some of the differences between the OLD and NEW models. In the former we used a constant roughness length over sea.

In quasi-geostrophic theory diabatic effects enter as differential heating in the omega equation. Since strong vertical velocity implies vortex stretching and hence intensification, we adjusted the sea surface temperature according to

(a) SST' = 2(SST − 273K) + 273K
(b) SST' = 2(SST − 273K) + 275K

and rerun case 1 with the 50-km resolution model. For experiment (a) we obtain a 2 mb improvement compared to the reference run (normal SST) while experiment (b) gave a 3-mb deeper low. A closer examination revealed, however, that the improvements were purely linear. Both experiments gave an overall warmer lower part of the atmosphere, and the surface pressure drop could be explained by simple hydrostatic calculations. Nonlinear feedbacks could not be detected.

2.4 Other Model Results

These cases present many challenges to the numerical modeler and particularly case 1 represents a unique opportunity to verify model forecast due to the detailed observations of Shapiro et al. (1987). To the authors' knowledge this case has, in addition to the Norwegian experiments, been run with the National Center for Atmospheric Research/Pennsylvania State University (NCAR/PSU) mesoscale model using both ECMWF and National Meteorological Center (NMC) analyses (S. Mullen, personal communication; Nordeng 1987, not published), with the Navy Operational Regional Atmospheric Prediction System (NORAPS), (Langland and Miller, 1988), and with the ECMWF limited area model (A. Simmons, personal communication). Case 2 has been run with the NCAR/PSU model (S. Mullen, personal communication). Figure 9 shows forecasted sea level pressure with the NCAR/PSU model (Nordeng, 1987, not published) using ECMWF initial fields for case 1. The result is characteristic for the numerical forecasts of this case; a disturbance is forecasted at approximately the right time and place, but one forecasts a broader scale cyclone than the intense, hurricane-like cyclone that was observed. Some improvements are reported when using higher horizontal resolution (Grønås et al., 1986; Langland and Miller, 1988), but the improvements are marginal.

3. A PROPER CONDENSATION DESCRIPTION

It is well known that the strongest developments are obtained by skipping all kind of parameterization of convection and retaining the grid-resolvable, "large-

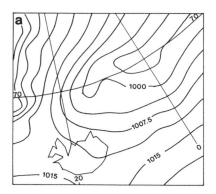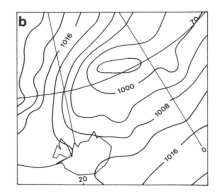

Figure 9. Thirty-nine-hour forecasts of mean sea level pressure and temperature at intervals of 5 K from 0000 GMT 26 February 1984 with the NCAR/PSU mesoscale model using a horizontal resolution of 50 km (a) with parameterization of convection (Anthes, 1977), (b) with explicit condensation.

scale or stable" condensation process only. This was demonstrated by Sardie and Warner (1985) for numerical simulations of two polar low cases and in Grønås et al. (1986), and Nordeng (1987) as well. The more intense development is easily understood dynamically. Without convection latent heat is released at low levels, which destabilizes the model atmosphere, giving rise to strong vertical velocity and hence increased circulation by vortex stretching. When using the NCAR/PSU model, a clear improvement is obtained by using the explicit condensation scheme of Hsie et al., (1984), rather than using a scheme that parameterizes convection (Anthes, 1977), (Figures 9a and 9b). In Nordeng (1987) it was argued from experiments with a 50-km resolution model that the method gave the strongest developments, but could not be justified since it gave unphysical (unstable) stratifications in the model.

However, since our present methods to parameterize convection are clearly unsatisfactory, in particular when using a horizontal resolution of the order of 25 km or less, we kept the NEW physics, but changed parameterization of condensation to the simplest possible (i.e., "large scale," resolvable, with an instantaneous fallout of rain). So far, we have tested the scheme on case 2. We started from a 12-hr forecast with the 50-km model (NEW physics), and used a horizontal resolution of 25 km for 18 hr. Due to limited computer resources we could only use 10 vertical layers. Figures 10a–10d show surface pressure for a section of the integration area after 6, 9, 12, and 15 hr of integration respectively, which correspond to 0600, 0900, 1200, and 1500 GMT 1 March

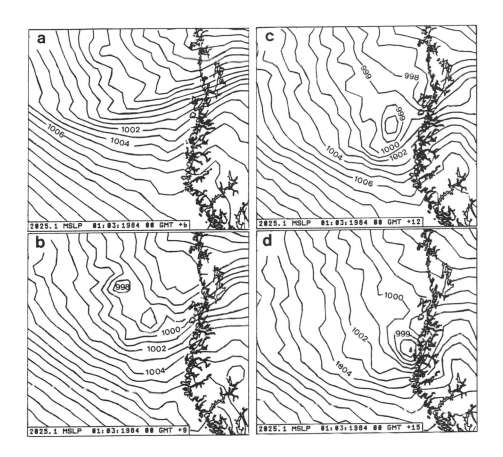

Figure 10. Forecasts with a horizontal resolution of 25 km from 0000 GMT 1 March 1984 of sea level pressure at intervals of 1 mb (a) 6-hr forecast, (b) 9-hr forecast, (c) 12-hr forecast, (d) 15-hr forecast.

1984. There is a lot of noise, but the noise does not destroy the evolution of a small-scale cyclone approaching the Norwegian coast. This cyclone does not develop when convection is parameterized (Figure 2b). Figure 11 shows a barograph from VIGRA airport close to where the cyclone made landfall. We noticed a strong drop in surface pressure between 1200 and 1400 GMT 1 March 1984 of 2 mb per hour. The barograph clearly demonstrates the mesoscale structure of the low. A corresponding plot taken from the model shows a similar pressure drop at exactly the right time. The warm core structure is demonstrated by a 20 °C temperature rise as the cyclone passes (Figure 12). To the authors' knowledge, this is the first time the intense inner core of a polar low has been

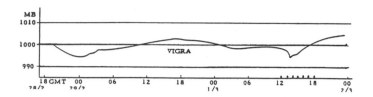

Figure 11. Pressure registration at Vigra Airport (62°N 6°E) from 1800 GMT 28 February to 0000 GMT 1 March 1984.

modeled. The scale of the cyclone is demonstrated in Figure 13, while the stratification in an area of strong ascending motion is shown in Figure 14. First of all we noticed the strong vertical velocity (~ 25 μbar s^{-1}) at the top of a conditionally (and saturated) unstable layer.

We obtained no further improvements, however, by using the NCAR/PSU model and explicit condensation with horizontal resolution of 25 km on case 1 (not shown). A similar experiment as described above has not been run on case 1 with the Norwegian model.

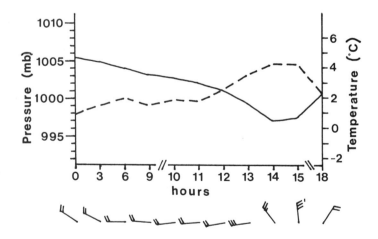

Figure 12. Pressure (full line) and screen (2 m) temperature (broken line) during integration with the 25-km resolution model from 0000 GMT 1 March 1984 for Vigra Airport (62°N 6°E). Winds are orientated relative to north. Full barbs correspond to 5 m s^{-1}, half barbs to 2.5 m s^{-1}. Note change in scale on time axis.

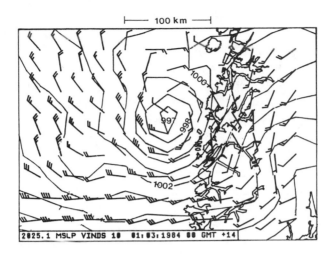

Figure 13. Pressure at intervals of 1 mb and winds from the lowest model layer (~ 40 m above the sea) for a 14-hr forecast from 0000 GMT 1 March 1984 with the 25-km resolution model, full barbs are 5 m s^{-1}, while half barbs correspond to 25 m s^{-1}.

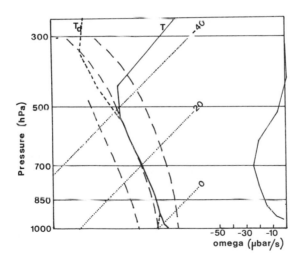

Figure 14. Temperature (full line) and dew point temperature soundings (short broken line) for a 14-hr forecast from 0000 GMT 1 March 1984. Moist adiabatic lapse rates are drawn as lines running upward to the left (long broken lines). Imbedded is vertical velocity (omega).

4. DISCUSSION

The following conclusions should be regarded as temporary since more experiments have to be run. We speculate, however, that the failure in forecasting the intense rapid growth, which is observed for at least some polar lows, comes from the way we parameterize convection in numerical models. Convection schemes are without exception constructed to work in the tropics and adopted for use in the middle latitudes. We are now adopting the same schemes for polar regions in high resolution models. It is not obvious that this can be justified. We have demonstrated that an almost perfect forecast of a polar low is obtained if we abandon parameterization of convection. In fact, case 2 was rerun with a simple moist adiabatic adjustment that operates for conditionally unstable and saturated stratifications only. We got rid of the noise, but also the development. The nature of the noise may be inspected by looking into the dispersion relation for free gravity waves (Orlanski, 1981).

$$\omega^2 = \frac{N^2\delta^2 + f^2\omega^2}{a\delta^2 + 1} \tag{1}$$

where ω is the frequency, N is the Väsäla-Brunt frequency, δ is the aspect ratio and the tracer a is 1 for nonhydrostatic equations and zero for hydrostatic equations. In the limit of large δ,

$$\omega^2 \sim \left[\begin{array}{l} N^2\delta^2, \text{ hydrostatic} \\ N^2, \text{ nonhydrostatic} \end{array} \right.$$

As the resolution increases so that smaller horizontal scales may be resolved (large δ), the hydrostatic growth rate is unbounded while the nonhydrostatic growth rate is bounded. The smallest resolved waves in a grid point model will therefore grow unlimited in a hydrostatic model if the stratification becomes unstable. We have tried to run with a 12.5-km resolution, but the noise level destroyed any evolution of the low.

Here we have to face the following controversy: The hydrostatic equations describe a kind of motion they are not constructed to work for! The only possible explanation is that we do more harm in specifying heating/moistening profiles and disregarding vertical advection of horizontal momentum (as most parameterization schemes do) than allowing the inadequate equations to take care of this explicitly. Nordeng (1987) proposed a scheme for parameterization of slantwise convection that is consistent for absolute momentum transport. The scheme efficiently removes symmetric unstable stratifications but was combined with a scheme for vertical convection to take care of vertical instability. The only reason that the forecast does not blow up has to come from the implicit horizontal diffusion inherent in the time integration scheme (Bratseth, 1983).

To solve the present problem we suggest that a scheme that allows the model equations to explicitly take care of all physical processes should be constructed. This could be done by combining a scheme like the explicit-condensation scheme of Hsie et al. (1984) with a nonhydrostatic correction method as proposed by Orlanski (1981). The explicit scheme of Hsie et al. (1984) uses cloud water and rain water as prognostic equations and adjusts for water loading in the hydrostatic equation. Orlanski's method is to adjust for vertical acceleration in the hydrostatic equation. Further work will continue along these lines.

ACKNOWLEDGMENTS

Some of the numerical simulations were sponsored and performed at the IBM Scientific Center in Bergen, Norway. Simulations with the NCAR/PSU model were performed while T.E.Nordeng was a visiting scientist at NCAR (National Center for Atmospheric Research) and NOAA/WPL, Boulder, Colorado, sponsored by The Office of Naval Research. We are grateful for their contributions, which made this work possible, to Y.-H Kuo (NCAR) for inspiring discussions, and to Evelyn Donald for help in setting up the experiments at NCAR.

REFERENCES

Anthes, R.A., 1977: A cumulus parameterization scheme utilizing a one-dimensional cloud model. *Mon. Wea. Rev., 105,* 270–286.
Blackadar, A.K., 1979: High resolution models of the planetary boundary layer. *Advances in Environment and Scientific Engineering.* I. Gordon and Branch, 276 pp.
Bratseth, A., 1983: Some economical, explicit finite-difference schemes for the primitive equations. *Mon. Wea. Rev., 111,* 663–668.
Emanuel, K.A., and R. Rotunno, 1989: Polar lows as arctic hurricanes. *Tellus, 41A,* 1–17.

Geernaert, G.L., S.E. Larsen, and F. Hansen, 1987: Measurement of the wind stress, heat flux, and turbulence intensity during storm conditions over the North Sea. *J. Geophys. Res.*, *92(C12)*, 13127–13139.

Geleyn, J.F., 1985: On a simple, parameter-free partition between moistening and precipitation in the Kuo scheme. *Mon. Wea. Rev.*, *113*, 405–407.

Grønås, S., A. Foss, and M. Lystad, 1986: *Numerical Simulations of Polar Lows in the Norwegian Sea*. Tech. Reps. Nos. 5 and 18, The Polar Lows Project, The Norwegian Meteorological Institute, P.O. Box 43, Blindern, 0313 Oslo 3, Norway.

Hoem, V., and S. Hoppestad, 1986: *Polar Low Case Studies III, 1982–1985*. Tech. Rep. No. 12, The Polar Lows Project, The Norwegian Meteorological Institute, P.O. Box 43–Blindern, 0313 Oslo 3, Norway

Hsie, E.-Y., R.A. Anthes, and D. Keyser: Numerical simulation of frontogenesis in a moist atmosphere. *J. Atmos. Sci.*, *41*, 2581–2594.

Iversen, T., and T.E. Nordeng, 1987: *A Numerical Model Suitable for a Broad Class of Circulation Systems on the Atmospheric Mesoscale*. NILU Rep. Tr: 2/87, Norwegian Institute for Air Research, Lillestrøm, Norway.

Langland, R., and R.I. Miller, 1988: Polar low sensitivity to sea surface temperature and horizontal grid resolution in a numerical model. Preprint, *of the Second Conference on Polar Meteorology and Oceanography*, American Meteorological Society, Boston, Mass. USA.

Large, W.G., and S. Pond, 1981: Open ocean momentum flux measurements in moderate to strong winds. *J. Phys. Oceanogr.*, *11*, 324–336.

Louis, J.F., 1979: A parametric model of the vertical eddy fluxes in the atmosphere. *Boundary Layer Meteor.*, *17*, 187–202.

Louis, J.F., M. Tiedtke, and J.F. Geleyn, 1981: A short history of the operational PBL-parameterization at ECMWF. *Workshop on Planetary Boundary Layer Parameterization*, ECMWF, 59–79.

Nordeng, T.E., 1986: *Parameterization of Physical Processes in a Three-Dimensional Numerical Weather Prediction Model*. Tech. Rep. No. 65, The Norwegian Meteorological Institute, P.O. Box 43–Blindern, 0313 Oslo 3, Norway.

Nordeng, T.E., 1987: The effect of vertical and slantwise convection on the simulation of polar lows. *Tellus*, *39A*, 354–376.

Nordeng, T.E., M. Reistad, and A.K. Magnusson, 1988: *An Optimum Use of Atmospheric Data in Wind Wave Modelling–Numerical Experiments*. Tech. Rep. No. 68, The Norwegian Meteorolgical Institute, P.O. Box 43, Blindern, 0313 Oslo 3, Norway.

Orlanski, I., 1981: The quasi-hydrostatic approximation. *J. Atmos. Sci.*, *38*, 572–582.

Rabbe, Å., 1986: *Analysis of a Polar Low in the Norwegian Sea. 29 February–1 March 1984*. Tech. Rep. No. 14, The Polar Lows Project. The Norwegian Meteorological Institute, P.O. Box 43–Blindern, 0313 Oslo 3, Norway.

Sardie, J.M., and T.T. Warner, 1985: A numerical study of the development mechanism of polar lows. *Tellus*, *37A*, 460–477.

Shapiro, M.A., L.S. Fedor, and T. Hampel, 1987: Research aircraft measurements of a polar low over the Norwegian Sea. *Tellus*, *39A*, 272–306.

ON THE EFFECT OF USING DIFFERENT FORMULATIONS FOR THE FORCING IN THE OMEGA EQUATION APPLIED TO POLAR LOWS[1] [2]

Torben Strunge Pedersen
Geophysical Institute, University of Copenhagen
Copenhagen, Denmark

ABSTRACT

Even though our knowledge about the structure and dynamics of polar lows has increased substantially in recent years, some uncertainty still remains concerning the primary driving mechanisms. This is in part due to the great variety of disturbances that classify as polar lows, and, further, that the relative importance of the different forcing mechanisms may change during the life cycle of the disturbances. To investigate this problem diagnostic studies have been made based on a polar low development in the Barents Sea. The vertical velocity is chosen as the primary diagnostic variable as it reacts to adiabatic as well as diabatic forcing. The vertical velocity is determined diagnostically by the use of an omega equation. The equation will be considered in a quasi- as well as a semi-geostrophic formulation. Traditionally, the adiabatic forcing terms in the omega equation have been split up into two components expressing differential advection of vorticity and the Laplacian of the thermal advection. These terms, however, partly cancel and alternative formulations have been proposed. These formulations and their relevance for the description of the adiabatic forcing terms during a polar low development are considered. The diabatic forcing through latent heat release, which is of major importance for some developments, is difficult to parameterize and especially the vertical distribution is not well known. However, to get an idea of the relative importance of the effect, a simple Ekman-pumping formulation together with a prescribed vertical distribution of the latent heat release has been applied in some experiments.

[1]This is a condensed version of an article published in *Tellus, 41A*, no. 1, January 1989
[2]Supported by the Office of Naval Research under Grant N00014-87-G-0232.

POLAR AND ARCTIC LOWS
Paul F. Twitchell, Erik A. Rasmussen,
and Kenneth L. Davidson (Eds.)

233

1. INTRODUCTION

In this work different versions of the omega equation are applied in a diagnostic study of a polar low development to investigate if it is feasible to identify which forcing mechanisms are of primary importance.

The synoptic situation during the polar low development in mid-December 1982 has been described in detail by Rasmussen (1985a, 1985b). He showed that the initial development was connected with an upper level cold core vortex moving southwards from the Svalbard region. Figure 1 shows an analysis of the situation at 0000 GMT on 13 December 1982. During the next 24 hr the polar low moves southeast and develops into a tight vortex with a surface pressure of 980 mb.

The data for this study are based on an initialized analysis for 13 December at 0000 GMT, 1982 produced by a high resolution limited area model (HIRLAM).

In Section 2 different formulations of the quasi- and semi-geostrophic model are described. In Section 3 we consider the diagnostics obtained with the

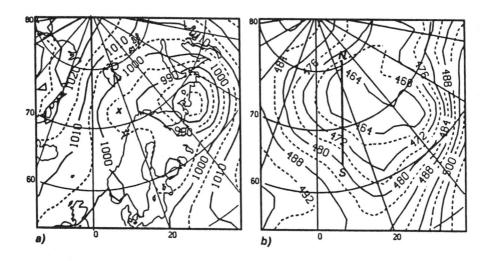

Figure 1. HIRLAM analysis for 0000 GMT 13 December 1982 (a) MSL pressure in millibars, the x indicates the position of the polar low; (b) 530-mb heights in dm. The thick black line indicates the orientation of the cross section shown in Figures 3 and 4.

quasi-geostrophic model, and in Section 4 the results from the semi-geostrophic model are given. Finally, Section 5 contains some concluding remarks.

2. THE MODELS

We consider a hydrostatic and Boussinesq atmosphere on a polar stereographic f-plane. As vertical coordinate the pseudo height

$$z = \left[1 - \left(\frac{p}{p_o} \right)^x \right] \frac{C_p \theta_o}{g} \tag{1}$$

introduced by Hoskins and Bretherton (1972) is applied.

The models use 13 vertical levels and a 21×21 horizontal grid using grid distances of 667 m and 200 km for the vertical and horizontal directions, respectively. The boundary conditions are given in Section 2.3.

2.1 The Quasi-Geostrophic Omega Equation

Traditionally the adiabatic forcing in the quasi-geostrophic omega equation, QGO, has been described by two terms. One is the vertical derivative of vorticity advection (AVO) and the other the Laplacian of the thermal advection (ATH). With the approximations mentioned above the adiabatic omega equation can be written as

$$N^2 \nabla^2 w + f^2 \frac{\partial^2 w}{\partial z^2} = f \frac{\partial}{\partial z} (\vec{V}_g \cdot \nabla \zeta_g) - \frac{g}{\theta_o} \nabla^2 (\vec{V}_g \cdot \nabla \theta) \tag{2}$$

$$\text{AVO} \qquad\qquad \text{ATH}$$

This formulation has the drawback that AVO and ATH have a common and possibly significant part, (ACA),

$$f \vec{V}_g \nabla \frac{\partial \zeta_g}{\partial z} \tag{3}$$

$$\text{ACA}$$

which cancels. Hoskins et al. (1978) showed that this problem is avoided if the total adiabatic forcing is expressed as the divergence of a vector $\vec{q}$ defined by

$$\vec{Q} = -\frac{g}{\theta_o}\left(\frac{\partial \vec{V}_g}{\partial x}\ \nabla\theta,\ \frac{\partial \vec{V}_g}{\partial y}\ \nabla\theta\right) \tag{4}$$

whereby Eq. (2) becomes

$$N^2 \nabla^2 w + f^2 \frac{\partial^2 w}{\partial z^2}\ =\ 2\nabla\cdot\vec{Q} \tag{5}$$

The solution to the omega equation can be split up into three parts associated with the lower boundary condition (w_{EQG}), the vorticity forcing (w_v), and the thermal forcing w_T), respectively.

$$N^2\nabla^2 w_{EQG} + f^2\ \frac{\partial^2 w_{EQG}}{\partial z^2}\ =\ 0$$

$$N^2\nabla^2 w_v + f^2\ \frac{\partial^2 w_v}{\partial z^2}\ =\ AVO \tag{6}$$

$$N^2\nabla^2 w_T + f^2\ \frac{\partial^2 w_T}{\partial z^2}\ =\ ATH$$

w_v and w_T vanish at the lower boundary where w_{EQG} is given by Ekman pumping (see Section 2.3). As the differential operator on the left-hand side of Eq. (2) is linear the total solution is

$$w\ =\ w_{EQG} + w_v + w_T \tag{7}$$

By partitioning the solution in this way it is possible to compare the significance of the different forcing terms as described in Section 3.

2.2 The Semi-Geostrophic Omega Equation

The semi-geostrophic omega equation (SGO) is derived in two steps. First the geostrophic momentum approximation (GMA) is made and then a coordinate and variable transformation as described by Hoskins (1975) and Hoskins and Draghici (1977) is performed.

The GMA is based on the assumption that the Rossby number

$$R_o = \frac{|\, d\vec{V} \,/\, dt \,|}{|\, f\vec{V} \,|} \tag{8}$$

is small. This is also the basic assumption behind the quasi-geostrophic approximation. However, the approach differs in the evaluation of the ratio between the rate of change of momentum and the Coriolis force. The quasi-geostrophic approximation applies horizontally uniform velocity and length scales giving the constraint $V/L < f$. In the GMA the ratio is considered separately for the direction along and perpendicular to the movement of an air parcel, giving rise to two criteria for the smallness of the Rossby number

$$\frac{1}{V}\,\frac{dV}{dt} < f, \; \frac{V}{r} < f \tag{9}$$

where V is the velocity and r the radius of curvature for an air parcel. These criteria, based on the Lagrangian view that the magnitude and direction of the momentum of an air parcel must change little during the time f, are weaker than the Eulerian criterion used for the quasi-geostrophic approximation.

A consequence of the less restrictive criteria in the GMA is that the horizontal advection by the ageostrophic velocity is retained. The advected quantities like momentum and vorticity, though, are still represented by their geostrophic values.

The geostrophic momentum equations are transformed to semi-geostrophic space by means of the coordinate transformations:

$$X = x + \frac{v_g}{f} \; ; \; Y = y - \frac{u_g}{f} \; ; \; Z = z; \tag{10}$$

Introducing a modified vertical velocity

$$w^* = \frac{f}{f + \zeta_g} \, w \tag{11}$$

The semi-geostrophic version of the omega equation becomes

$$\vec{V}^2(q_g w^*) + f2 \frac{\partial^2 w^*}{\partial Z^2} = 2\nabla \cdot \vec{Q} \tag{12}$$

with the geostrophic potential vorticity, q_g, and $\vec{Q}$ given by

$$q_g = (f + \zeta_g) \, \frac{\partial \theta}{\partial Z} \, ; \, \vec{Q} = - \, \frac{g}{\theta_o} \left(\frac{\partial V_g}{\partial X} \, \nabla \theta, \, \frac{\partial \vec{V}_g}{\partial Y} \, \nabla \theta \right) \tag{13}$$

The only difference between Eq. (12) and the corresponding quasi-geostrophic formulation, Eq. (5), is the replacement of the Brunt-Väisälä frequency with the geostrophic potential vorticity, and that the partial derivatives are related to the semi-geostrophic space.

Results from the semi-geostrophic model presented in the following are obtained as follows. With potential temperature, and geopotential and geostrophic wind given in the physical space, a transformation to the semi-geostrophic space is performed. Using a modified Barnes procedure as described by Maddox (1980) the data are interpolated to a regular grid. The omega equation is solved in the semi-geostrophic space and the results presented in physical space.

2.3 Boundary Conditions

To solve the omega equation as given by Eqs. (2), (5) or (12) the vertical velocity must be specified at the boundaries of the model domain. The lower boundary is assumed to coincide with the top of the planetary boundary layer, (PBL), and w is given by the Ekman pumping. At the top and the lateral boundaries $w = 0$ is assumed.

The classic Ekman pumping is obtained by matching quasi-geostrophic dynamics with that of the Ekman layer. This yields

$$w_{EQG} = \frac{1}{2} \left(\frac{2K}{f} \right)^{1/2} \zeta_g \tag{14}$$

where ζ_g is the geostrophic vorticity at the top of the PBL and K is constant. For the QGO Eq. (14) is used as boundary condition.

Wu and Blumen (1982) derived an Ekman-pumping formula consistent with the GMA by incorporating this approximation in the Ekman-layer dynamics:

$$
w_{EGM} = l\frac{1}{2}\left(\frac{2K}{f}\right)^{1/2}\left\{-\underbrace{\left(2 - 1/(1+\zeta_g/f)^{1/4}\right)\cdot\nabla\cdot\vec{v}_t}_{w_{DIV}} + \underbrace{\zeta_g}_{w_{EQC}}\right.
$$

$$
\left. -\underbrace{\frac{1}{4f}\vec{v}_g\cdot\nabla\zeta_g}_{w_{ISA}} - \underbrace{\frac{3}{4f}\,\vec{k}\cdot\nabla\times(\vec{v}_g\zeta_g)}_{w_{CRO}}\right\} \tag{15}
$$

where $\vec{v}_t$ is the horizontal velocity at the top of the PBL.

Following Wu and Blumen (1982) the individual terms in Eq. (15) can be interpreted as the effect of different physical mechanisms. The first term, w_{DIV}, is related to the ageostrophic divergence at the top of the PBL. Here as in Wu and Blumen (1982) this divergence is assumed to be negligible. The second term, w_{EQG}, corresponds to the quasi-geostrophic Ekman pumping, Eq. (14), and is related to the divergence of the viscous stress. The advection of geostrophic vorticity expressed by the third term, w_{ISA}, accounts for the effect of the isallobar wind in the boundary layer. Finally, the last term, w_{CRO}, is associated with the cross isobar flow which is needed to balance the viscous stress.

Ignoring w_{DIV} Eq. (15) becomes

$$
w_{EGM} = w_{EQC} + w_{ISA} + w_{CRO} \tag{15a}
$$

which is used as boundary condition for the SGO.

3. FORCING IN THE QUASI–GEOSTROPHIC OMEGA EQUATION

In connection with the study of polar lows in the region near Bear Island it is appealing to express the forcing in the conventional way as shown in Eq. (2). The contrast between the cold ice pack and the relatively warm Norwegian current gives rise to marked differences in the temperature at low levels (see Figure 2). Thus, the ATH term in Eq. (2) can be expected to be significant at the lower levels. At upper levels the AVO term can be expected to be dominant as the development is associated with a southward moving upper level vortex.

Figure 3 shows a cross section of the vertical velocity fields associated with the different terms expressing the adiabatic forcing. In the vicinity of the polar low ATH dominates AVO practically everywhere. The figure shows that ACA can be significant giving contributions that equal or even exceed the total magnitude of the ATH and AVO terms.

That the thermal forcing dominates vorticity forcing at upper levels in the vicinity of the low is somewhat surprising. Generally it is expected that ATH is most significant at low levels and AVO at higher levels, see, for example, Reed (1985).

The fact that ATH does not produce significant velocities at low levels is probably connected with the position of the polar low. It is at this time already situated somewhat to the south of the shallow baroclinic zone connected with the ice/sea transition in the HIRLAM analysis. Further, it should be noted that the lowest level where ATH is evaluated in the model corresponds to 1.334 km.

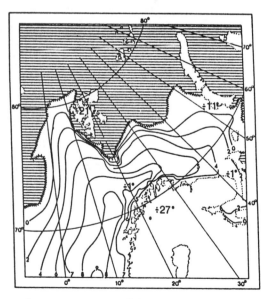

Figure 2. Sea surface temperatures (°C) and sea-ice extent (hatched) (after Rasmussen, 1985b). Representative 2 m temperatures have been added.

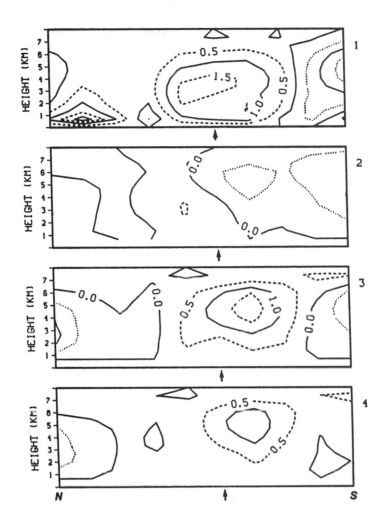

Figure 3. Vertical velocity (cm s⁻¹) from the QGO with adiabatic forcing, along the cross section indicated in Figure 1, 0000 GMT 13 December 1982. Dash/dot indicates ascent/descent. The arrow identifies the location of the polar low (1) total forcing, (2) AVO, (3) ATH, (4) ACA.

To get an idea of the effect of the thermal forcing in the lowest part of the atmosphere a version of the model without Ekman pumping has been tested. In this case the lower boundary is placed at the surface with $w = 0$ so that the AVO and ATH forcing can be included at the 667 m level. The result (not shown) does not vary much from that obtained previously, except, of course, for the lack of the Ekman-pumping contribution. Along the ice edge north of the polar low where most of the baroclinicity is confined below 1 km ATH still does not give rise to the largest vertical velocities at low levels, partly because the geostrophic wind is nearly aligned with the isotherms at this time. Satellite images and observations, however, indicate the presence of shallow frontal structures over the sea, which may yield ATH forcing. These features are not present in this HIRLAM analysis.

Anyway, it is somewhat misleading to talk of the forcing as being dominated by thermal or vorticity effects in this case. As noted earlier the cancellation term, which contributes to both the thermal and the vorticity forcing, constitutes a major part of both of them.

4. SEMI–GEOSTROPHIC DIAGNOSTICS

The polar low considered here has a scale of ~ 500 km and application of the quasi-geostrophic approximation may not be suitable. Especially if the ageostrophic horizontal advection is of importance, use of the more general GMA should give results that differ.

The solution to the SGO is shown in Figure 4(1) for the same cross section as in Figure 3. A comparison with the quasi-geostrophic solution [Figure 3(1)] reveals a few qualitative differences. Most notably around the polar low the SGO yields a stronger updraft at lower levels and a weaker updraft at higher levels.

As noted in Section 2.3 the Ekman pumping used for the SGO, Eq. (15a), includes two additional terms. Wu and Blumen (1982) found for an idealised, steady, axisymmetric circular vortex that the magnitude of w_{EGM} was larger in anticyclones and smaller in cyclones compared to w_{EQG}. The difference, according to their Figure 3, ranges from 0 to about 60% depending on the Rossby number and distance from the centre of the vortex. This tendency of a reduced upward motion in a cyclonic region is also found here. However, the magnitude is less than 0.1% near the polar low. Actually, the largest values found in this case for w_{ISA} and w_{CRO} are $-6 \, 10^{-5}$ cm s^{-1} and $-4 \, 10^{-4}$ cm s^{-1}, respectively. They are associated with the major low east of the Kola Peninsula (see Figure 1) where $w_{EQG} \simeq 4$ cm s^{-1}. Consequently the difference between the SGO and QGO solutions cannot be ascribed to the change in the Ekman-pumping formulation but must primarily be associated with the effect of the ageostrophic horizontal advection.

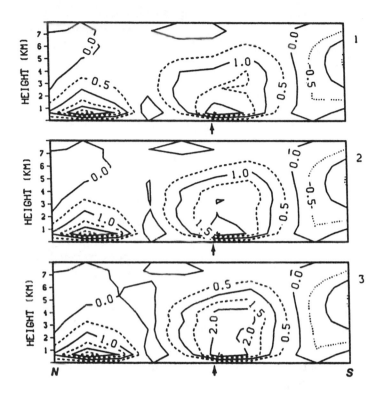

Figure 4. Vertical velocity from the SGO with adiabatic forcing (1), two types of diabatic forcing as described in the text (2) and (3). Conventions and units as in Figure 3.

Blumen and Wu (1983) studied the effect of an Ekman layer on baroclinic instability including w_{DIV} in the Ekman pumping. They found that w_{DIV} had an amplitude of approximately half the value of w_{ISA} and w_{CRO}. As these terms yield insignificant contributions to w_{EGM}, in this case our negligence of w_{DIV} seems justified.

Figure 4(1) shows that the adiabatic forcing produces ascent in the region of the incipient polar low, but only of a modest intensity. However, this forced ascent may help to organise the convection. Already at 0250 GMT a satellite image [see Rasmussen (1985b) Figure 4] shows intense convection and it is during the next 12 to 21 hr that the polar low develops into a tight vortex at the surface.

To get an estimate of the potential influence of diabatic heating through latent heat release some preliminary tests have been made. Diabatic forcing is incorporated in the omega equation through an additional term on the right-hand side of Eq. (12), giving

$$\nabla^2(q_g'w*) + f^2 \frac{\partial^2 w*}{\partial Z^2} = 2\nabla \cdot \vec{Q} + \frac{g}{\theta_o} \nabla^2\left(\frac{H\theta}{C_p T}\right) \qquad (16)$$

where H is the diabatic heating rate.

To get an estimate of H we assume that the diabatic forcing is controlled by the Ekman pumping. If $w_{EGM} \leq 0$ no diabatic forcing is applied. In areas with $w_{EGM} > 0$ it is assumed that all the water vapour pumped from the PBL into the free atmosphere is condensed within a vertical column heating it. The last assumption is not strictly valid as some of the heat released will be dispersed by gravity waves (e.g., Frank, 1983). Observations at ship AMI, a little to the southeast of the polar low at 0000 GMT 13 December, show mixing ratios of 2 g kg^{-1} increasing to 3 g kg^{-1} as the polar low passes later on. A value of 2 g kg^{-1} is used here. Over land and/or ice this value may be too large and a land/sea mask is included in the calculation of H so the mixing ratio can be reduced there.

The vertical distribution of latent heat release in polar lows is not well known. Analysis of an explosively deepening cyclone by Liou and Elsberry (1987) shows that strong diabatic heating at 600–700 mb is of major importance for that development. Though polar lows and the type of cyclone analysed by Liou and Elsberry differ in many ways, we choose a similar distribution with 75% in the 502–737 mb layer and the rest distributed evenly in the remainder of the troposphere, i.e., in the two layers 880–737 mb and 502–407 mb. Further we assume H is zero over land.

Figure 4(2) shows the result. It is seen that the upper part of the ascent region in the polar low has intensified. The region of ascent to the north of the polar low has practically not changed as no diabatic forcing is applied over the ice. Additional tests have been made including diabatic forcing over land. The polar low, situated over water, does not exhibit any changes, whereas the northern ascent region intensifies with the local diabatic forcing.

Increasing the water vapour content to 3 g kg^{-1} yields an increase in the middle and upper part of the polar low updraft. This is seen in Figure 4(3) showing that the solution is rather sensitive to the strength of the diabatic forcing.

Testing different vertical distributions of H yields only small changes in the updraft associated with the polar low. Generally the updraft intensifies most at the level of maximum heating. This intensification, though, is less pronounced with the maximum in the lowest part of the troposphere.

5. CONCLUDING REMARKS

Preliminary diagnostic studies of a polar low have been made based on an analysis produced by a high resolution limited area model.

In the QGO the thermal forcing dominates the vorticity forcing at upper levels in the vicinity of the low. At this stage of the polar low development the thermal forcing at low levels is not strong. The latter may be due to inadequacies in the treatment of the lowest part of the atmosphere in the analysis used here.

The part that cancels between the ATH and AVO terms can be large and locally exceed the individual terms. This makes the interpretation of the forcing in terms of thermal and vorticity effects uncertain and calls for caution when trying to interpret the ATH and AVO terms as the effect of individual mechanisms.

Application of the SGO yields results that differ from the QGO solution. The SGO produces an updraft that is stronger at lower levels and weaker at upper levels in the polar low region. The difference must primarily be ascribed to the inclusion of ageostrophic horizontal advection in the SGO.

The difference between the Ekman pumping based on quasi-geostrophic theory and that based on the GMA is negligible in this case. The difference never exceeds the order of $^0/_{00}$.

The total adiabatic forcing produces only modest ascent in the polar low region. This may, however, be conductive for the organization of the convection and the subsequent development of the polar low.

Diabatic forcing controlled by the Ekman pumping and the mixing ratio in the PBL intensifies the vertical velocity field in the polar low region.

Preliminary tests show that the intensification at the polar low is more sensitive to variations in the mixing ratio than the vertical distribution of H.

The diabatic forcing has a localised effect on the vertical velocity field in the sense that inclusion of diabatic forcing over land affects the field there, but for practical purposes not in the polar low region over the sea.

ACKNOWLEDGMENTS

I would like to thank the HIRLAM group for providing the analysis used for this study and Wayne Schubert for bringing the works by R. Wu and W. Blumen to my attention. The computations were performed at the Danish Meteorological Institute.

REFERENCES

Blumen, W., and R. Wu, 1983: Baroclinic instability and frontogenesis with Ekman bounday layer dynamics incorporating the geostrophic momentum approximation. *J. Atmos. Sci.*, *40*, 2630–2637.

Frank, W.M., 1983: The cumulus parameterization problem. *Mon. Wea. Rev.*, *111*, 1859–1871.

Hoskins, B.J., and F. Bretherton, 1972: The geostrophic momentum approximation and the semi-geostrophic equations. *J. Atmos. Sci.*, *32*, 233–242.

Hoskins, B.J., and F. Bretherton, 1972: Atmospheric frontogenesis models: Mathematical formulation and solution. *J. Atmos. Sci.*, *29*, 11–37.

Hoskins, B.J., and I. Draghici, 1977: The forcing of ageostrophic motion according to the semi-geostrophic equations in an isotropic coordinate model. *J. Atmos. Sci.*, *34*, 1859–1867.

Hoskins, B.J., I. Draghici, and H.C. Davies, 1978: A new look at the omega equation. *Quart. J. Roy. Meteor. Soc.*, *104*, 31–38.

Liou, C.-S., and R.L. Elsberry, 1987: Heat budgets of analyses and forecasts of an explosively deepening maritime cyclone. *Mon. Wea. Rev.*, *115*, 1809–1824.

Maddox, R.A., 1980: An objective technique for separating macroscale and mesoscale features in meteorological data. *Mon. Wea. Rev.*, *108*, 1108–1121.

Rasmussen, E., 1985a: *A Polar Low Development Over the Barents Sea.* Tech. Rep. No. 7, The Norwegian Pola Lows Project, 42 pp.

Rasmussen, E., 1985b: A case study of a polar low development over the Barents Sea. *Tellus*, *37A*, 407–418.

Reed. R.J., 1985: Baroclinic instability as a mechanism for polar low formation. *Proceedings of the International Conference on Polar Lows*, Oslo, Norway, 141–149.

Wu, R., and W. Blumen, 1982: An analysis of Ekman boundary layer dynamics incorporating the geostrophic momentum approximation. *J. Atmos. Sci.*, *40*, 1774–1782.

POLAR LOW SENSITIVITY TO SEA SURFACE TEMPERATURE AND HORIZONTAL GRID RESOLUTION IN A NUMERICAL MODEL

Rolf H. Langland and Ronald J. Miller
Naval Environmental Prediction Research Facility
Monterey, California, U.S.A.

ABSTRACT

A numerical model simulates the development of a polar low observed by instrumented aircraft. The forecast of the polar low central pressure and wind pattern is significantly improved by changing the model grid spacing from 80 km to 30 km. Providing the model with higher accuracy sea surface temperature and ice coverage fields contributes to minor improvements in the polar low forecast.

1. INTRODUCTION

Accurate forecasting of polar lows presents many challenges to the numerical modeler. The relatively small horizontal extent of most polar lows requires correspondingly small grid resolution in the model, perhaps on the order of 50 km or less. With a primitive equation model, great demands are placed on computer resources, since for numerical stability the model time step is reduced, while the number of grid points must be sufficient to cover a domain of adequate size and avoid lateral boundary contamination.

In addition, it seems reasonable to assume that accurate polar low forecasts require fairly detailed depictions of sea surface temperature (SST) and sea ice coverage. Changes in SST of only one or two degrees Celsius could, in some situations, be enough to alter stability in the lower troposphere and affect the vertical distribution of convective heating and moistening, as well as latent and sensible heat fluxes from the sea surface. Of course, other considerations of

Copyright © 1989 A. DEEPAK Publishing
ISBN 0-937194-19-0

model physics and parameterization are also quite important. For example, the effects of various convective parameterization schemes on polar low forecasts are discussed by Nordeng (1987).

At the present time, relatively few numerical simulations of polar lows have appeared in the literature. This is due in part to a lack of detailed observational studies with which model forecasts might be verified. A summary of recent theoretical and modeling studies of polar lows is given by Rasmussen and Lystad (1987).

This study presents some model forecasts and sensitivity tests applied to one well-documented polar low case study (Shapiro et al., 1987). Our objective is to analyze in a fairly qualitative sense the effects of changing model grid resolution and incorporate alternative SST and ice edge analyses.

2. DESCRIPTION OF THE NORAPS MODEL

The model used is the Navy Operational Regional Atmospheric Prediction System (NORAPS). It is run on a 109×82 horizontal grid with 30-km or 80-km resolution. There are 12 sigma levels corresponding to 0.962, 0.887, 0.812, 0.737, 0.650, 0.550, 0.450, 0.362, 0.287, 0.218, 0.137 and 0.050. A time step of 100 s is used (with DX=30 km), on an Arakawa Scheme-C grid.

A modified Kuo approach is used for cumulus parameterization, where moisture convergence greater than 7.055×10^{-10} kg m^{-2} s^{-1} and conditional instability are used to trigger convection. The PBL is assumed to be well-mixed and as deep as the lowest layer in the model (about 40 mb). The moisture analysis and boundary conditions are obtained from the forecast fields of NOGAPS (Navy Operational Global Atmospheric Prediction System).

Surface fluxes of heat, moisture, and momentum are computed as in Deardorff (1972). The model uses monthly climatological values of albedo, ground wetness, and surface roughness over land. Over ice, the albedo is 0.6, and surface roughness is equal to 0.0002435 m. Over water, the albedo is 0.09 and the surface roughness is computed according to Charnock's relation. Short wave and long wave radiation are parameterized in the model.

Before the actual forecast run is made, NORAPS is brought through a 48-hr sequence to allow development of mesoscale circulation features. At 12-hr intervals, the boundary conditions are updated from the analyzed fields of NOGAPS. The update cycle is found to significantly improve forecasts of cyclogenesis in NORAPS (Hodur, 1987).

3. POLAR LOW SIMULATIONS: 25–27 February 1984

The polar low chosen for our test simulations was observed by instrumented aircraft during the 1984 Arctic Cyclone Expedition. A detailed description and analysis of the observations is presented in Shapiro et al. (1987). This case represents a unique opportunity to verify model forecast capability of polar low development.

The forecast results will be used to compare four simulations of the Shapiro polar low (Table 1). Tau-zero forecast time for the final forecast in each sequence is 0000 GMT 27 February 1984. The model domains are depicted in Figure 1.

Figure 2a depicts the NORAPS run 1 (DX − 80 km) forecast of surface wind at 1200 GMT 27 February 1984. Here, "surface" wind refers to the 0.962 sigma level in the model. A semi-closed circulation is centered near 70°N 7°W, somewhat to the north and west of the location observed by Shapiro. A trough extends from north of Iceland into the Barents Sea. Lowest forecast pressure in the trough is 1001 mb, while the observed lowest pressure at 1340 GMT was 982 mb, near 69°N 3°W (Figure 3).

The surface wind field from NORAPS run 2 (DX=30 km) is shown in Figure 2b for 1500 GMT. In this case, a well-defined closed circulation appears closer to the observed location. Wind velocities up to 50 knots are forecast to

TABLE 1. SPECIFICATIONS FOR MODEL SIMULATIONS OF THE POLAR LOW*

DX	SST Resolution/Ice Edge	Analysis Type
Run 1: 80 km	low (380 km resolution)	type 1
Run 2: 30 km	low (380 km resolution)	type 1
Run 3: 30 km	high (50 km resolution)	type 1
Run 4: 30 km	high (50 km resolution)	type 2

*Ice edge analysis types are discussed in text.

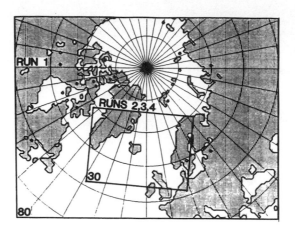

Figure 1. NORAPS model domain. Simulations with horizontal grid resolution of 80 km and 30 km on a 109×82 mesh.

the northwest of the low, in better agreement with observations. The forecast underestimates wind velocities to the south of the polar low. Minimum pressure in run 2 at 1500 GMT is 994 mb.

Figure 4 depicts the SST distribution and ice edge patterns used in runs 2, 3, and 4. The dashed contours represent the SST field used operationally in

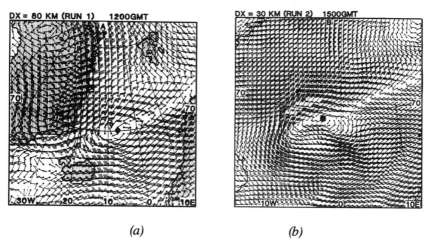

(a)   (b)

Figure 2. NORAPS forecast surface wind, 27 February 1984,. Flag equals 50 knots, solid dot is location of lowest surface pressure. Note: figure depicts a portion of model domain. (a) Run 1 (DX=80 km), 1200 GMT. (b) Run 2 (DX=30 km), 1500 GMT.

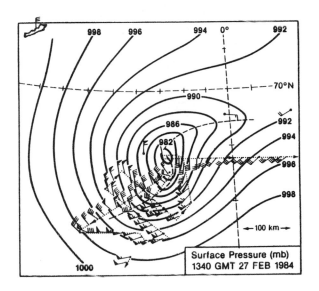

Figure 3. Surface pressure analysis (mb) at 1340 GMT 27 February 1984, from Shapiro et al. (1987). (Reprinted with permission of Tellus).

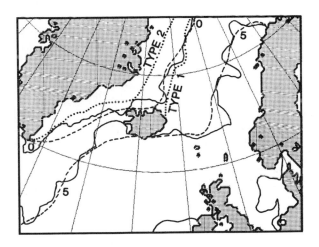

Figure 4. Sea surface temperature (SST) valid 27 February 1984. Contour interval is 5°C. Dotted lines are ice edge boundaries. SST and ice analysis discussed in text.

NORAPS. This is created by interpolating an observational field with a resolution of approximately 380 km onto the model grid. The dotted line labeled "type 1" in Figure 4 is the boundary between open sea and ice-covered sea as defined for NORAPS operational use. This is obtained from a climatological field with an approximate 2.5° resolution, which is interpolated onto the model grid.

The solid contours in Figure 4 represent the SST field based on a high resolution analysis for the Norwegian Sea produced by the Navy Fleet Numerical Oceanography Center. This analysis is derived from satellite and surface observations and has a resolution of approximately 50 km. Inspection of the two SST analyses reveals that to the north and west of Iceland, there are regions where the high resolution field indicates warmer SST readings. The dotted line labeled "type 2" is an alternative analysis of the ice edge boundary based on an observational analysis of solid ice coverage (Naval Polar Oceanography Center, 1984). In the type 2 analysis, the ice edge moves to a more realistic position closer to Greenland.

Model run 3 can be compared with run 2 to determine the effect of providing NORAPS with high resolution SST fields while leaving the ice edge unchanged from its operational configuration (type 1 line in Figure 4). The surface wind field from the high resolution SST forecast is depicted in Figure 5a. As in run 2,

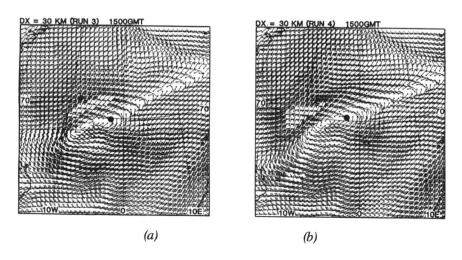

(a)   (b)

Figure 5. As in Figure 2b. (a) For NORAPS run 3 (DX=30-km, high resolution SST field). (b) For NORAPS run 4 (DX=30-km, high resolution SST and ice edge fields).

maximum wind speeds are about 50 knots on the northwest side of the low. Minimum pressure at 1500 GMT in run 3 is 993 mb. The center of the low is narrower than in run 2, and aligned more closely to the main southwest-northeast trough axis.

Finally, to determine the effect of adjusting the ice edge boundary from type 1 to type 2, run 4 was made, in which the high resolution SST field used in run 3 is retained. Figure 5b depicts the run 4 forecast of surface wind. The trough where polar low cyclogenesis takes place has migrated with the ice edge back towards Greenland and is no longer in the position found in run 3. The lowest central pressure at 1500 GMT in run 4 is 992 mb, near 70°N 4°W.

4. SUMMARY

The NORAPS model demonstrates a high degree of skill in forecasting the polar low of 27 February 1984. A decrease in horizontal grid spacing from 80 km to 30 km results in an improvement in forecast quality, as measured by the positioning of the polar low and deepening of the central pressure (Figure 6). The 30-km resolution forecast (run 2) is clearly better than the 80-km forecast (run 1). The inclusion of high resolution SST fields (run 3) does not, in these

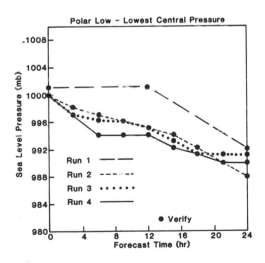

Figure 6. Model forecast of polar low central pressure (mb) during 24-hr forecasts starting at 0000 GMT 27 February 1984; verifying pressure at 1340 GMT shown as solid dot.

simulations, markedly change the forecast. Relocating the ice edge boundary (run 4) causes a more rapid deepening of the polar low, as shown in Figure 6, and a shift in position of the trough in which the polar low develops.

5. REFERENCES

Deardorff, J.W., 1972: Parameterization of the planetary boundary layer for use in general circulation models. *Mon. Wea. Rev., 100,* 93–106.

Hodur, R., 1987: Evaluation of a regional model with an update cycle. *Mon. Wea. Rev., 115,* 2707–2718.

Naval Polar Oceanography Center, 1984: Eastern-Western Arctic Sea Ice Analysis (unclassified). Published by Naval Oceanography Command, NSTL Station, Bay St. Louis, MS 39529.

Nordeng, T.E., 1987: The effect of vertical and slantwise convection on the simulation of polar lows. *Tellus, 39A,* 4, 354–375.

Rasmussen, E., and M. Lystad, 1987: The Norwegian polar lows project: A summary of the International Conference on Polar Lows, 20–23 May 1986, Oslo, Norway. *Bull. Am. Meteor. Soc., 68,* 801–816.

Shapiro, M.A., L.S. Fedor, and T. Hampel, 1987: Research aircraft measurements of a polar low over the Norwegian Sea. *Tellus, 39A,* 4, 272–306.

CHAPTER 4 – CASE STUDIES

Introduction

Case study papers were sought for and hold a separate place in this book because they define the scientific questions and provide the statistics in emerging topical areas of mesoscale vortices (polar/arctic lows) in the polar regions. We believe case studies and statistics in this chapter represent the third multi-authored set on polar lows. The first set was the proceedings for the August 1984 Conference on Polar Lows in Copenhagen, Denmark. The second set appeared as a series of reports arising from the multi-year Norwegian Polar Low Project. Many of these appeared later in the special issues of Tellus referenced in the preface of this book.

Case studies are critical to increase understandings of, and formulating models for, largely unknown features of mesoscale cyclones in both arctic and antarctic regions. As stated by Kellogg and Twitchell as recently as 1986, the cyclones have just become recognized as distinct meteorological phenomena with regard to structure as well as to geographical area. A few case studies from different platforms led to organized efforts to gain knowledge of their intensity and scale. With the capabilities and efforts to perform extensive observational programs, there still remains a scarcity of case studies for the polar low when compared with other oceanic region mesoscale systems such as hurricanes. Those case studies that exist have provided considerable insights and challenges to serve as fuel for discussion and for concerted cooperative efforts to answer scientific questions.

The chapter's first two papers, by Shapiro and coauthors, are based on aircraft-derived data as well as supporting synoptic data and illustrate the nature of both the scientific questions and partial answers that a *few case studies can generate. An unpredictable and relatively short-lived polar low event is only observed in situ with some luck involved. This is why the Norwegian Sea aircraft (NOAA P-3) data and its existing analyses and interpretations are valuable. In both papers included here, the baroclinicity and vorticity, versus convective, features receive the attention. The first of these two papers emphasizes the role of middle and upper level features as essential contributing factors. The second paper clearly emphasizes that observed structures do not differ significantly from previously described cold frontal structures in midlatitude continental regions.*

The third paper in this chapter, by Douglas and Shapiro, reviews observational studies of polar lows with detailed accounts of research aircraft flights into a storm over the Norwegian Seas and other flights into a Gulf of Alaska storm system. The similarity of the structure of these storms suggests that analytical and numerical simulations discussed in other papers can be verified.

In the fourth paper, Fett illustrates with meteorological satellite data that many polar lows in the Greenland, Norwegian, and Barents Seas are associated with boundary layer fronts presumably due to vector wind induced convergence field as well as baroclinic zones. He states further that cold air at upper middle levels is essential to their formation. Hence, the importance of both surface dynamic forcing and upper level instability mechanisms is demonstrated on the basis of the case studies.

Parker's paper provides this chapter's documentation of polar low occurrences in the northwestern Arctic (Beaufort Sea) and based on reconstructed data, describes features that are similar to those described more often for the Greenland, Norwegian, and Barents Seas.

The last two papers, by Bromwich and by Turner and Row, present and describe mesoscale cyclones occurring in the Antarctic. Their descriptions differ significantly from those of the arctic case studies because the role of the high Antarctic Plateau to the formation of katabatic winds and steering of synoptic scale flow is emphasized. Similar factors associated with the Greenland Plateau, or Svalbard Island, or the Norwegian North Cape were not in discussions of arctic polar lows. However, the scale and intensity of the antarctic cyclones are similar to the arctic ones.

A CASE STUDY OF AN ICE–EDGE BOUNDARY LAYER FRONT AND POLAR LOW DEVELOPMENT OVER THE NORWEGIAN AND BARENTS SEAS

M.A. Shapiro and L.S. Fedor

NOAA/ERL/Wave Propagation Laboratory
Boulder, Colorado, U.S.A.

ABSTRACT

This study presents the analysis of the structure of selected mesoscale weather systems observed over the Norwegian and Barents Seas during the Arctic Cyclone Expedition, 1984. Observations taken with the NOAA P-3 research aircraft and its dropwindsonde system, high-resolution AVHRR images from the polar orbiter satellites, and conventional surface and upper air stations were used to describe an ice-edge boundary layer front and associated jet stream along the west coast of Spitsbergen, an arctic cold-air outbreak and arctic cold front, and the development of two polar lows. Results show that the ice-edge front and jet formed in response to horizontal gradients in sensible heating as cold air ($\sim -20°C$) flowed off the arctic ice pack over the warm ($0°C$) underlying ocean along the west coast of ice-snow-covered Spitsbergen. Polar low development occurred within the shallow (~ 1 km) frontal baroclinicity at the leading edge of the arctic cold-air outbreak.

1. INTRODUCTION

During January and February 1984, the Arctic Cyclone Expedition (ACE) flew research flights with a NOAA WP-3D (P-3) research aircraft, exploring the mesoscale structure of oceanic and atmospheric weather systems over the North Atlantic Ocean and its adjacent Arctic seas. Observations were taken of atmospheric and ocean fronts; ocean wave characteristics (sea state); atmospheric boundary layer fluxes; and polar lows, the mesoscale (~ 500 km) cyclones that form within or at the leading edge of arctic air streams, as described by Harrold

POLAR AND ARCTIC LOWS
Paul F. Twitchell, Erik A. Rasmussen, and Kenneth L. Davidson (Eds.)

257

and Browning (1969), Rabbe (1975), Mullen (1979), and Rasmussen (1979, 1985). This study reports on the 14 February 1984 NOAA P-3 flight from Bodφ, Norway, north to Spitsbergen, which described a boundary-layer front and its associated low-level (~ 850 mb) jet stream that formed along the ice edge west of Spitsbergen under conditions of northerly wind flow. This flight also documented the structure of an arctic cold-air outbreak and its associated arctic front. The NOAA satellite images captured the evolution of two polar-low cloud circulations along this front. The geography of this area is shown in Figure 1.

Our initial interest in the atmospheric boundary layer structure along the west coast of Spitsbergen and its adjacent ice edge was aroused during a September 1983 visit by the lead author to The Norwegian Meteorological Institute (DNMI), Oslo, Norway. At that time, an examination of satellite cloud images revealed a discontinuity in cloud structure along the west coast of Spitsbergen, extending southward into the Norwegian Sea, as shown in Figure 2a. The discontinuity was marked by a band of enhanced cumulus activity. This cloud structure was of special interest, since as shown in Figure 2b, a polar low formed on the discontinuity. In the ensuing discussions, it was hypothesized that this discontinuity in cloud structure formed during northerly low-level flow, approximately parallel to the ice edge. The north-south-oriented shallow cumulus cloud "streets" over the ice-free sea (Figure 2a) were interpreted as the signature for longitudinal convective roll vortices (Kuettner, 1959; LeMone, 1973) within the neutrally

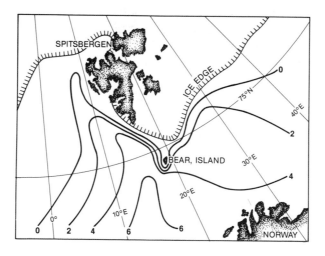

Figure 1. Regional geography of the Arctic Cyclone Expedition. Solid lines indicate sea surface temperature in °C. The ice edge is for 20 February 1984.

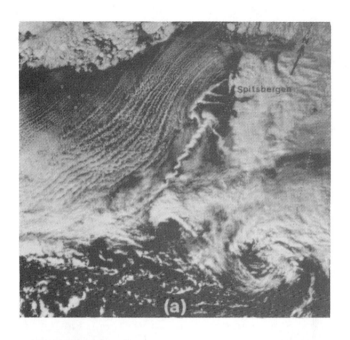

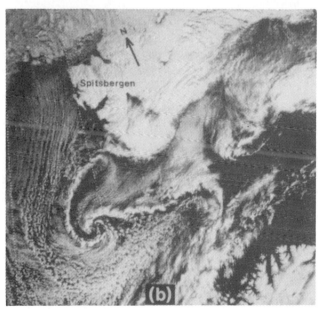

Figure 2. NOAA satellite infrared images at (a) 1220 GMT 18 April 1985, and (b) 0741 GMT 19 April 1985 (from Hoem et al., 1984).

stable ($\partial\theta/\partial p = 0$) surface boundary layer, downwind from the east-west-oriented ice edge west of Spitsbergen between $0°$ and $10°$E longitude. It was further hypothesized that a boundary layer front could form between two distinct boundary layer flows: (1) the flow that was heated by the upward flux of sensible heat as initially cold air ($\leq 20°$C) passed over warmer water ($>0°$C) after exiting the northern ice cap to the west of Spitsbergen, and (2) the adjacent nonheated northerly flow over snow-covered Spitsbergen and the sea ice to its south. Additional forcing for ice-edge boundary layer frontogenesis was provided by the narrow sea-surface temperature frontal structure (Figure 1) along the ice edge where a branch of the Gulf Stream passes northward along the west coast of Spitsbergen. This air-sea interactive regime was the focus of the NOAA aircraft mission on 14 February 1984.

2. INSTRUMENTATION

Standard navigational and meteorological parameters were measured and recorded by the P-3 onboard data acquisition system. Omega dropwindsondes were released to obtain vertical profiles of wind, temperature, and moisture; an airborne downward-looking infrared radiometer measured sea and ice temperatures; high-resolution (~ 1 km) NOAA-7 and NOAA-8 infrared satellite images described the cloud systems associated with the atmospheric flows under investigation; and the NOAA gust-probe system measured vertical fluxes of heat, momentum, and moisture.

3. SYNOPTIC PERSPECTIVE AND FLIGHT SUMMARY

The synoptic setting for the 14 February 1984 flight is shown in the surface analyses for the period 0600 to 1500 GMT (Figure 3). These analyses show the development of >15 m s^{-1} northerly flow in the area west of two polar lows that were traveling eastward parallel to the arctic ice edge at $\sim 74°$N (see ice-edge position, Figure 1). The leading edge of the cold-air outbreak (the arctic front) reached the northern coast of Norway between 1200 and 1500 GMT (Figures 3c and 3d). Pressure rises exceeding 10 mb in 3 hr were observed in the cold air behing the western polar low and the arctic front (Figures 3c and 3d) as the low traveled rapidly eastward at 17 m s^{-1} from the Norwegian Sea into the Barents Sea. As the western low passed south of the southern tip of Spitsbergen, $\leq 20°$C, arctic air was ejected off the ice cap between Spitsbergen

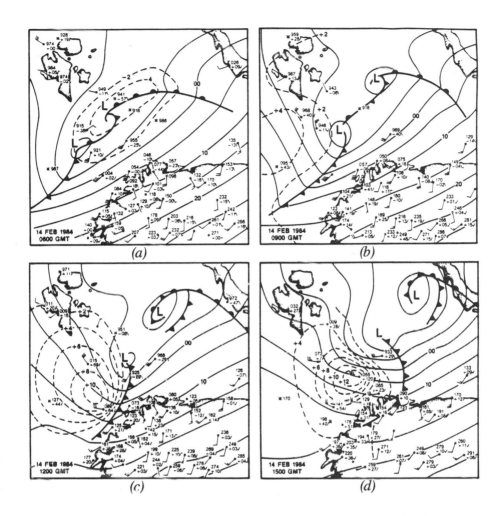

Figure 3. Surface pressure and pressure tendency analyses for 14 February 1984. Solid lines indicate surface pressure in millibars; dashed lines show 3 hr pressure tendency in millibars (3 hr)⁻¹; heavy lines are surface fronts. Boxes with an inner x show drifting ocean buoys. Wind vectors at full barb =5 m s⁻¹, and at half barb=2.5 m s⁻¹.

and Greenland and flowed southward out over the warm water (>0 °C) to the south of the ice edge. The arctic outbreak created the desired meteorological conditions west of Spitsbergen toward which the NOAA research aircraft was dispatched at ~1000 GMT 14 February 1984. Figure 4 presents the flight track and the position of the dropwindsonde profiles taken during the flight.

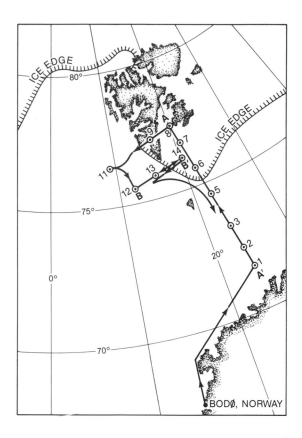

Figure 4. NOAA research aircraft flight track on 14 February 1984. Dotted circles show the locations and identifying numbers of dropwindsonde deployments. AA' and BB' are the projection lines for the cross sections shown in Figures 5 and 6, respectively.

   The first portion of the flight charted vertical structure of the arctic front and the variation in height of the boundary layer beneath the front as a function of distance from the ice edge by deploying the dropwindsondes numbered 1, 2, 3, 4 (failed), 5, 6, 7, and 8 at an altitude of ~5 km along the line AA' of Figure 5 (and Figure 7). The aircraft then headed westward and released dropwindsondes 9, 10 (failed), and 11 to the west of Spitsbergen. The onboard computer data reduction of dropwindsondes 8, 9, and 11 provided the first in-flight glimpse of the ice-edge boundary layer front below. The aircraft then turned southward and descended to 3 km where dropwindsondes 12, 13, and 14 were

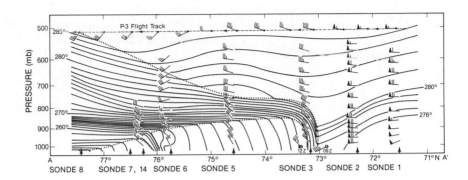

Figure 5. Cross-section analysis of potential temperature (K, solid lines) through the arctic front on 14 February 1984 along the line AA' of Figure 4. Dropwindsonde locations are indicated by heavy arrows with identifying numbers from Figure 4 plotted below. The dashed line with flight direction arrows and selected flight-level wind vectors shows the research aircraft flight track. Wind vectors without dotted heads indicate dropwindsonde wind profiles; wind vectors with flag=25 m s^{-1}, at full barb=5 m s^{-1}, and at half barb=2.5 m s^{-1}. Dotted lines show frontal boundaries.

released along the line BB' of Figure 4 (and Figure 7), normal to the boundary layer front. The final portion of the flight consisted of three aircraft penetrations of the front and its low-level jet stream along the line BB' at 1500, 800, 232, and 116 m above the sea.

4. VERTICAL STRUCTURE

Our discussion of the vertical structure of the mesoscale systems under investigation begins with the cross section through the arctic front along the line AA' of Figures 4 and 7. The analysis was based upon the sequential dropwindsondes discussed in Section 3, the 5-km flight-level data, and surface observations from the Norwegian research ship AMI, stationed off the northern coast of Norway. The cross-section analysis (Figure 5) shows the leading edge of the arctic cold-air outbreak (the arctic front) at 73 °N. Note the increase in the height of the neutrally stable subfrontal boundary layer downwind from the ice edge, resulting from the transfer of sensible heat from the warm ocean to the cold air above. The arctic front was a sharp inversion (lid) over the near adiabatic boundary layer beneath. Surface wind speeds exceeding 25 m s^{-1} were observed in the west-southwesterly flow in advance of the front, and 6-m-high sea swells and wind-blown ocean-spray foam streaks were seen from the aircraft. The second frontal structure in the cold air between sondes 6 and 7 was the ice-edge boundary layer front discussed in the following paragraphs.

One of the most illuminating analyses of this study was the wind, potential temperature, and potential vorticity cross section (Figure 6) prepared from the

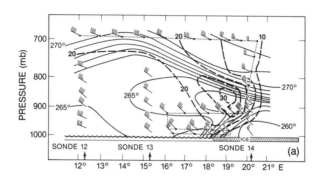

Figure 6a. Cross-section analysis of potential temperature (K, solid lines) and wind speed (m s⁻¹, dashed lines) along the projection line BB' (Figure 4) on 14 February 1984. Aircraft flight track follows the dotted line. Vectors with dotted heads show selected aircraft winds; vectors without heads indicate dropwindsonde winds. Numbered arrows locate dropwindsondes.

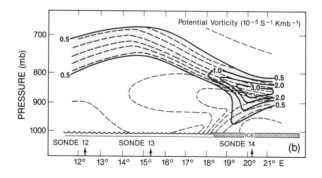

Figure 6b. Cross-section analysis of potential vorticity (10⁻⁵ s⁻¹ K mb⁻¹) for Figure 6a. Solid lines indicate potential vorticity and dashed lines indicate potential temperature. Stippled area shows potential vorticities in the range (0.5 − 2.0) × 10⁻⁵ s⁻¹ K mb⁻¹.

dropwindsonde and aircraft traverses along the line BB′ of Figures 4 and 7. This analysis illustrates the structure of the boundary layer front and its associated low-level jet stream above the ice edge west of Spitsbergen. Figure 6a shows the front sloping upward from the sea surface to 860 mb. The front was 80-km wide and contained a 5-K thermal gradient across that distance. The thermal stratification above the sea surface to the warm side (west) of the front was near adiabatic west of 18 °E and superadiabatic over the zone of highest sea surface temperature associated with the warm ocean current that flows northward along the ice edge (Figure 1). The barotropic level for this system (level of zero thermal gradient) was found at 870 mb. It was there that the dropwindsonde and aircraft observations charted the ~ 30 m s⁻¹ low-level northwesterly jet stream about the baroclinicity of the boundary layer front. The horizontal thermal gradient above the jet was opposite to that found in the front beneath. The boundary layer inversion between 12 °E and 17 °E was elevated by the upward flux of sensible heat on air parcels traveling within the convectively overturning boundary layer.

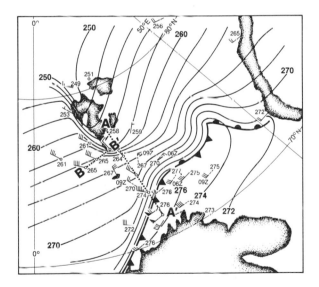

Figure 7. Surface potential temperature (K, solid lines) analysis at 1200 GMT 14 February 1984, prepared from a composite of land and ship observations (wind vectors without circle heads), and dropwindsondes (wind vectors with circle heads). Cross-section projection lines AA' and BB' (dotted lines) are for Figures 5 and 6, respectively. The observations were taken between 0400 and 1200 GMT and were space-time adjusted to 1200 GMT. Heavy solid lines indicate the arctic front.

Cumulus clouds within the boundary layer were observed from both the aircraft and the satellite, which gave further evidence of the convective overturning. This ice-open-water differentially heated boundary layer also contained large relative vorticity (10^{-4} s^{-1}) and potential vorticity ($\sim 10^{-5}$ s^{-1} K mb^{-1}) within the front and to the eastern side of the jet stream core as shown in Figure 6.

The physical structure of this boundary-layer jet-front system (Figure 6) is analogous to that of upper tropospheric jet-front tropopause folds (e.g., Shapiro, 1980), except that in the present case, the high potential vorticity is found within the stably-stratified arctic boundary layer inversion. We suggest that the folded potential vorticity structure may be attributed to the following processes. Radiative heat loss from the arctic sea-ice pack and adjacent snow- and ice-covered land generates high potential vorticity within the surface-based radiative inversion, whose depth may range from a few hundred meters up to >2 km as in dropwindsonde 8 (Figure 5). This radiative heat loss and the resulting inversion are strongest during the winter months when no direct solar radiation reaches the Arctic latitudes. When the thermally stable, high-potential-vorticity air flows off of the ice cap and over the relatively warm open water, the heating through the vertical gradient of sensible heat flux destroys the low-level potential vorticity by decreasing the thermal stratification on air parcels. The boundary layer jet-stream frontal structure and associated potential vorticity fold (Figure 6b) are formed in the zone separating the heated (sensible heating) ocean boundary layer from the radiatively cooled land and sea-ice boundary layer.

The formation of differentially heated boundary layer fronts is not restricted to the ice edge west of Spitsbergen. Similar structures may form wherever the wind flow is parallel to (or very weak at) adjacent snow-ice-covered and open-water surfaces. When ice-edge boundary layer fronts are driven out to sea by changes in the synoptic wind flow, they maintain their identity in spite of the modifying effect of the diabatic heating by the warm underlying sea. Oceanic arctic fronts such as those shown in Figure 5 can often be traced from their origin at the ice edge and propagate as far away as the coast of Norway and the United Kingdom.

5. SENSIBLE HEAT FLUX

The heating of near-surface air parcels after they exited the northern ice pack and traveled southward over the warm sea surface was calculated from (1) aircraft measurements of the vertical gradient of the vertical heat flux ($\overline{w\theta'}$), and (2) the temporal change of potential temperature on trajectories within the surface boundary layer. When we applied procedure (1), the aircraft gust probe measurements gave the vertical heat fluxes along the flight legs to the warm side of the front (Figure 6). The measured fluxes at 232 and 871 m were 0.051 (upward) K m s^{-1} and -0.001 (downward), respectively. The sea surface heat flux was calculated from the bulk aerodynamic technique,

$$(\overline{w\theta'})_{\text{surface}} = -C_d V \Delta\theta \tag{1}$$

where C_d is the surface drag coefficient ($C_d = 1.6 \times 10^{-3}$), V is the surface wind speed ($V \sim 15$ m s^{-1}, determined from aircraft and dropwindsonde measurements), and $\Delta\theta$ is the air-minus-sea potential temperature difference ($\Delta\theta = -5$ K, determined from the aircraft downward-looking radiometer sea surface temperatures and dropwindsonde profiles). These measurements gave $(\overline{w\theta'})_{\text{surface}} = 0.12$ K m s^{-1} (upward). From these fluxes, the sensible heating by the vertical gradient of the heat flux in the layer extending from the surface to 232 m was

$$\frac{d\theta}{dt} = -\frac{\partial(\overline{w\theta'})}{\partial z} = (30 \times 10^{-5})\text{K s}^{-1} = 26 \text{ K day}^{-1} \tag{2}$$

The trajectory method, procedure (2), for calculating surface boundary layer heating makes use of the temporal and spatial characteristics of the wind and potential temperature fields. The surface thermal gradient in the cold northwesterly arctic flow west and south of Spitsbergen was 10 K (500 km)$^{-1}$ (Figure 7). The surface wind was directed normal to the thermal gradient at a speed of ~ 15 m s^{-1}. The local tendency of potential temperature was negligible, since Bear Island (Figure 1) reported no surface temperature changes during the period 0900 to 1500 GMT. From these measurements, the trajectory estimate for surface air parcel for negligible local changes ($\partial\theta/\partial t \approx 0$) was

$$\frac{d\theta}{dt} = \vec{V} \cdot \nabla\theta = (30 \times 10^{-5})\text{K s}^{-1} = 26 \text{ K day}^{-1} \tag{3}$$

a coincidence in agreement with the results of Eq. (2). These heating rates are of comparable magnitude to summertime diurnal surface temperature rises over the central U.S. prairie lands.

6. HORIZONTAL STRUCTURE

The horizontal structure of the boundary layer front, arctic cold-air outbreak and associated arctic front, and the identification of two polar lows were derived from a composite analysis of land- and sea-surface synoptic observations, aircraft dropwindsonde soundings, aircraft flight-level data, and NOAA-7 and NOAA-8 polar-orbiting satellite infrared images. Figures 7 and 8 present the composite analysis of surface potential temperature, and the streamline analysis of the surface wind, respectively, at 1200 GMT 14 February 1984. The asynoptic (off-time) observations from ships and key land stations (i.e., Bear Island) and the 116 m flight-level data and dropwindsondes were space-time adjusted to 1200 GMT using the phase velocity of the two polar lows as determined from their movement in the satellite images.

The potential temperature analysis and wind vectors (Figure 7) show the cyclonic wind shear and baroclinicity of the arctic front at the leading edge of the northwesterly flow of the arctic cold-air outbreak. The 20 to 35 m^{-1} southwesterly flow in advance of the arctic front was associated with a narrow zone of warm potential temperature ($\sim$275 K) along the Norwegian coast. The

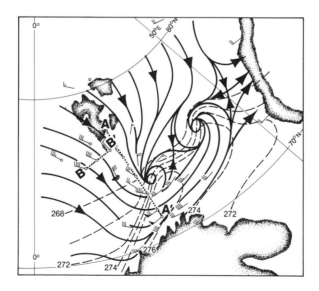

Figure 8. Surface streamline analysis (heavy lines with directional arrows) for Figure 7. Selected isopleths of potential temperature (dashed lines) and wind vectors are the same as in Figure 7.

relatively dense data set defined the ice-edge boundary layer front along the west coast of Spitsbergen, and the wind vectors on both sides of the front showed cross-frontal shears of ~ 14 m s^{-1} (100 km)$^{-1}$, in agreement with the cross-section analysis (Figure 6a). The largest baroclinicity (Figure 7) occurred at the merger of the boundary layer and arctic fronts. The second wave in the arctic front was drawn on the basis of the satellite images (discussed in Section 7); without supporting temperature or wind observations in the vicinity of 75 °N 40 °E.

The composite surface streamline analysis for 1200 GMT 14 February 1984, along with selected isotherms from Figure 7 is presented in Figure 8. The observations clearly define the cyclonic circulation of the weak polar low that formed within the baroclinicity of the merging of the boundary layer and arctic fronts.

7. NOAA-7 AND NOAA-8 INFRARED SATELLITE IMAGES

Some of the most intriguing observations of this study were the infrared images from the NOAA-7 and NOAA-8 polar-orbiting satellites. The images at 0440, 0830, 0945, and 1622 GMT 14 February were received and recorded at Tromsφ, Norway, and processed at the NOAA Wave Propagation Laboratory to highlight the fine-scale structure contained in the cloud-top radiative temperatures. Note that the following discussion is our interpretation of the imagery. Without the companion three-dimensional wind, temperature, moisture, and radar reflectivity data sets, the discussion of such images is qualitative and should be treated as so by the reader. The images represent only cloud structure and cloud-top temperatures, and their quantitative relationship to mesoscale circulations and intensity of meso-convective processes is only speculative.

The first image for this case was taken at 0440 GMT 14 February (Figure 9). The warm (dark) cloud tops southwest of Spitsbergen (Figure 9a) were associated with a shallow stratocumulus (SC) layer cloud in the weakly stratified boundary layer beneath the arctic front (Figures 5 and 7). The radiative cloud-top temperatures of the SC clouds over the lower-left portion of the image were between 260 and 270 K. Reference to the cross sections in Figures 5 and 6a shows that the temperatures at the top of the boundary layer (850 mb) were also within this temperature range, suggesting that the vertical development of these clouds was suppressed by the large static stability of the arctic front. In the center of the image, a high-level, east-west-oriented cirrus shield (CS) had a sharp southern edge and a "fingered-feathery" northern edge. Radiative cloud-top temperatures of the shield were ~ 210 K, similar to the 400-mb tropopause

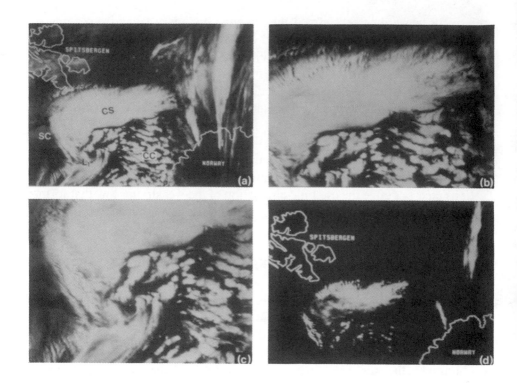

Figure 9. NOAA satellite infrared image at 0440 GMT 14 February 1984. Gray-scale temperature shadings range from ~280 K for the warmest (darkest) sea surface temperatures to ~210 K for the coldest (brightest) cirrus clouds near the tropopause. Added notations for selected cloud features are: stratocumulus cloud layers (SC), cirrus cloud shield (CS), deep cellular convective clouds (CC), and polar lows (L). (a) Two-kilometer-resolution image; (b) One-kilometer-resolution 1× enlargement centered on the polar low L_1 of Figure 8a; (c) One-kilometer-resolution 1× enlargement centered on the cirrus shield (CS) of Figure 8a; (d) Two-kilometer-resolution image, enhanced to define the "over-shooting" most active deep cumulus convection within the cirrus cloud shield of Figure 9a.

temperature in the adjacent Bear Island 0000 GMT 14 February 1984 rawin-sonde sounding. The lower center of the image contains deep cumulus convection (CC), which was bounded by the leading edge of the low-level arctic outbreak to the west, the high-level cirrus shield to the north, and the northern coast of Norway to the south. These convective clouds developed within the warm, moist, and convectively unstable (to ~400 mb) southwesterly flow in advance of the two cyclonic disturbances and frontal structure shown in Figures 3a and 7. South of the western edge of the cirrus cloud shield, the low-level stratocumulus of the arctic outbreak (partially shielded by broken high-level cirrus) and the deep cellular convective clouds in the warm air intertwined cyclonically to produce a cloud circulation signature of polar low L_1, the western polar low in Figures 3 and 8. This circulation was defined in the later satellite images. Figures 8b and 8c show the enlargements of the cirrus cloud shield, and polar low respectively of Figure 9a. Note the area of coldest (highest) cloud tops in the western portion of the cloud shield (Figure 9c). Figure 9d shows the enhanced image for Figure 9a in which the most convective portion of the cloud shield is marked by the cold cloud-top temperatures associated with deep cumulus convection extending above 400 mb. It was this convective portion of the cirrus shield that took on the structure of a mature polar low, L_2 during the following 6 hr.

The image at 0830 GMT (Figure 10) showed the eastward movement of polar low L_1 to directly south of the southern tip of Spitsbergen, with the deep cellular convection wrapping cyclonically halfway around its nearly cloud-free inner eye. This image first suggested the formation of a second polar low, L_2, which appeared in the area that previously contained the highest cloud tops of the cirrus shield in Figures 9c and 9d. Figures 10b and 10c present the enlargement images of these two polar-low cloud systems. Note that the most active and deepest convection for polar low L_1 (Figure 10b) was north of its relatively cloud-free eye.

The following image at 0945 GMT (Figure 11) showed only slight changes from that at 0830 GMT (Figure 10) as the two polar lows continued their eastward track into the Barents Sea. The enlargements showed the merging of the cirrus anvil outflow from the convective clouds north of polar low L_1 (Figure 11b) and a further development of the eye structure for polar low L_2 (Figure 11c) as the high-level cirrus shield wrapped further about its cloud-free inner core.

The final image at 1622 GMT (Figure 12) shows polar low L_1 almost void of deep convective clouds, and that polar low L_2 had reached the stage of a mature polar low, with the high-level cirrus clouds completely encircling the cloud-free inner core. The sharp discontinuity in the shallow low-level cumulus streaks west and south of Spitsbergen (Figure 12a) occurred at the boundary layer front described above. The enlargement (Figure 12b) showed cyclonic structure in the low-level boundary layer cumulus of L_1 with only few remaining penetrating cumulonimbi.

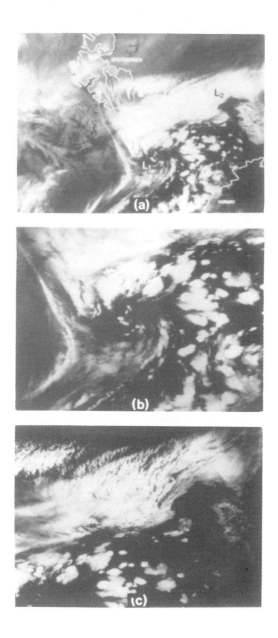

Figure 10. NOAA satellite infrared image at 0830 GMT 14 February. The gray-scale temperatures and cloud-type notations are the same as in Figure 9a. (a) Two-kilometer-resolution image; (b) One-kilometer-resolution 2× enlargement centered on the polar low L_1 of Figure 10a; (c) One-kilometer-resolution 2× enlargement centered on the polar low L_2 of Figure 10a.

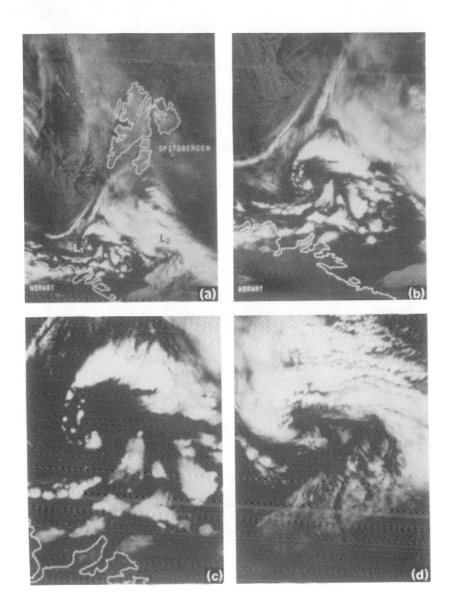

Figure 11. NOAA satellite infrared image at 0945 GMT 14 February 1984. The gray-scale and cloud-type notations are the same as in Figure 9. (a) Two-kilometer-resolution image; (b) One-kilometer-resolution 1× enlargement centered over polar low L_1 of Figure 10a; (c) One-kilometer-resolution 2× enlargement of polar low L_1 of Figure 10a; (d) Enlargement of polar low L_2 of Figure 10a.

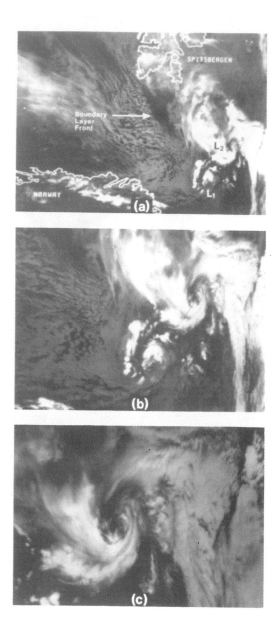

Figure 12. NOAA satellite infrared image at 1622 GMT 14 February 1984. The gray-scale and cloud-type notations are the same as in Figure 9. (a) One-kilometer-resolution image; (b) Enlargement centered over polar lows L_1 and L_2 of Figure 12a; (c) Enlargement of polar low L_2 of Figure 12a.

Another interesting example of mesoscale cloud structure under northerly flow over and south of Spitsbergen is shown in Figure 13. The image at 0415 GMT 2 March 1984 (Figure 13a) shows the boundary layer cumulus cloud streets characteristic of arctic outflows over the Norwegian Sea. The leading edge of the outbreak (the arctic front) was at the developed cumulus cloud band just west of the Norwegian coast. A second band of enhanced cumulus extended southward in the lee of the ice edge west and south of Spitsbergen, and merged with deeper convective arctic frontal clouds. The enhanced cloud structure at the merger of the two cloud bands was interpreted as a polar low by DNMI meteorologists. We suggest that the cloud band that extends southward from west of Spitsbergen is tied to a southerly extension of an ice-edge boundary layer front similar to the one described in Section 4. Note the series of periodic cumulus clusters of ~50-km scale that lie just east of the suspected boundary layer cloud

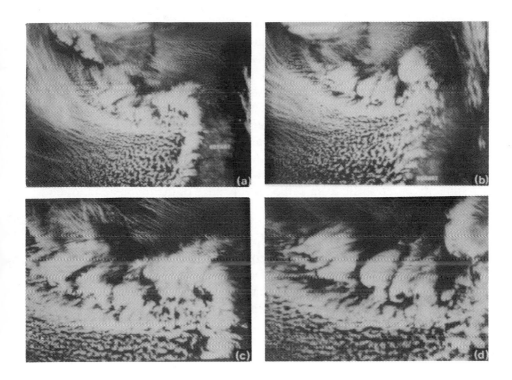

Figure 13. NOAA satellite 1-km-resolution infrared images at (a) 0415 GMT 2 March 1984, and (b) 0845 GMT 2 March 1984; (c) Enlargement of polar low L_1 and trailing convective clusters from Figure 13a; (d) Enlargement of comma-shaped convective clusters in Figure 13b.

band. The next image at 0845 GMT (Figure 12b) shows that the polar low and arctic frontal clouds had struck the northern coast of Norway, and that the trailing cumulus clusters in the vicinity of the suspected boundary layer front had developed into 50-km-diameter cyclonic comma-shaped cloud systems. The enlargement (Figure 12c) shows the fine-scale structure of the polar low and periodic cumulus clusters of Figure 12a. The enlargement of the periodic comma clouds (Figure 12d) shows one comma with a cloud-free inner eye. Without the supporting meteorological observations, one can only imagine what type of instability mechanism (barotropic, baroclinic, moist convective, or a combination thereof) could be responsible for the formation of such fascinating cloud structures.

8. SUMMARY AND CONCLUSIONS

This study presents our analysis, interpretations, and speculations of the structure and origin of selected mesoscale weather systems over the Norwegian and Barents Seas. The unique observations, taken with the NOAA research aircraft and its dropwindsonde system, described an ice-edge boundary layer front and its associated low-level jet stream along the west coast of Spitsbergen, and a arctic cold-air outbreak and its associated arctic cold front, which propagated from the arctic ice cap into the Norwegian Sea. When these special aircraft observations were combined with conventional surface observations and the NOAA-7 and NOAA-8 satellite infrared images, the resulting analyses and satellite interpretations suggested that two polar lows evolved and propagated within the baroclinicity of the arctic front at the leading edge of the outbreak of arctic air from off the polar ice cap. It should be noted that the baroclinicity and vorticity at the boundary layer front and arctic cold front were confined to a shallow layer (~ 1.5 km) near the sea surface. Mansfield (1974) suggested that cyclonic perturbations within the middle and upper troposphere are most likely a contributing factor in the spin-up of deep (~ 7 km) polar lows. For the present study, there were insufficient upper air observations to establish the presence of upper level influences upon the polar low developments discussed in this study.

REFERENCES

Harrold, P.W., and K.A. Browning, 1969: The polar low as a baroclinic disturbance. *Quart. J. Roy. Meteor. Soc.*, *95*, 710-723.

Hoem, V., S. Hoppestad, and A. Rabbe, 1984: *Polar Low Case Studies, I, 1982-1985.* Tech. Rep. No. 3, Section 3, Polar Lows Project, Norwegian Meteorological Institute, Oslo.

Kuettner, J., 1959: The band structure of the atmosphere. *Tellus, 11,* 267-294.

LeMone, M.A., 1973: The structure and dynamics of horizontal roll vortices in the planetary boundary layer. *J. Atmos. Sci., 30,* 1077-1091.

Mansfield, D.A., 1974: Polar lows: The development of baroclinic disturbances in cold air outbreaks. *Quart. J. Roy. Meteor. Soc., 100,* 541-554.

Mullen, S.L., 1979: An investigation of small synoptic-scale cyclones in polar air streams. *Mon. Wea. Rev., 107,* 1636-1647.

Rabbe, A., 1975: Arctic instability lows. *Meteorologiske Annaler, 6,* Norwegian Meteorological Institute, Oslo.

Rasmussen, E., 1979: The polar low as an extratropical CISK disturbance. *Quart. J. Roy. Meteor. Soc., 105,* 531-549.

Rasmussen, E., 1985: A case study of a polar low development over the Barents Sea. *Tellus, 5,* 407-418.

Shapiro, M.A., 1980: Turbulent mixing within tropopause folds as a mechanism for the exchange of chemical constituents between the stratosphere and troposphere. *J. Atmos. Sci., 37,* 994-1004.

RESEARCH AIRCRAFT OBSERVATIONS OF AN ARCTIC FRONT OVER THE BARENTS SEA

M.A. Shapiro, T. Hampel, and L.S. Fedor
NOAA/ERL/Wave Propagation Laboratory
Boulder, Colorado, U.S.A.

ABSTRACT

Research aircraft flight-level and dropwindsonde observations documented the structure of an arctic front that formed over the Barents Sea south of Spitsbergen and the Arctic Sea ice pack. The front formed in the confluence of warm southwesterly open ocean flow and cold northeasterly off-ice flow. The front was extremely shallow in slope, not extending above 800 mb in the 400-km extent of its observation. Analyses of front-normal and parallel wind components were used to diagnose the vorticity and divergence fields for the front. Airborne weather radar observations document a mesoscale (~ 100 km) precipitation feature that formed within the cyclonic vorticity and convergence of the leading edge of the frontal zone.

1. INTRODUCTION

During January and February 1984, the Arctic Cyclone Expedition flew research flights with the NOAA P-3 Orion research aircraft, exploring the mesoscale structure of weather systems over the North Atlantic Ocean and the adjacent arctic seas. This study presents the analysis of the observations taken on 18 February 1984, when measurements were made of a strong arctic front that was positioned off and over the arctic ice cap south of Spitsbergen over the Barents Sea. Figure 1 shows the regional geography, position of the arctic ice edge, aircraft flight track, and location of the aircraft dropwindsonde deployments on 18 February 1984.

POLAR AND ARCTIC LOWS
Paul F. Twitchell, Erik A. Rasmussen,
and Kenneth L. Davidson (Eds.)

279

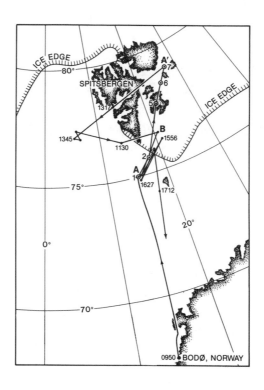

Figure 1. Regional geography and NOAA research aircraft track on 18 February 1984. Selected time marks (GMT) are entered along the flight tract. Dotted circles show the location and identifying numbers of dropwindsonde deployments. AA' is the cross-section projection for Figure 3; AB is the cross-section projection for Figures 5 and 6.

The original scientific objectives for this day's flight were twofold. The first was the documentation by dropwindsonde of the vertical structure of the arctic front off and over the ice along the east coast of Spitsbergen. The second objective was to investigate the cold air northerly arctic outflow from off the ice edge to warm open water west of Spitsbergen and document the associated boundary-layer transformation processes. Unfortunately, the low-level flow west of Spitsbergen was found to be southerly, and so the remainder of this research flight focused upon describing the structure of the arctic front to the east, using direct flight-level observations from the aircraft.

2. INSTRUMENTATION

Standard navigational and meteorological parameters were measured and recorded by the aircraft onboard data acquisition system. Omega-navigation dropwindsondes were released to obtain vertical profiles of wind, temperature, and moisture; an airborne downward-looking infrared radiometer measured sea and ice temperatures; and the NOAA gust-probe system measured vertical fluxes of heat, momentum, and moisture.

3. THE ARCTIC FRONT OF 18 FEBRUARY 1984

The first indication of arctic frontal structure south of Spitsbergen came from the dropwindsonde wind and temperature profiles (Figure 2) taken between 1141 and 1238 GMT along the northward, 465-mb flight leg, along the line AA′ of Figure 1. The onboard readout from sondes 2, 5, 6, and 7 (3 and 4 failed;

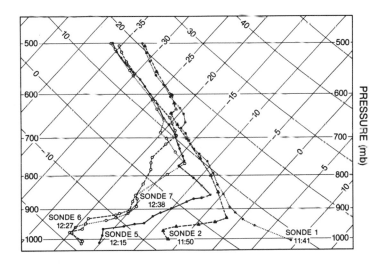

Figure 2. Dropwindsonde temperature profiles (°C) between 1141 and 1238 GMT 18 February 1984. The geographical positions of the profiles are indicated in Figures 1 and 3.

1 was in the warm air) profiled a sharp low-level frontal inversion below 800 mb
(Figure 2). The winds recorded by the dropwindsondes were easterly below and
southwesterly above the frontal inversion layer. A cross-section analysis of potential temperature (Figure 3) was prepared along the line AA' of Figure 1 (and
Figure 4), from the five dropwindsondes (Figure 2) and the aircraft 465-mb flight-
level observations. Measurements along the low-level flight tracks (Figure 3),
between ~ 1500 to 1700 GMT, described the detailed structure of the leading
edge of the front discussed later in this article. From Figure 3, we note that the
frontal layer had little slope between 75.5° and 77.5°N, and increased in depth
north of 78°N. The frontal inversion was quite extreme, having a 10°C increase
of temperature between 970 and 850 mb during the profile taken by sonde 5
(Figure 2). The leading edge of the front was located south of the ice edge over
the open water. The position of the ice edge shown in Figure 3 (and Figure 5)
was determined from the downward-looking infrared radiometer temperature
measurements.

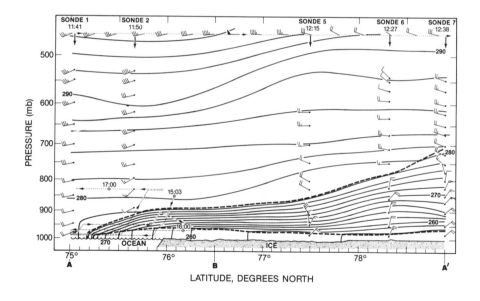

*Figure 3. Cross-section analysis of potential temperature (K, thin solid lines)
for the arctic front on 18 February 1984 along the AA' of Figure 1. Heavy
dashed lines indicate frontal boundaries; the NOAA flight track follows dotted
line with GMT time marks. Wind vectors along the upper flight track indicate
465-mb flight-level winds: full barb=5 m s⁻¹, half barb=2.5 m s⁻¹. Wind
vectors with black circle heads indicate dropwindsonde winds (m s⁻¹).*

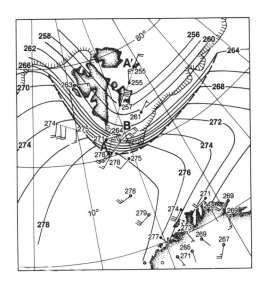

Figure 4. Surface potential temperature analysis (K, thin solid lines) at 1200 GMT 18 February 1984. The heavy dashed line indicates the warm side of the arctic front. Dropwindsonde near-surface winds, potential temperatures, and dotted-head wind vectors with plotted temperatures are located along flight track AA'. Wind vectors without dotted heads indicate near-surface aircraft observations; wind vectors with cloud-cover circles indicate ship and land surface observations. Wind barb values are the same as in Figure 3; the ice edge is the same as in Figure 1. Lines AA' and AB are cross-section projection lines for Figures 3, 5, and 6.

The horizontal structure of the arctic front was analyzed from a composite of conventional surface observations, the aircraft dropwindsonde measurements, and near-surface flight-level observations. The surface potential temperature analysis (Figure 4) shows the baroclinicity of the front that extended southward from the Fram Strait west of Spitsbergen and then turned eastward just south of the southern extent of the ice edge south of Spitsbergen. Substantial cyclonic shear was found across the arctic front, as shown in Figure 4, and in Figures 5 and 6 discussed below.

The detailed structure of the leading edge of the arctic front along the line AB of Figures 1 and 4 was analyzed from dropwindsondes 1 and 2 (Figures 2 and 3); flight-level measurements at 830, 950, 970, and 990 mb; and two aircraft

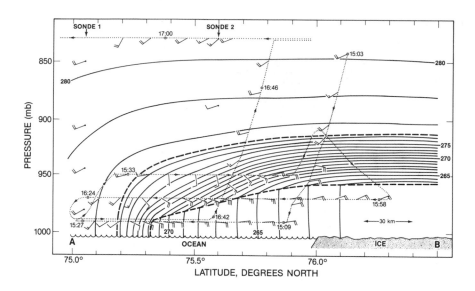

Figure 5. Frontal potential temperature analysis (K, thin solid lines) derived from the aircraft flight-level measurements and dropwindsondes along the line AB of Figure 1. The dotted line shows the flight track with selected wind vectors; GMT time marks follow the dotted line. Vectors with dotted heads indicate dropwindsonde winds; wind vectors are the same as in Figure 3.

soundings between ∼ 940 and 990 mb. The analysis of potential temperature through the leading edge of the front (Figure 5) shows the front sloping downward from the ice edge to the sea. The frontal width was ∼ 120 km at 960 mb and decreased to ∼ 20 km at the sea surface. The plotted wind vectors show the horizontal and vertical directional wind shear across the front: easterly below the frontal inversion and southwesterly above and in advance of the leading edge. The upper boundary of the front was isentropic, whereas its lower boundary had a 10 K potential temperature increase in the 100-km distance between the ice edge and the trailing edge of the front at the sea surface. The baroclinicity within the neutrally stable ($\partial\theta/\partial p \approx 0$) subfrontal boundary layer reflected the upward flux of sensible heat from the sea surface acting upon air parcels whose temperatures before exiting the ice edge were ∼ 10 °C colder than that of the underlying open water. The reader should note that the ice edge (Figure 1) was oriented northwest to southeast in the vicinity of the cross section (Figures 4 and 5), thus, the easterly flow in the subfrontal boundary layer was clearly directed off ice.

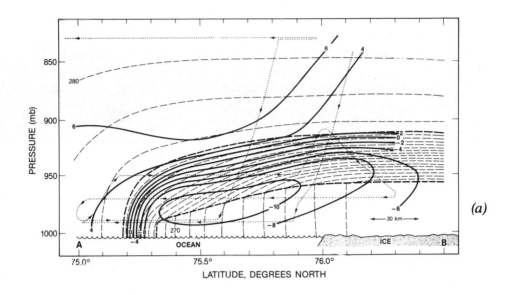

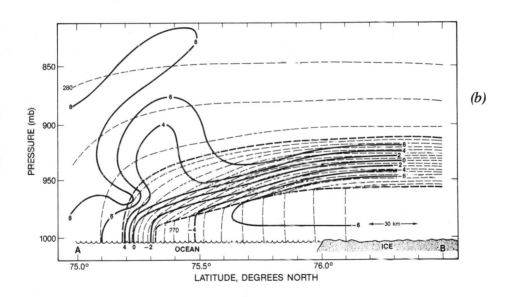

Figure 6. Wind velocity analysis for Figure 5. Solid lines indicate (a) front-parallel (300°) u-component wind speed (m s⁻¹) and (b) front-normal (210°) v-component wind speed (m s⁻¹). Potential temperature (thin dashed lines) and frontal boundaries (heavy dashed lines) are the same as in Figure 5.

The analysis of wind velocity was performed on its front-parallel and front-normal components (Figures 6a and 6b, respectively). The orientation of the front was taken from the surface potential temperature analysis (Figure 4). The analysis of the front-parallel (300°) wind component (u, positive toward the southeast), shown in Figure 6a, contains a low-level jet that was strongest (~ 15 m s^{-1}) at the top of the adiabatic subfrontal boundary layer. The vertical shear in u across the frontal layer at 75.4°N was 15 m s^{-1} km^{-1}. The front-parallel geostrophic vertical wind shear, $\partial U/\partial z$, across the frontal layer at 75.4°N was calculated from the horizontal thermal gradient in Figure 6.

$$\frac{\partial U}{\partial z} = \frac{g}{f\theta}\frac{\partial \theta}{\partial n} = 13 \text{ m s}^{-1}\text{ km}^{-1}, \tag{1}$$

where $g=9.8$ m s^{-1} is gravity; $f=1.5\times 10^{-4}$ s^{-1} is the Coriolis parameter; $\theta=270$ K is potential temperature in the middle of the frontal layer at 75.4°N; $\partial\theta/\partial n=20$ K (100 km)$^{-1}$ is the front-normal potential temperature gradient; and Eq. (1) is the geostrophic thermal-wind equation. A comparison between Eq. (1) and the observed shear (15 m s^{-1}) in Figure 6a shows that the u-component wind was in near-geostrophic thermal wind balance with the cross-frontal thermal gradient. The vertical shear of u in the neutrally stratified baroclinic boundary layer beneath the front (Figure 6a) was not in the geostrophic thermal wind balance owing to the first-order importance of the boundary-layer momentum flux processes, which are not included in the geostrophic thermal wind approximation.

The analysis of the front normal (210°) wind component (v, positive toward the northeast) in Figure 6b shows positive values of $v>8$ m s^{-1} in the warm air in advance of the front, and negative values >6 m s^{-1} in the cold boundary layer beneath the front. The vertical shear in v across the front layer at 75.4°N was comparable to that in u, reaching 10 m s^{-1} km^{-1} in the lower portion of the frontal layer. The vertical shear in v was not geostrophically balanced by the analyzed along-front thermal gradient shown in Figure 4, in which

$$\frac{\partial v}{\partial z} \neq \frac{\partial V}{\partial z} = \frac{g}{f\theta}\frac{\partial \theta}{\partial x}, \tag{2}$$

where $\partial v/\partial z$ and $\partial V/\partial z$ are the actual and geostrophic front-normal vertical shear, and $\partial\theta/\partial x$ is the along-front thermal gradient, respectively. We suggest one of the following: (1) the analysis of the front-parallel temperature gradient in Figure 4 is incorrect; (2) the neglected turbulent vertical fluxes contributed to the imbalance in Eq. (2); or (3) the distribution of v (Figure 7b) is the ageostrophic response to geostrophically forced secondary circulations (Sawyer, 1956; Eliassen, 1962), i.e., the cross-front ageostrophic wind. The data base used to develop Figure 4 was of insufficient density to resolve this issue. However, considering the magnitude and spatial distribution of the cross-front divergence and the location of the zone of heaviest precipitation (discussed next), it is most likely that suggestion 3 is the answer.

The vorticity and divergence (convergence) of the arctic front were calculated from the front-parallel and front-normal wind components, respectively. The results show that the vorticity (Figure 7a) and divergence (Figure 7b) were concentrated within a narrow zone of $\simeq 20$-km width at the leading edge of the front. The largest values of cyclonic relative vorticity ($\simeq 10^{-3}$ s^{-1}) were found near the sea surface (1000 mb) and diminished to $\simeq 10^{-4}$ s^{-1} at 940 mb. The convergence (Figure 7b) was of comparable magnitude to the vorticity ($\simeq 10^{-3}$ s^{-1}), and was colocated with the area of maximum relative vorticity (Figure 7a). During the aircraft penetrations of the front, the heaviest precipitation was encountered at the leading edge of the front in the region of maximum cross-frontal convergence.

Of special interest was the aircraft radar reflectivity image taken during the return flight to Bodo, Norway. The aircraft had just passed over the leading edge of the front on a southward heading, when a 100-km-scale cyclonic precipitation echo was noted on the radar screen (Figure 8). This image was located within the cyclonic vorticity and convergence of the front along the southern portion of the line AB (Figures 4 and 5). We suggest that mesoscale vortices of this scale could form within the baroclinicity and cyclonic vorticity of such shallow frontal layers. Unfortunately, there were no satellite images available to indicate whether the observed precipitation radar echo was tied to a polar-low cloud feature.

4. SUMMARY AND CONCLUSIONS

This study presents analyses describing the structure of a shallow outbreak of cold air from off the arctic ice pack. Though the described arctic frontal structure is unique with regard to the remote geographic region within which it was observed, it does not differ significantly from the cold-frontal structure described by Sanders (1955) over the central United States. It is significant that studies

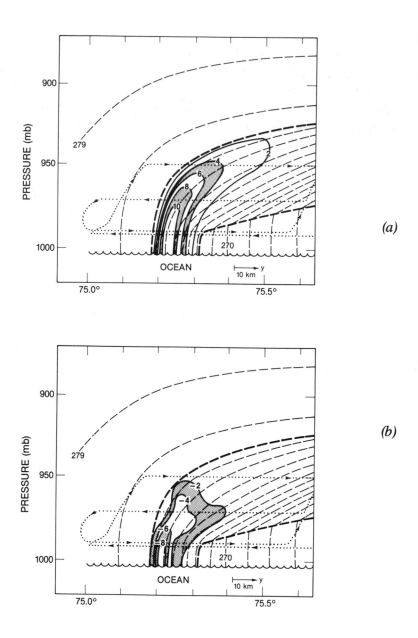

Figure 7. (a) Relative vorticity ∂v/∂y (10⁻⁴ s⁻¹, solid lines) and (b) divergence ∂v/∂y (10⁻⁴ s⁻¹, solid lines) at the leading edge of the arctic front shown in Figures 5 and 6. Potential temperature (thin dashed lines), frontal boundaries (heavy dashed lines), and flight track (dotted lines) are the same as in Figures 5 and 6.

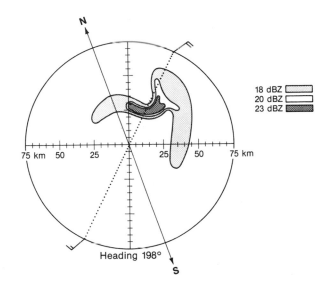

Figure 8. NOAA aircraft belly-radar precipitation echo at 1657 GMT 18 February 1984. The aircraft was positioned at the center of the circle and flying southward at a heading of 198°. The dotted line is the flight path AB of Figure 4, across the front; selected wind vectors at 940 mb from Figure 5 are plotted at the end points. Radar reflectivity contours are 18, 20, and 23 dBZ.

such as this are just beginning to describe such mesoscale systems over the arctic ice and adjacent seas.

REFERENCES

Eliassen, A., 1962: On the vertical circulation in frontal zones. *Geofys. Publ.*, *24*, 147–160.

Sanders, F., 1955: An investigation of the structure and dynamics of an intense surface frontal zone. *J. Meteor.*, *12*, 542–552.

Sawyer. J.S., 1956: The vertical circulation at meteorological fronts and its relationship to frontogenesis. *Proceedings of the Roy. Soc. London*, *A234*, 346–362.

A COMPARISON OF THE STRUCTURE OF TWO POLAR LOWS OBSERVED BY RESEARCH AIRCRAFT

Michael W. Douglas
Cooperative Institute for Research in the Environmental Sciences (CIRES)
University of Colorado
Boulder, Colorado, U.S.A

M.A. Shapiro
NOAA/ERL/Wave Propagation Laboratory
Boulder, Colorado, U.S.A.

ABSTRACT

The analysis of observations collected from research aircraft flights into two polar lows are presented and compared in order to identify those features common to both. A polar low development northeast of Iceland on 27 February 1984 is compared with a polar low observed on 4 and 5 March 1987 over the northern Gulf of Alaska. Close similarities in the low-level wind fields and thermal structures are reflected in similar satellite cloud signatures. Some previous observational and theoretical studies have attempted to relate polar lows to tropical cyclones; here the structure of the polar lows is briefly compared with a tropical monsoon depression, which shows similarities in both satellite imagery and in aspects of the temperature field, such as the presence of a warm "eye."

1. INTRODUCTION

The study of small synoptic-scale (less than ~ 1000-km wavelength) or large mesoscale cyclones in high latitudes, hereafter referred to as polar lows, has followed several paths in recent years. Many observational studies (e.g., Reed, 1979; Mullen, 1979; Rasmussen, 1981, 1985; Businger, 1985; Forbes and Lottes, 1985) have used satellite imagery to assist in the synoptic and subsynoptic scale

POLAR AND ARCTIC LOWS
Paul F. Twitchell, Erik A. Rasmussen,
and Kenneth L. Davidson (Eds.)

analysis of sparse surface and upper air observations. Fewer observational studies have been based exclusively on satellite imagery (e.g., Zick, 1983). Modeling studies, using both analytical (e.g., Mansfield, 1974) and numerical (e.g., Sardie and Warner, 1983) models, have mostly used idealized initial states, because specification of the initial state for real polar lows has not been possible with the widely spaced observations. The modeling results suggest that both baroclinic instability and conditional instability of the second kind (CISK) may be important during the intensification of polar lows, though the former appears to have more support.

Although observational studies using frequent surface observations at a single station (e.g., Rasmussen, 1985) can provide an estimate of the structure of polar lows when the assumptions of steady state, uniform translation, and radial symmetry are made, these assumptions are restrictive. The only way to relax these assumptions is to provide mesoscale observations over a short time interval (approximately several hours) and over the domain of the polar low. Currently, only instrumented aircraft can come close to satisfying these observational requirements. The intent of this paper is to summarize the results of research flights into two polar lows made during the past 4 years, and to describe characteristics common to both. Results from a flight into a polar low over the Norwegian Sea, reported in detail by Shapiro et al. (1987) (hereafter referred to as S87), are compared with results from a more recent polar low investigation over the Gulf of Alaska, during the Alaska Storms Program in March 1987. In the last section of this paper we briefly discuss the suitability of comparing polar lows with certain tropical depressions (also investigated by research aircraft), as has been done in some recent theoretical and observational papers on polar lows.

2. SUMMARY OF THE AIRCRAFT FLIGHTS

A detailed account of the flight into the 27 February 1984 polar low over the Norwegian Sea was described in S87. For this case a NOAA P-3 aircraft spent nearly 4 hr, mostly at 300 m, in the vicinity of the polar low, collecting flight-level, radar, and omega dropwindsonde (ODWS) data. From its appearance in satellite imagery the storm was near maximum intensity during the 300-m flight segment. The flight track at 300 m is shown in Figure 1a.

The Gulf of Alaska polar low was flown into twice with a NOAA P-3, with the midpoint of the first flight ~2200 UTC 4 March 1987 and that of the second flight ~0000 UTC 6 March 1987. The slow propagation of the Alaskan polar

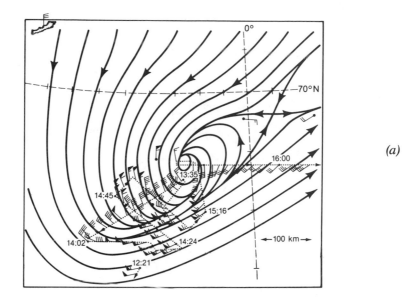

(a)

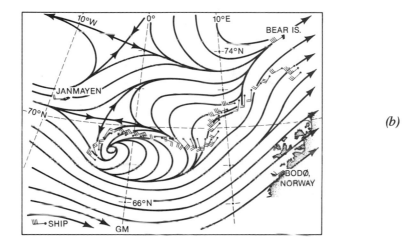

(b)

Figure 1. Streamline analysis for Norwegian Sea and Gulf of Alaska polar lows shows: (a) The 956-mb streamlines of data adjusted to 1340 UTC 27 February 1984 (from Shapiro et al., 1987), (b) The 580-mb streamlines of data adjusted to 1340 UTC 27 February 1984 (from Shapiro et al., 1987). The wind plotting convention is the same as in Figure 3; note differing scales on the figures.

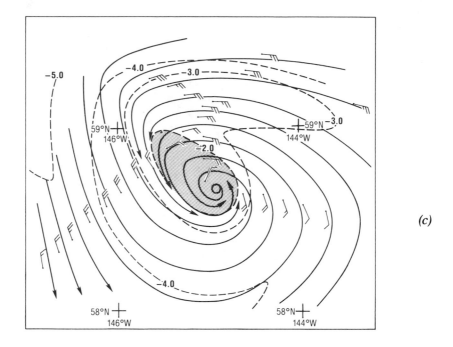

(c)

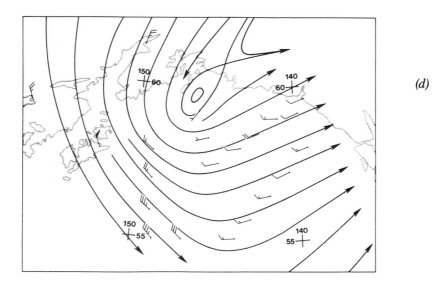

(d)

Figure 1 (continued). (c) The 960-mb streamlines (and isotherms at 1°C interval with air warmer than −2°C shaded) using aircraft data adjusted to 2000 UTC 4 March 1987, (d) The 580-mb streamlines for 0000 UTC 5 March 1987.

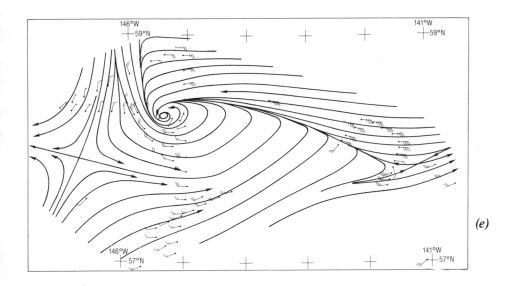

(e)

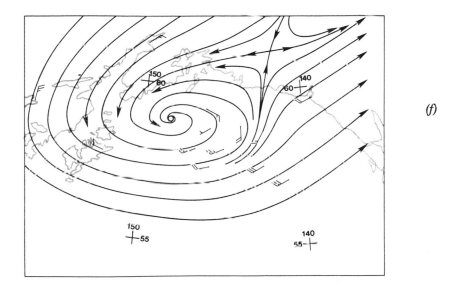

(f)

Figure 1 (concluded). (e) The 956-mb streamlines using aircraft data adjusted to 2300 UTC 5 March 1987, (f) The 580-mb streamlines using ODWS data obtained between 0142 and 0322 UTC 6 March 1987.

low and its close proximity to the aircraft staging base, Anchorage, Alaska allowed for more time at both low level and at ODWS level than was possible in the Norwegian Sea polar low flight. The first flight into the Gulf of Alaska polar low consisted of both a low-level and a high-level flight segment. During the low-level segment the P-3 located and crossed the polar low circulation center twice at 960 mb (see Figure 1c). This pattern documented the central pressure, radius of maximum winds, and thermal field in and around the eye. The aircraft then climbed to the highest level possible to begin an ODWS deployment; the first 2 hr were at 460 mb, the last 1.75 hr at 388 mb. Eighteen ODWS's were successfully deployed. It should be pointed out that on this flight no radar data were recorded because of a recording system malfunction, hence the emphasis on an ODWS mission.

The second flight, centered near 0000 UTC 6 March, focused upon the mesoscale structure of the polar low. After an initial flight pattern about the center at 956 mb (see Figure 1e), the aircraft climbed to 666 mb for another center-oriented flight pattern (not shown). Then the aircraft headed east, in search of a second center suggested by satellite imagery, which had been obtained just before takeoff. The aircraft descended to 956 mb again, ultimately finding a cyclonic shear line (east side of Figure 1e) that was subsequently crossed four times. The aircraft then ascended for a 2-hr ODWS deployment pattern before returning to Anchorage. Nine sondes were released (all but one from 388 mb), of which only one provided less than completely satisfactory data.

3. SYNOPTIC–SCALE PERSPECTIVE OF THE POLAR LOWS

The synoptic-scale flow regimes within which polar lows develop are important because many studies to date (e.g., Mansfield, 1974; Reed and Duncan, 1987) suggest that the stability (both static stability and baroclinic stability) characteristics of the synoptic-scale flow strongly influence the growth of disturbances on the scale of polar lows. Also, since only synoptic-scale circulations can be resolved with operationally available data, numerical forecasting of the development of polar lows must be closely tied to the synoptic-scale meteorological flow.

3.1 The Norwegian Sea Polar Low

The rapid subsynoptic development of the Norwegian Sea polar low on 27 February 1984 occurred in advance of a synoptic-scale short wave (Figure 2),

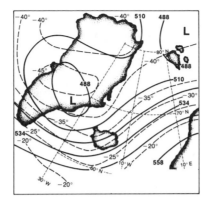

Figure 2. The 500-mb height (solid lines, m×10) and temperature (dashed lines, °C) analysis is shown 1 day prior to the most intense stage of the Norwegian Sea polar low. (After Figure 2b in Shapiro et al., 1987.)

which was propagating northeastward around a larger scale quasi-stationary low center over southcentral Greenland. A series of cloud vortices, evident in satellite imagery (Figures 27 and 29 in S87), developed as the rapidly moving short wave appeared to initiate several successive lower tropospheric cyclones, which propagated more slowly than the speed of the short wave.

3.2 The Gulf of Alaska Polar Low

In contrast to the rapid development of the Norwegian Sea polar low, the Gulf of Alaska polar low developed slowly. During the period of polar low development (4–6 March 1987), the upper tropospheric flow over the Gulf of Alaska was dominated by a slowly retrograding trough (Figure 3); this trough extended from northern Alaska across the Gulf of Alaska and southward along the west coast of North America to near 30 °N. Beneath this trough axis, northerly near-surface flow of cold air over the Gulf of Alaska on 3 and 4 March 1987 (not shown), together with upward heat fluxes from the warmer sea surface, produced a boundary layer of near-zero dry static stability. Thus, a near-neutral boundary layer was overlain by a deep layer of very cold air of low static stability beneath the upper tropospheric trough axis. In this shallow, offshore flow, the polar low slowly developed. The polar low was initially very shallow, being

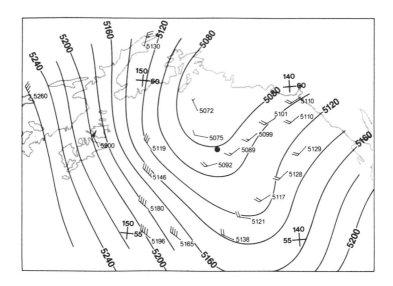

Figure 3. The 500-mb height analysis (20-m interval) over the Gulf of Alaska and vicinity at 0000 UTC 5 March 1987 is shown 1 day prior to the most intense stage of the Gulf of Alaska polar low. ODWS-derived heights (m) and winds (full barb, 5 m s⁻¹, half barb, 2.5 m s⁻¹) are also shown. The surface position of the low is marked by the solid dot. Latitude and longitude are indicated at the cross marks.

confined to below 800 mb at 0000 UTC 5 March. The vertical extent of the polar low slowly increased and by 0000 UTC 6 March extended to above 600 mb, when it lay east of the 500-mb trough axis, a favorable position for synoptic-scale upward motions (inferred from quasi-geostrophic reasoning). On 7 March, a southerly flow, developing in response to synoptic-scale cyclogenesis to the southwest of the polar low, began to advect warmer air over the northern Gulf of Alaska. Although undocumented, it is likely that this southerly flow quickly eliminated the boundary layer baroclinic zone, which had developed in the off-shore flow, and also increased the boundary layer stability by reducing the air-sea temperature contrast. With these synoptic-scale conditions the polar low quickly dissipated. Whereas the Alaskan polar low could be described as slowly deepening (3.5 mb over 24 hr), since central pressures were measured accurately on two successive days by the P-3 it was difficult to quantify the deepening rate for the Norwegian Sea polar low. The horizontal scale of the mesoscale vortex on 27 February was so small (Figure 9 of S87) that routine observations were not available near the lowest pressure, and since the low was only observed at one time, the deepening rate is uncertain. The European Center for Medium-Range Weather Forecasts (ECMWF) synoptic-scale analysis at 1200 UTC 27 February (Figure 4d in S87) did not capture the tight gradients near the polar low center. The analysis depicted the synoptic-scale low center with a minimum pressure of near 995 mb 3 hr before the P-3 reported a 979-mb central pressure and thus cannot be used to accurately estimate deepening rates. The limitations of the operational upper air observation network in resolving the polar low circulations and the need for supplementary data can be appreciated by attempting a 580-mb wind analysis using only the winds from the four rawinsonde stations shown in Figure 1b.

4. MESOSCALE STRUCTURE OF THE POLAR LOWS

4.1 Wind Fields

Comparison of the structure of the polar lows was carried out at two levels, 956 mb (300 m pressure altitude) and 580 mb, because analyses at these levels are shown in S87. Figures 1a–1f show the 956-mb and 580-mb streamline analyses for 1340 UTC 27 February 1984 and for 1200 UTC on both 5 and 6 March 1987. Although the 956-mb aircraft data on the 2 days in March were position adjusted for the translation of the vortex, the 580-mb ODWS positions were not adjusted, because evolution and translation of features could not be distinguished. The uncertainty affects the analysis only slightly, however, because the translation of the synoptic-scale trough was very slow, about 3 m s^{-1} towards the west.

At low levels (956 to 960 mb, Figures 1a, 1c, 1e) the streamline analyses show two zones of confluence; the confluence line southwest of the center on 27 February (Figure 1a) was better documented by aircraft traverses than the corresponding line (extending north-northwest from the center) at 2000 UTC on 5 March (Figure 1c) and 2300 UTC 6 March (Figure 1e). The shear line extending eastward from the center on 6 March is better described than the corresponding feature on 27 February, which was suggested from only two ODWS observations. The eastern shear line on 27 February may be underestimated in intensity, considering the satellite observed convective cloud cover (Figure 4) along this line.

Figure 4. NOAA polar orbiting infrared imagery is shown at 1340 UTC 27 February 1984 (from Shapiro et al., 1987).

At 580 mb the cyclonic circulation shown near the center of the Norwegian polar low in Figure 1b was more intense than the Alaskan polar low's circulation at 0000 UTC 5 March (Figure 1d). However, between 5 and 6 March the midtropospheric circulation intensified, and by 0000 UTC 6 March (Figure 1f) the circulation had approached the intensity of the Norwegian Sea low.

The shallow depth of the Alaskan polar low circulation at 0000 UTC 5 March is apparent from the vertical variation of relative vorticity above the low center (Figure 5). The vorticity was obtained analytically by fitting a plane to the u and v components of the wind from the three ODWS observations nearest to, and surrounding the low-level circulation center. The cyclonic vorticity decreased to near zero by 800 mb, increased above 600 mb, and had a second maximum

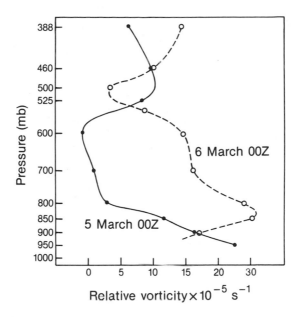

Figure 5. Vertical profiles of relative vorticity (10^{-5} s^{-1}) are determined from the three ODWS wind observations nearest to, and surrounding, the Gulf of Alaska polar low at 0000 UTC on 5 and 6 March 1987.

near 500 mb. Thus, the vortex was confined to the lowest 200 mb on 5 March. The low increased in vertical extent considerably over the next 24 hr, and by 0000 UTC 6 March the vorticity decreased much more slowly with height.

4.2 Thermal Structure

The low-level (956-mb) thermal structure of both polar lows was similar (Figures 1c, 6a–6b), having coldest temperatures to the north and west of the center and warmest air very close to the center. Cold air advection was a maximum from the west to south of each vortex; in association with a region of general subsidence as inferred from satellite imagery (see Section 5). The warmest air in the Norwegian Sea low (Figure 6a) was to the southwest of the circulation center, contrasting with the pattern on 5 and 6 March (Figures 1c and 6b), where the warmest air was centered on the center. In both cases, a strong thermal gradient bounded the inner eye. A sharp thermal contrast extended along the shear line east from the center on 6 March, whereas the contrast was not as sharp on

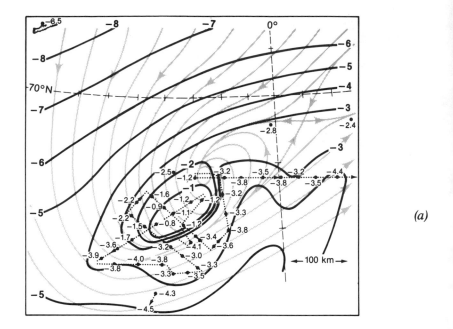

(a)

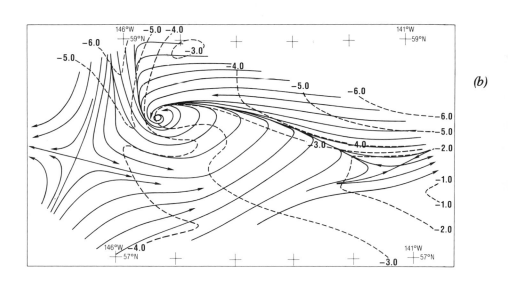

(b)

Figure 6. The 956-mb temperature analysis with streamlines superimposed is shown for (a) 1340 UTC 27 February 1984 (after Shapiro et al., 1987), (b) 0000 UTC 6 March 1987. Isotherm contour interval is 1°C. Isotherms solid in (a), dashed in (b).

5 March and 27 February. This may have been due to the lack of sufficient high-spatial-resolution aircraft data northeast of the low center on 27 February. However, the strong thermal gradient at low levels in this area is supported by data from ODWS 9 on 27 February, which showed that the wind shifts (see Figures 8 and 13 in S87) from easterly at 7 m s^{-1} at 950 mb to southwesterly at 15 m s^{-1} at 850 mb.

To obtain the layer mean thermal structure of the polar lows, ODWS-derived thicknesses can either be analyzed directly or the aircraft radar altimeter observations at two flight levels can be subtracted to find the intervening layer thickness. The latter procedure was used by S87 to obtain the 1013- to 580-mb thickness distribution for the Norwegian Sea polar low. In their analysis (Figure 7), the warm core of the vortex is apparent, with an axis of warm air extending northeast from the center. The vertical extent of this warm core was greater than that

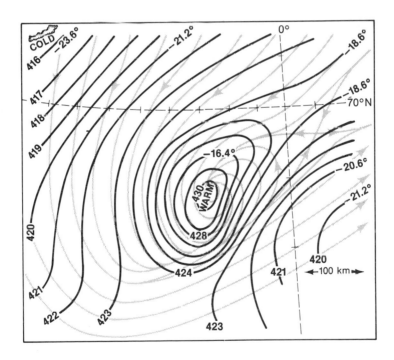

Figure 7. The 1013- to 580-mb thickness analysis (m) is shown at 1340 UTC 27 February 1984 (from Shapiro et al., 1987). The 956-mb streamlines are superimposed.

associated with the Gulf of Alaska polar low at 0000 UTC 5 March. At this time the 1000- to 850-mb layer thickness analysis, based on ODWS-derived thicknesses (Figure 8a), shows a small, warm region coincident with the circulation center suggested by the 925-mb ODWS winds. A wedge of warm air extends southeastward from this center, associated with air that had wrapped around the vortex on the south side and had experienced the combined effects of subsidence warming (suggested from scant clouds in satellite imagery) and upward sensible heat fluxes (ODWS surface temperatures were near 0 °C whereas the sea surface temperature varied between 4 °C and 5 °C.) The main baroclinic zone parallels the coastline except far to the southeast, where a synoptic-scale low center and frontal zone distort the pattern. In the 500- to 700-mb layer thickness analysis (Figure 8b), a synoptic-scale trough in the thickness field dominates the entire Gulf of Alaska, showing the warm core of the polar low to be very shallow. This agrees with the vertical variation of vorticity (Figure 5), which also suggested the very shallow extent of the polar low.

Flight-level radar altitude data at 956 mb and 666 mb during the second flight into the Gulf of Alaska polar low permitted derivation of the thickness field (Figure 9) in the same manner as for the Norwegian Sea polar low. In the 956- to 666-mb layer, the Gulf of Alaska polar low center was clearly warmer than the surroundings, though the difference was less than at the 956-mb level (Figure 6b), indicating that the temperature excess of the warm core decreased with height. As with the Norwegian Sea polar low, the thickness pattern in Figure 9 shows an elongation toward the east, paralleling the low-level shear line (Figure 1e).

5. SATELLITE IMAGERY AND A CONCEPTUAL MODEL OF THE POLAR LOWS

Comparison of satellite imagery of both polar lows shows that while the Gulf of Alaska polar low exhibited the growth and decay of circulation centers (Figures 10a–10d) during its more than 2-day existence, its general appearance was close to that of the Norwegian Sea polar low at 1340 UTC 27 February (Figure 4), except for the larger size of the latter's cloud "comma" (compare Figures 4 and 10b). At this time (see Figure 6a) cold air had wrapped around the south side of the Norwegian Sea low and was approaching the main convective cloud band from the south and southwest. This cloud band was most developed east and northeast of the vortex center, and less clearly defined to

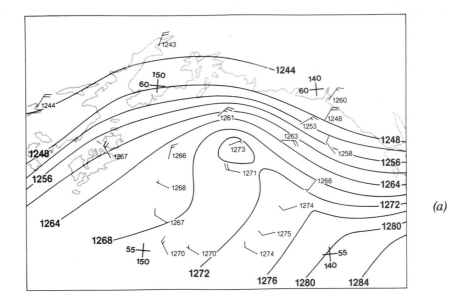

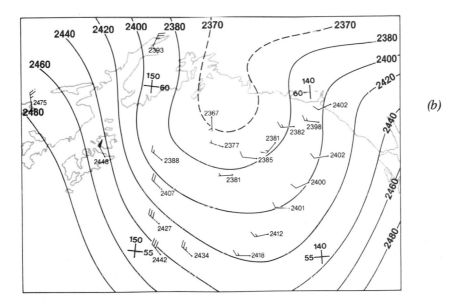

Figure 8. The thickness (m) analysis based on ODWS data at 0000 UTC 5 March 1987 is shown for (a) the 850- to 1000-mb layer with 925-mb ODWS winds included and (b) the 500- to 700-mb layer with 600-mb ODWS winds included. Contour interval for (a) 4 m, and (b) 20 m.

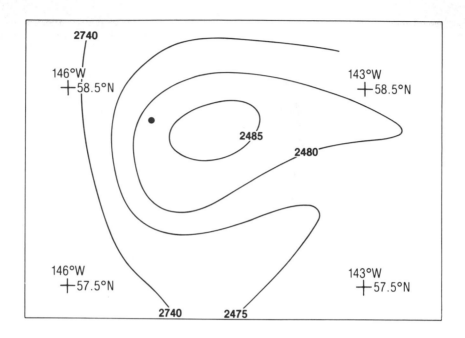

Figure 9. The thickness analysis (m) is shown for layer 956 to 666 mb at 0000 UTC 6 March 1987. Contour interval is 5 m, which corresponds to a layer mean temperature difference of 0.47°C. Polar low center at 956 mb is indicated by solid dot.

the south and east of the center. The Gulf of Alaska polar low showed a very similar distribution of temperature (Figure 1c) and cloudiness (Figures 10a–10b) near 0000 UTC 5 March.

A conceptual model of the cloud and thermal fields of the polar lows at low levels, based on the imagery and analyses described above, is shown in Figure 11. Cold air advection is continuous around the south side of the storm, though the advection becomes very weak as the cloud band, extending east from the center, is approached. Beneath the cloud band, the surface winds are easterly but the vertical shear is westerly. The strongest horizontal and vertical shear about the polar low is associated with this "warm frontal zone." West of the low center, strong cyclonic shear is associated with the strong thermal gradient between the warmest air in the center of the polar low, and the colder air to the west, which has been advected southward.

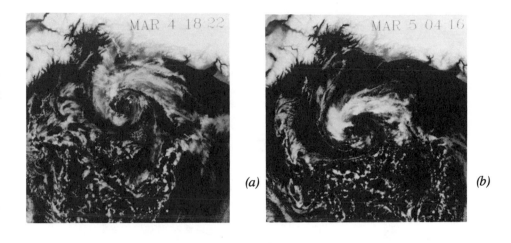

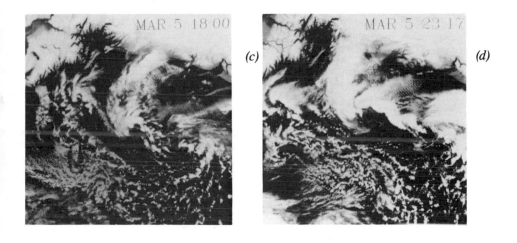

Figure 10. Infrared imagery from the Defense Meteorological Satellite Program (DMSP) satellite for the Gulf of Alaska polar low shows the same geographical area for (a) 1822 UTC 4 March 1987, (b) 0416 UTC 5 March 1987, (c) 1800 UTC 5 March 1987, (d) 2317 UTC 5 March 1987. The distance across the image is ~ 720 km east-west.

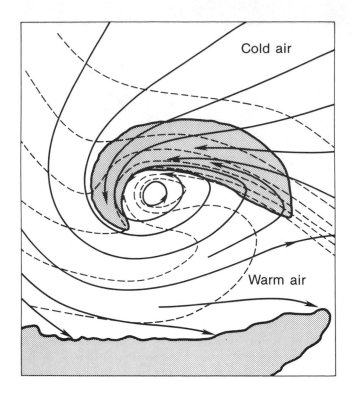

Figure 11. A conceptual model of a polar low that exhibits a satellite cloud signature similar to those of the Norwegian Sea and Gulf of Alaska polar lows is described in the text. Streamlines are solid and isotherms are dashed. High cloud cover is schematically shown by the irregular heavy solid line.

6. COMPARISONS WITH A TROPICAL DEPRESSION

Polar lows have been compared with tropical cyclones, both in their observed structure (e.g., Rasmussen, 1983) and dynamical aspects (e.g., Emanuel and Rotunno, 1989). The similarity of the satellite imagery of the two polar lows discussed above to some tropical disturbances, such as depressions that develop during the southwest monsoon of south Asia, suggests that a comparison would be of interest. Research aircraft data collected in one such monsoon depression over the Bay of Bengal on 7 and 8 July 1979 (during the Summer Monsoon Experiment) can be used for comparison (on these days the best data set to date was obtained in a tropical depression displaying major cloud field asymmetries) with the polar lows described above.

The monsoon depression developed in a basic state of moderate easterly shear and maximum cloudiness (Figure 12) was found to the south of a lower tropospheric confluence line that extended westward from the depression center (Figure 13). This area of cloudiness was coincident with the region of maximum lower tropospheric warm air advection (Douglas, 1987) and in this sense resembles the main polar low cloud band. Another polar low feature, an eye with a minimum in cloudiness and relative humidity and a maximum in temperature, was also found at the monsoon depression center. Although the thermal advection in the boundary layer near the center was quite small (Figure 13), the region of maximum cold air advection at 945 mb, east and north-east of the center, was also the region of minimum cloud cover, as with the polar lows.

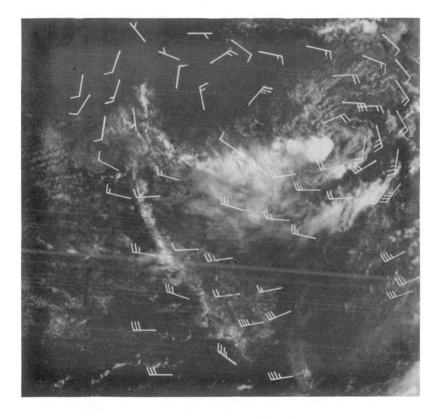

Figure 12. A DMSP visible image at 0600 UTC 8 July 1979 shows cloudiness associated with a monsoon depression over the northern Bay of Bengal (center of depression is ~20°N, 85°E). Wind observations at 850 mb are super-imposed. Image is ~2600 km across (east-west).

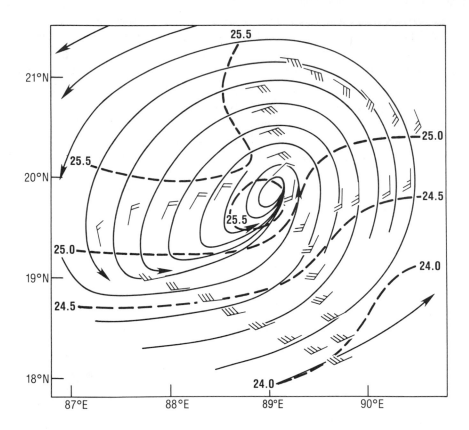

Figure 13. Streamline (solid) and isotherm (dashed, °C) analysis of 945-mb NOAA P-3 data is adjusted to 0757 UTC 7 July 1979. Contour interval is 0.5°C.

Some differences between the monsoon depression and the polar lows should, however, be stated. The vertical extent of the warm core above the surface center of the monsoon depression, extending to above 400 mb (not shown), was greater than that of either polar low. Also, the horizontal scale of the monsoon depression, if measured by the distance between successive depressions, is about 2000 km, considerably larger than polar lows. However, this may be misleading, as the east-west dimension of significant cloudiness associated with the monsoon depression, about 700 km (Figure 12), is not much different than that of the Norwegian Sea polar low (Figure 4). Finally, the basic-state thermal gradient is reversed in the case of the monsoon depression, with warmest air to the north at all levels. However, by inverting Figures 12 and 13, their similarity to Figure 11 becomes apparent.

Despite some similarities in their observed structure, the important dynamics involved in the intensification of polar lows and monsoon depressions may not be similar. The development of any wind field circulation in a baroclinic basic state will produce thermal advections whose associated vertical motions will modify the cloud field. The relative importance of condensation heating and baroclinic and barotropic conversion processes of the synoptic-scale shear flow to the spin-up of a vortex can only be determined from energy budgets based on either good observational or numerically simulated data over the life cycle of the vortex.

7. DISCUSSION

The results presented herein indicate that despite differences in the associated synoptic-scale circulations, the two polar lows that were observed in detail by research aircraft, had similar wind and thermal field structure, especially in the lower troposphere, and that the cloud field of each polar low was correspondingly similar. These similarities appear to be closely tied to the lower tropospheric baroclinicity, suggesting that it plays an important role in the development of the polar lows. This is an encouraging finding, since for operational forecasting purposes satellite imagery often provides the only available information on polar lows. If a workable conceptual model can be developed to describe the three-dimensional structure of polar lows from satellite imagery, it might be possible not only to make nowcasts of the surface winds about polar lows but also to introduce, or "bogus," idealized circulations into operational forecast models. Such "conceptual model bogusing" has been suggested by Krishnamurti (1985) for tropical analyses to improve the specification of tropical wave structure. Polar lows fall into a similar, poorly observed class of phenomena.

Another positive aspect of the observed similarity of structure of the two polar lows is that analytical or numerical simulations of polar lows now can be verified against a conceptual model that has some (albeit limited) generality. However, this conceptual model applies only to the mature polar low, as mesoscale in situ observations on which to base a conceptual model of the life cycle of polar lows are presently too limited. Only in the Gulf of Alaska polar low were mesoscale observations made on two successive days and this particular polar low exhibited only slow deepening. It remains a goal to collect mesoscale observations at successive times to document the evolution of a rapidly intensifying polar low.

ACKNOWLEDGMENTS

Thanks are extended to NOAA's Office of Aircraft Operations for providing the aircraft from which these observations were taken and to the Office of Naval Research/Applied Research Directorate for supporting the results reported herein. Thanks are due to Bernie Walters of the Pacific Marine Environmental Laboratory/NOAA for supplying the Gulf of Alaska polar low satellite images.

REFERENCES

Businger, S., 1985: The synoptic climatology of polar low outbreaks. *Tellus, 37A,* 419–432.

Douglas, M.W., 1987: *The Structure and Dynamics of Monsoon Depressions.* Rep. No. 86-15. Available from the Deptartment of Meteorology, Florida State Univ., Tallahassee, Florida.

Emanuel, K.A., and R. Rotunno, 1989: Polar lows as arctic hurricanes. *Tellus, 41A,* 1–17.

Forbes, G.S., and W.D. Lottes, 1985: Classification of mesoscale vortices in polar air streams and the influence of the large-scale environment on their evolutions. *Tellus, 37A,* 132–155.

Krishnamurti, T.N., 1985: Tropical upper tropospheric motion field. In *Proceedings of the NASA Symposium on Global Wind Measurements,* A. Deepak Publishing, Hampton, Virginia, pp. 15–20.

Mansfield, D.A., 1974: Polar lows: The development of baroclinic disturbances in cold air outbreaks. *Quart. J. Roy. Meteor. Soc., 100,* 541–554.

Mullen, S.L., 1979: An investigation of small synoptic cyclones in polar air streams. *Mon. Wea. Rev., 107,* 1636–1647.

Rasmussen, E., 1981: An investigation of a polar low with a spiral cloud structure. *J. Atmos. Sci., 38,* 1785–1792.

Rasmussen, E., 1983: A review of mesoscale disturbances in cold air masses. In D.K.Lily and T. Gal-Chen (Eds.), *Mesoscale Meteorology—Theories, Observations and Models,* Ridel, Boston, 247–283.

Rasmussen, E., 1985: A case study of a polar low development over the Barents Sea. *Tellus, 37A,* 407–418.

Reed, R.J., 1979: Cyclogenesis in polar airstreams. *Mon. Wea. Rev., 107,* 38–52.

Reed, R.J., and C.N. Duncan, 1987: Baroclinic instability as a mechanism for the serial development of polar lows: A case study. *Tellus, 39A,* 376–384.

Sardie, J.M., and T.T. Warner, 1983: On the mechanisms for the development of polar lows. *J. Atmos. Sci., 40,* 869–881.

Shapiro, M.A., L.S. Fedor, and T. Hampel, 1987: Research aircraft measurements of a polar low over the Norwegian Sea. *Tellus, 39A,* 272–306.

Zick, C., 1983: Method and results of an analysis of comma cloud developments by means of vorticity fields from upper tropospheric satellite wind data. *Meteor. Rdsch., 36,* 69–84.

POLAR LOW DEVELOPMENT ASSOCIATED WITH BOUNDARY LAYER FRONTS IN THE GREENLAND, NORWEGIAN, AND BARENTS SEAS

Robert W. Fett
Naval Environmental Prediction Research Facility
Monterey, California, U.S.A.

ABSTRACT

During the Marginal Ice Zone Experiment (MIZEX), in the spring of 1987, documentation was obtained of westward moving boundary layer fronts crossing the Fram Strait between Spitsbergen and Greenland. The frontal position was distinguished by an enhanced north/south-oriented band of clouds. Cumulus cloud streets embedded in moderate southeasterly flow led up to this feature. Analysis of surface observations indicated that the frontal position was located along the axis of an inverted low-level trough in which vorticity was also maximized. The frontal structure was confined to low levels below 2000 m, separating colder air to the east from warmer air to the west, and was capped by an inversion that tended to be higher east of the front. Multiple cloud vortices were noted, forming along the frontal cloud band, with the largest of the vortices appearing at the south end of the front.

The examples observed during MIZEX 1987 did not undergo further intensification. It was noted, however, that similar-appearing fronts, at times, had associated vortices that became polar lows. It is suggested that upper level support, in the form of cold 500-mb troughs or lows, distinguished developing from non-developing cases. It has been noted that low-level baroclinicity and vorticity are required conditions prior to polar low evolution. The occurrence of boundary layer fronts bringing cold arctic air from off the ice pack over relatively warm water may be one of the mechanisms leading to the low-level pre-conditioning necessary to establish the environment in which polar low development can occur. Examples of boundary layer frontal movement across the Fram Strait and also the Barents Sea, and polar low development in conjunction with such movement are the subject of this paper.

POLAR AND ARCTIC LOWS
Paul F. Twitchell, Erik A. Rasmussen,
and Kenneth L. Davidson (Eds.)

313

1. INTRODUCTION

During the MIZEX-87 experiment a boundary layer front (BLF) was observed propagating in a westerly direction through the Fram Strait (Davidson, Schultz and Fett, 1988). Ship-launched rawinsonde data and NOAA AVHRR HRPT data provided intensive documentation of the frontal feature. The front was apparent as an enhanced north/south-oriented cloud band that separated colder air to the east, which had just come off the ice in fresh southeasterly flow, from warmer air to the west over the Fram Strait in light northerly flow. Figure 1 is a NOAA-9, HRPT infrared (channel 4) view, which shows the BLF just off the west coast of Spitsbergen in the Fram Strait at 1110 GMT 25 March 1987. The Fram Strait region, between Spitsbergen and Greenland, is ice free due to northerly flow of a branch of the Gulf Stream in this region. Three research ships (Polar Circle, Hakon Mosby, and Valdivia) were operating within the strait at the time of these data. They documented that the frontal structure was confined to the boundary layer that was about 1000-m deep west of the front and 2000-m deep to the east.

Multiple mesoscale vortices were observed by satellite to form along the frontal boundary, as in Figure 1, and particularly at the southern end of the front

Figure 1. A NOAA-9 AVHRR HRPT infrared view over Svalbard acquired at 1110 GMT 25 March 1987.

or base of the associated inverted trough. Figure 2 shows a NOAA-10 HRPT infrared (channel 4) view of the system at 1017 GMT on the following day, 26 March. Anticyclonically curved cloud streets clearly delineate the southeasterly flow west of the frontal boundary. Valdivia, operating in the region of the cloud streets, reported 25–30 knots southeasterly flow with snow showers near the time of the NOAA-10 data. Ships to the west of the front were reporting light northerly winds. Note the mesoscale cloud vortices at the southern end of the BLF. These vortex examples did not undergo further intensification although similar appearing vortices have sometimes been noted to develop into polar lows. A reason for this may have been the fact that upper levels at the time showed high pressure and relatively warm temperatures. Polar low development has been shown, in a statistical sense, to occur most frequently when negative height anomalies and cold temperatures are present aloft (Businger, 1985).

The role of BLF's in polar low development has not previously been described, probably because of a lack of documentation concerning their nature.

Figure 2. A NOAA-10 AVHRR HRPT infrared view over Svalbard acquired at 1017 GMT 26 March 1987.

It is clear from the MIZEX-87 experiment that these features can effectively precondition the lower levels to optimize the possibility of polar low development. Rasmussen (1985a) provided the rationale for the necessity of a preexistent low-level conditioning in his statement that "a basic mechanism for all polar low development" is that "first low-level vorticity is formed in some way or other." He then postulated that it would be possible for deep convection to proceed and "then CISK can take over and through a rapid development concentrate the already available vorticity into a polar low."

This paper focuses on the probability that two mechanisms, the BLF and an upper cold low or trough, unite to stimulate polar low production—at least as a dominant mechanism in the northern Greenland, Norwegian, and Barents Seas area.

2. THE POLAR LOW OF 13 DECEMBER 1985

A case study of this development was described and documented by Rasmussen (1985b). The storm in question developed into an intense-appearing polar low by 0250 GMT 13 December 1982 when its image was captured by NOAA-7 just southwest of Bear Island. Rasmussen found that this development was linked to the movement of an upper cold low from near Novaya Zemlya on 11 December to a position near Bear Island on 12 and 13 December. Low-level analyses were remarkably free of any evidence of low formation near the satellite-observed vortex center until late on 13 and early on 14 December.

Figure 3 shows one of the earliest available satellite views, a NOAA-7 infrared image acquired at 1257 GMT 12 December 1982. The vortex center is clearly evident west of Bear Island (shown by an "x"). However, also apparent on this image (and not discussed by Rasmussen) is a north/south-oriented band of clouds extending from the region just west of the vortex center. This band of clouds is remarkably similar to the BLF examples shown in Figures 1 and 2. A small mesoscale vortex is even apparent near the northern limit of the band. Although not as clearly shown as in Figure 2, thin cumulus cloud streets appear to converge into the leading edges of the band from the east to southeast.

Figure 4 is a copy of the Fleet Naval Oceanography Center's (FNOC) 1200 GMT surface analysis on 10 December 1982. The outline of the satellite-observed storm and a suggested streamline analysis has been superimposed upon the analysis. As usual data are so sparse around incipient polar low developments

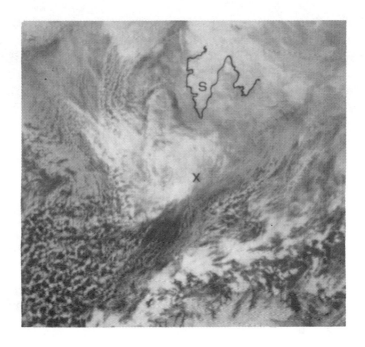

Figure 3. A NOAA-7 AVHRR HRPT infrared IMAGE over Svalbard acquired at 1257 GMT 12 December 1982. (Courtesy of E. Rasmussen.)

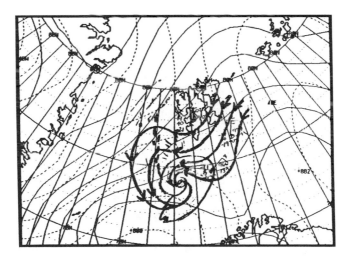

Figure 4. The Fleet Numerical Oceanography Center's surface analysis at 1819 GMT 10 December 1982. (Storm outline and streamlines have been superimposed.)

that surface analyses normally offer no clue that such a development is about to take place. The important point of this example is that the polar low development appeared to be linked to the nearby proximity of a preexistent BLF as well as an upper cold low. From a forecast perspective, monitoring the movement of both systems would have been very helpful in predicting the particular polar low evolution.

3. POLAR LOW EVOLUTION ON BLF'S WEST OF NOVAYA ZEMLYA

Figure 5 is a Defense Meteorological Satellite Program (DMSP) infrared view showing the Novaya Zemlya region at 1819 GMT 12 December 1983. A BLF is evident extending in a north/south-orientation just west of the island. The BLF is distinguished by having a series of developed mesoscale vortices near its southern end—similar, but better developed than in Figure 1. The vortices, in themselves, imply that low-level vorticity is being maximized in this region.

An examination of 500-mb data (not shown) reveals, additionally, that an upper cold low existed over northern Novaya Zemlya on 2 December. This system

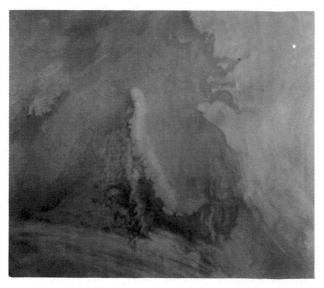

Figure 5. A DMSP infrared image centered on Novaya Zemlya 1819 GMT 12 December 1983.

moved southwest on the following day in tandem with a cold pool of air aloft having temperatures of less than $-45\,°C$. Major ingredients of polar low genesis appeared to exist.

Figure 6 is from a DMSP infrared mosaic at about 0840 GMT on the following day, 3 December 1983. Two larger scale vortices are now apparent and candidates for polar low status.

An individual DMSP infrared image (Figure 7) at 1616 GMT 3 December shows finer details of vortex evolution. These systems, however, failed to intensify beyond this stage.

4. POLAR LOW DEVELOPMENT FROM SOUTHWARD MOVING BLF'S IN THE FRAM STRAIT

BLF's in the Fram Strait and other locations move in the direction of the low-level flow or surge. Quite often such a surge occurs from north to south between Spitsbergen and Greenland.

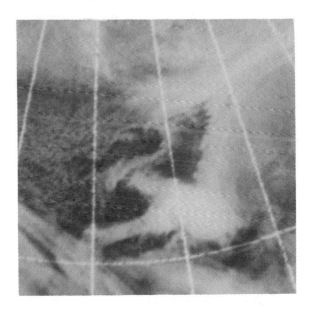

Figure 6. A DMSP infrared mosaic at 0840 GMT 3 December 1983. Novaya Zemlya is in center of mosaic.

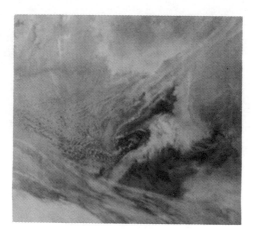

Figure 7. A DMSP infrared image acquired at 1616 GMT 3 December 1983. Novaya Zemlya is in right center of image.

Figure 8 shows such a surge through the Fram Strait in a DMSP infrared view at 2318 GMT 19 December 1983. An enhanced cloud line looking much like a rope cloud delineates the frontal position. Note that open-celled cumulus in the strait do not exist all the way up to the frontal boundary but changed form to overcast stratus 40–60 nm behind the front. Winds are believed to turn strongly anticyclonically in this region to a northeasterly direction. This causes a change from open to closed cell or overcast stratus and helps create the cyclonic vorticity Maximized at the frontal location. (See Figure 2 as an example of strong anti-cyclonic turning of the wind near the frontal boundary.)

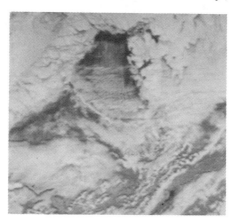

Figure 8. A DMSP nighttime visual image acquired at 2318 GMT 19 December 1983. Svalbard is in upper right corner of image.

Figure 9. A DMSP nighttime visual image acquired at 2035 GMT 20 December 1983. Svalbard is in upper right corner of image.

At 2035 GMT 20 December (Figure 9), a cloud vortex is observed in DMSP nighttime visual data. In this image the pronounced anticyclonic turning of the winds from northerly to northeasterly, in the flow leading to the vortex, is revealed by the change in alignment of cloud streets in that area; 500-mb data (not shown) reveal a cold trough aloft over the vortex region.

By 1913 GMT 21 December, as revealed by another DMSP nighttime visual image (Figure 10), a significant polar low had evolved.

Figure 10. A DMSP nighttime visual image acquired at 1913 GMT 21 December 1983. Spitsbergen is in upper right corner of image.

5. CONCLUSIONS

The results of this study (abbreviated in this presentation) indicate that BLF's in the northern Greenland, Norwegian, and Barents Seas region are important factors in polar low evolution. Vortex generation is an integral part in the life cycle of a BLF. Polar low generation, however, does not appear to evolve from such vortices unless a cold upper level low moves into the BLF region. A coupling of the features then appears to permit rapid cyclogenesis. Details of this life cycle interrelationship are lacking. From a forecast perspective, however, knowing the role of BLF's as one of the precursor indications of polar low development is important in itself. Armed with frequent high quality satellite views, the knowledgeable forecaster is unlikely to be surprised when development occurs.

REFERENCES

Businger, S., 1985: The synoptic climatology of polar low outbreaks. *Tellus, 37A,* 419–432.

Davidson, K.L., R.R. Schultz, and R.W. Fett, 1988: Observations of a Fram Strait ice edge westward propagating wave. Preprint, *Second Conference on Polar Meteorology and Oceanography,* 29–31 March 1988, Madison, WI. American Meteorological Society, 45 Beacon St., Boston, Mass. 02108, p. 15.

Rasmussen, E., 1985a: *A Polar Low Development Over the Barents Sea.* Polar Low Project, Tech. Report No. 7, Norwegian Meteorological Institute, Oslo, Norway, 42 pp.

Rasmussen, E., 1985b: A case study of polar low development over the Barents Sea, *Tellus, 37A,* 407–418.

POLAR LOWS IN THE BEAUFORT SEA

M.N. Parker
Atmospheric Environment Service
Edmonton, Alberta, Canada

ABSTRACT

The frequency of polar lows in the southern Beaufort Sea is investigated by examining detailed surface weather charts available for the period 1976 to 1985. One system, which developed over the Chukchi Sea and subsequently affected the study area, is briefly examined. Reasons for the lack of polar low development observed in the Canadian Beaufort Sea are presented.

1. INTRODUCTION

Forecasters in Scandinavia and Britain have long been aware that small-scale vortices frequently develop deep within cold air masses streaming over warm water surfaces. Since these weather systems, commonly known as polar lows, most frequently occur over data sparse ocean areas, are small-scale features, develop rapidly, frequently produce gale or near gale force winds, and are not adequately handled by the numerical models, they present a major challenge to the operational forecaster.

Synoptic conditions similar to those that lead to the development of polar lows in the Norwegian Sea frequently exist over the Bering Sea and Gulf of Alaska. However, it was not until Businger (1987) that a detailed investigation of the occurrence of polar lows in this area appeared in the literature.

Offshore exploration has been active in the Beaufort Sea since the mid-seventies. Although the exploration area is ice bound for much of the year, there is a period during the late summer and fall months when cold arctic air can stream across the warmer water surfaces of the southern Beaufort Sea. It is only

POLAR AND ARCTIC LOWS
Paul F. Twitchell, Erik A. Rasmussen,
and Kenneth L. Davidson (Eds.)

323

natural then that the question of the occurrence of polar lows in the Beaufort Sea should arise. This paper addresses that question.

2. METHODOLOGY

A geographical reference map and map of the study area are presented in Figure 1. To support offshore hydrocarbon exploration in the Canadian Beaufort Sea, Atmospheric Environment Service, under contract to industry, operated a meteorological office at Tuktoyaktuk, NWT during the open water seasons of 1976 to 1985. During this time three hourly surface analyses were produced. These charts, drawn at a scale of 1:5,000,000 incorporated meteorological data from the offshore drill sites, support vessels, satellite imagery, and sea level

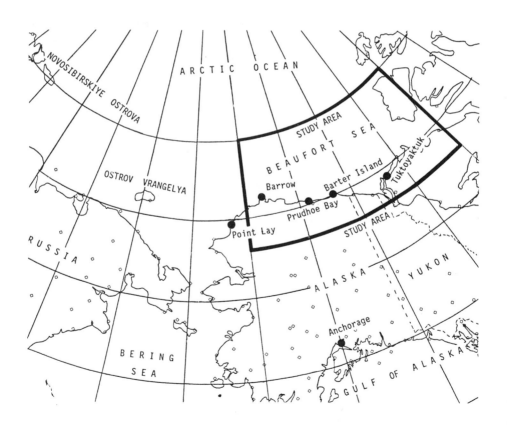

Figure 1. Geographic reference map and study area.

pressures from buoys on the polar pack ice. Maps for September, October, and November were examined for indications of polar lows. NOAA satellite imagery were also examined with particular attention to those periods when surface charts indicated cold air was moving over open water in the southern Beaufort Sea. Defense Meteorological Satellite Program (DMSP) imagery was obtained to aid in the study of one particular system.

3. RESULTS

During the period examined, no instances of polar lows developing over the southern Beaufort Sea were found. One case of polar low type development over the Chukchi Sea, which subsequently moved across northwestern Alaska into the study area, was detected. This particular system can be seen in Figure 2 as it approached Point Lay, Alaska.

Figure 2. Infrared satellite. 0620 UTC 14 October 1985. Produced from USAF DMSP film transparencies archived for NOAA/NESDIS at the University of Colorado, CIRES/National Snow and Ice Data Center, Campus Box 449, Boulder, CO 80309.

Prior to development of this system a baroclinic low and trough had crossed northwestern Alaska leaving the central and eastern sections of the Chukchi Sea under a northwesterly flow of arctic air. Surface temperatures at Ostrov Vrangelya were between $-12°$ and $-16°C$ while water temperatures would have been near or slightly above $0°C$. Looking further back in time to 0000 UTC 11 October 1985 an intense 50 kPa vortex and surface low were positioned near Novosi-birskiye Ostrova. The surface feature weakened to the point where it could no longer be analysed on the surface analyses. However, the 50 kPa centre moved southeast and then eastward to pass just south of Ostrova Vrangelya at 1200 UTC on 13 October 1985. During passage of this system 50 kPa temperatures of $-50°$ were recorded at Ostrov Vrangelya.

Satellite imagery at 2323 UTC 14 October 1985 indicated a low-level cyclonic circulation and heavy convection occurring near $70°N$ $170°W$ that cor-responds with the projected position of the 50 kPa centre. This may have represented a new surface low, or more likely represented enhancement of a weak surface or near surface feature that was present under the 50 kPa centre but could not be detected due to lack of surface data. In either case development was ac-companied by strong convection made possible by very cold upper temperatures moving over an area already conditionally unstable in the lower levels due to strong surface heating provided by an area of open water. Figure 3 shows the

Figure 3. Analysis for 50 kPa for 0000 UTC 15 October 1985.

50 kPa chart at time of surface development while Figure 4 shows the polar low as it passed just north of Barter Island, Alaska. During passage of the low mean surface winds at Barter Island reached 82 km/hr and gusts to 107 km/hr were recorded. The maximum reported gust during the event was 139 km/hr from Prudhoe Bay, Alaska. Only patchy cloudiness was reported over northern Alaska during passage of the system, and although some snow flurry activity was reported the main obstruction to visibility was blowing snow.

4. DISCUSSION OF RESULTS

During the period 1976 to 1985 only one case of a polar low affecting the study area could be identified. No cases of polar lows forming over the open waters of the southeastern Beaufort Sea could be identified. It is postulated that the main reason for this observation is the lack of a sufficient area of open water, especially during the latter portion of the year when the supply of cold arctic air increases. Figure 5 indicates the median area of six-tenths or less of ice cover on 17 September and 15 October. This amounts to a median northern fetch during

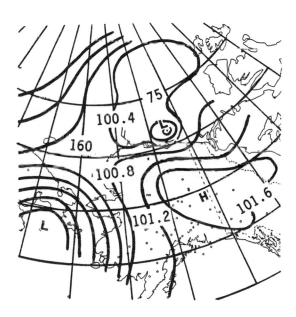

Figure 4. Surface analysis for 0000 UTC 15 October 1985.

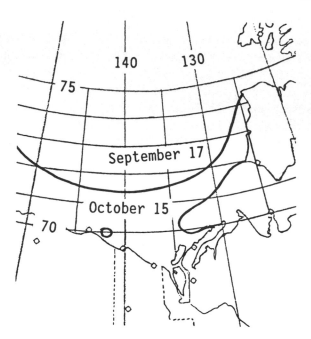

Figure 5. Median position of six-tenths or less of all ice. Data base 1953–1978 (after Markham, 1981).

mid-September of about 185–275 km with a substantially larger east-west fetch. The northern fetch is then about the same scale as the expected feature, leaving little area for development of low-level conditional instability over the water surface. Since development is likely to occur during times of moderate flow, it is reasonable to assume that if a polar low did begin to form the upper support would move over land or the polar pack before maximum surface development was attained. An equally important consideration is the lack of conditionally unstable arctic air masses over the Beaufort Sea during the time of minimum ice cover. During September only 0.9% of the upper air assents from Inuvik show an 85 kPa temperature equal to or less than −15°C, a value that would eventually result in conditionally unstable air over the open water of the Beaufort Sea. This only increases to 2.5% at Sachs Harbour. By October the frequency of arctic outbreaks begins to increase and these figures increase to 19.4% at Inuvik and to 29.5% at Sachs Harbour. These figures neither account for direction of flow nor do they include any information about the presence of a 50 kPa trough. The event of October 1985 demonstrates that vigorous polar lows can affect the southwestern Beaufort Sea and that they can strike with little warning. With different synoptic and ice conditions such a low could affect the exploration area of the Canadian Beaufort. During such a scenario it should be possible to provide some advance warning due to distances involved from the development area.

5. CONCLUSIONS

This study concludes that occurrences of true polar lows in the Canadian Beaufort Sea are rare events. However, this conclusion is based on the examination of only 10 years of data and use of a larger data set could alter the results.

Additionally, because of the limited data base over the Chukchi Sea, it is possible that other polar lows may have developed but were not detected.

It has been postulated that the main reason for the lack of polar low activity over the southeastern Beaufort Sea is the limited area of open water. A catch 22 being that as the supply of cold arctic air increases during the fall months the area of open water diminishes. Despite this one cannot rule out the unusual or rare event. Given optimum synoptic conditions during a year when ice cover is at a minimum it should be possible for a polar low to develop over the southeastern Beaufort Sea. For this reason forecasters must be aware of the factors that contribute to the development of these disturbances, otherwise early detection and advanced warning is unlikely. By the same token an understanding of polar lows will assist the forecaster in identifying and correctly handling events such as the October 1985 disturbance.

ACKNOWLEDGMENTS

This paper is a product of Scientific Services Division, Atmospheric Environment Service, Western Region. As such I wish to thank my colleagues for their advise and support during this study. In particular I wish to thank Mr. Hume, Chief, Scientific Services Division for providing the draft reviews and Ms. H. Gutzmann for her advice and assistance during preparation of the final copy. A special thanks is also due Mr. E. Hudson who, as officer in charge of the Beaufort Weather Office in 1985, recognized the significance of the October 1985 event and ensured that all charts and satellite imagery were saved.

REFERENCES

Businger, S., 1987: The Synoptic Climatology of Polar Low Outbreaks Over the Gulf of Alaska and the Bering Sea. *Tellus, 39A,* 307–325.

Markham, W.E., 1981: *Ice Atlaş Canadian Arctic Waterways.* Canadian Government Publishing Centre, Supply and Services Canada, Hull, P.Q., Canada, K1A-0S9, 198 pp.

BIBLIOGRAPHY

Environment Canada, 1982–1985: *Beaufort Weather Office Reports.* Proprietary, Atmospheric Environment Service, Edmonton, Alberta.

Forbes, G.S., and W.D. Lottes, 1985: Mesoscale vorticies in polar airstreams. *Tellus, 37A,* 132–135.

Lewis, P.J., 1987: *Severe Storms Over the Canadian Western High Arctic, A Catalogue Summary for the Period 1957–1983.* Atmospheric Environment Service, Canadian Climate Centre Report 87-2, 306 pp.

Moffett, R., 1987: Polar lows in eastern Canada. Atmospheric Environment Service, Professional Training Branch, Technical Note 87-4, 19 pp.

Økland, H.O., 1986: Heating by organized convection as a source of polar low amplification. *Proceedings of the International Conference on Polar Lows,* Oslo 1986, The Norwegian Meteorological Institute, 161–171.

Rasmussen, E., 1983: A Review of Mesoscale Disturbances in Cold Air Masses. *Mesoscale Meteorology—Theories, Observations and Models,* Reidel Publishing Co., 247–283.

Rasmussen, E., 1985: A case study of a polar low development over the Barents Sea. *Tellus, 37A,* 407–418.

Rasmussen, E., 1986: Different types of polar lows affecting Scandinavia. *Proceedings of the International Conference on Polar Lows,* Oslo 1986, The Norwegian Meteorological Institute, 17–29.

Reed, R.J., 1979: Cyclogenesis in polar air streams. *Mon. Wea. Rev., 107,* 38–52.

Wilhelmsen, K., 1986: Climatological study of gale producing polar lows near Norway. *Proceedings of the International Conference on Polar Lows,* Oslo 1986, The Norwegian Meteorological Institute, 31–39.

SUBSYNOPTIC–SCALE CYCLONE DEVELOPMENTS IN THE ROSS SEA SECTOR OF THE ANTARCTIC[1]

D. H. Bromwich
Byrd Polar Research Center
The Ohio State University
Columbus, Ohio, U.S.A.

ABSTRACT

Wintertime case studies of subsynoptic-scale cyclone developments are presented for those parts of the Ross Sea/Ross Ice Shelf area that are subject to strong and persistent flows of cold katabatic air from the East Antarctic ice sheet. Katabatic-induced boundary layer baroclinicity and a weak surface trough appear to be necessary conditions for cyclogenesis. Upper level support often accompanies storm formation, and governs subsequent storm evolution. Lower to midtropospheric data at much higher spatial resolution are clearly needed before definitive analyses can be conducted.

1. INTRODUCTION

Hemispheric satellite studies (Streten and Troup, 1973; Carleton, 1979) have shown that the circumpolar low pressure trough around Antarctica is generally a region of decay for synoptic-scale cyclones. Based upon recent work discussed below, it appears that cyclone development at the smaller subsynoptic-scale (hundreds of kilometers in spatial extent; Barry and Chorley, 1987, p. 202) is a frequent occurrence between the northern limit of the southern sea ice zone and the coast of Antarctica.

---

[1]This research was supported by National Science Foundation grants DPP-8314613 and DPP-8519977.

POLAR AND ARCTIC LOWS
Paul F. Twitchell, Erik A. Rasmussen,
and Kenneth L. Davidson (Eds.)

331

Carleton and Carpenter (1989, in review) studied daily thermal infrared satellite pictures during the winter (June to September) for the period 1977–1983; the spatial resolution of the Defense Meteorological Satellite Program (DMSP) images was 5.4 km. Spiraliform-type polar lows were found to develop with some frequency over the sea ice. Up to 40% of the subsynoptic-scale cyclogeneses poleward of 60°S were comprised of these forms with the majority being the inverted-comma type polar lows that predominate over the open ocean. The spirals preferentially formed near the sea ice limit and had a lifetime of 12–36 hr.

In a study to be reported in detail elsewhere, the present author used twice-daily automatic weather station (AWS) observations and daily DMSP satellite images (2.7-km resolution) for 1984 and 1985 to investigate cyclogenesis along the Victoria Land coast between Terra Nova Bay and Minna Bluff (see Figure 1). This area was found to be a very active region for subsynoptic cyclone formation with between one and two new lows forming each week throughout the year. About half the analyzed storms showed significant development before either moving out of the AWS domain or dissipating in situ, and nearly 70% of them formed without a maritime cyclone being present in the Ross Sea area. Apart from the open water season, most cyclones initially had little or no cloud signature, and were resolved only by the AWS pressure and wind observations. This result indicates the importance of low-level baroclinicity for their formation. Furthermore, the absence of distinct satellite signatures associated with these cyclones during at least their formative stages means that they would only be detected by a satellite survey of the type conducted by Carleton and Carpenter (1989, in review) at some later stage of their evolution.

In 1985 the AWS array expanded slightly over the northern Ross Ice Shelf. In addition to the genesis area near Franklin Island, a less active cyclogenetic area (approximately 40% as frequent) was found near Byrd Glacier. Apart from their smaller frequency of occurrence, the Byrd Glacier storms behaved in much the same fashion as those that formed near Franklin Island. Only 10%–20% of these subsynoptic storms were resolved at some stage by the Australian Bureau of Meteorology's hemispheric synoptic analyses. The proximity of cyclogenetic areas adjacent to regions of intense cold air transport from East Antarctica (Terra Nova Bay and Byrd Glacier; Parish and Bromwich, 1987) again suggests that low-level baroclinicity is a key ingredient for formation of these storms.

Case studies of subsynoptic-scale cyclone formation during the open water season at the Antarctic coast have been carried out by Bromwich (1987) and Turner and Warren (1988). In the first example the intense katabatic outflow

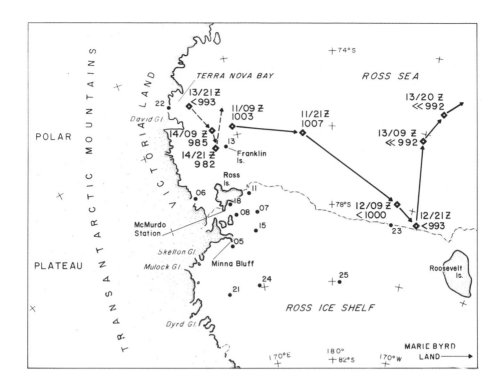

Figure 1. Surface trajectories at 12-hr intervals of subsynoptic cyclones that formed near Franklin Island, 11–15 May 1985. Notation: 11/09Z 1003 means that at 0900 UTC 11 May 1985 the cyclone was centered at the diamond location with a central pressure of 1003 hPa. Locations mentioned in the text are also noted. Filled circles with numbers attached are AWS sites; Stations 24 and 25 only operated during 1985, and Station 21 ran for the first half of 1984.

from Terra Nova Bay (Bromwich, 1989a) appeared to play a key role in the development and maintenance of a stationary storm system over the southwestern Ross Sea. It was inferred that boundary layer baroclinicity was responsible for the formation, but that large transfers of heat and water vapor into the katabatic airstream from the open Ross Sea governed its subsequent evolution. The complex eventually assumed an inverted comma shape on satellite imagery. Turner and Warren (1988) used NOAA satellite imagery and sounder (TOVS) data to show that a relatively shallow vortex formed in an intense lower tropospheric, baroclinic zone near Halley Station. Once again, it was inferred that heat fluxes from the open ocean into cold continental air played an important role in the development and maintenance of the storm that dissipated rapidly once it moved inland.

Here two case study periods of subsynoptic-scale cyclogenesis in the Ross Sea sector during winter are presented. Section 2 analyzes a subsynoptic development sequence near Franklin Island and Section 3, a similar series of events near Byrd Glacier; both regions are influenced by persistent flows of cold katabatic air from East Antarctica. The primary data base is comprised of at least twice-daily manual analyses of observations from AWS sites close to sea level (Sievers et al., 1986), and is supplemented by an average of two satellite images for each day. Direct wind measurements of the intense katabatic airstreams that typically blow from Terra Nova Bay and Byrd Glacier were not available, but proxy evidence of their presence is often provided by thermal infrared satellite pictures (Bromwich, 1989b). Since the ocean is ice covered, usually there is little heat energy to power the subsynoptic storms, and inadequate condensed water to serve as an effective tracer of their air motions on satellite images. Twice-daily surface and 500-hPa hemispheric maps produced by the Australian Bureau of Meteorology were examined to determine qualitatively the large scale and upper air conditions accompanying surface cyclogenesis. In addition the complete set of rawinsonde data from McMurdo Station was consulted. The Australian 500-hPa charts were modified to be consistent with omitted McMurdo observations and available satellite imagery.

2. A CYCLOGENETIC SEQUENCE OVER THE SOUTHWESTERN ROSS SEA

Prior to 0900 UTC 11 May 1985 a subsynoptic-scale low formed near Franklin Island (compare Figure 2). The hemispheric sea level chart for 3 hr later produced by the Australian Bureau of Meteorology, showed a deep but weakening synoptic-scale cyclone with a central pressure of 955 hPa to the north-west of the Ross Sea at 65 °S 157 °E. Surface winds of 25 m s^{-1} were recorded at Leningradskaya Station. The subsynoptic low appears to have formed in warm maritime air ahead of a front that was oriented nearly east-west across northern Victoria Land and was part of the synoptic-scale circulation. The validity of a closed center near Franklin Island in Figure 2 is strongly supported by subsequent regional analyses. A weak low similar to those discussed in the next section was present near Byrd Glacier, but dissipated near Station 24 by 1800 UTC 11 May.

Nine hours prior to the analysis in Figure 2 a rawinsonde ascent from McMurdo Station revealed a 4-km-deep stably-stratified cloud layer extending between the 900- and 500-hPa levels. Winds were from the north-northwest and the speed increased with height; strong speed shear was present above the 600-hPa

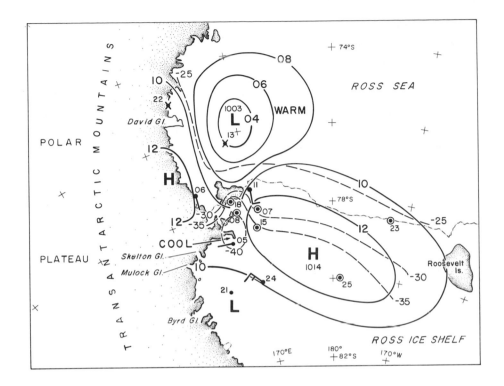

Figure 2. Subsynoptic-scale cyclogenesis near Franklin Island, 0900 UTC 11 May 1985. Sea level isobars (solid in hPa: 12=1012 hPa) and surface isotherms in °C (dashed) were constructed from AWS data. Wind speed notation: a circle concentric with the filled location circle represents calm conditions, no symbol attached to the direction line means less than 1.3 m s⁻¹, half a barb is 1.3–3.8 m s⁻¹, and a full barb denotes 3.9–6.4 m s⁻¹. AWS sites with crosses had no wind data, but did collect pressure and temperature observations. An AWS site with no symbol means that no data were available for that location.

level. The strongest speed of 52 m s⁻¹ was measured at the highest point of the wind sounding (471 hPa, 5.7 km). The modified Australian hemispheric 500 hPa-maps showed that the cyclone developed beneath an upper level ridge, probably with an associated jet streak; thus upper level divergence could have contributed to surface cyclogenesis (Uccellini and Kocin, 1987). Katabatic-induced boundary layer baroclinicity may also have played a role although, because of cloud cover, no satellite evidence is available to confirm this.

The cyclone moved to the east-southeast at an average speed of 6.4 m s⁻¹ to reach the northeastern edge of the Ross Ice Shelf (Figure 3) by 0900 UTC 12 May 1985. During this period the storm developed only slightly with weakly divergent flow shown by the 500-hPa charts; these analyses are uncertain because no rawinsonde ascents from McMurdo are available. On the Australian sea level chart for 1200 UTC 12 May, a weak ridge was shown covering the southern Ross Sea and Ross Ice Shelf. As a result the cyclone shown in Figure 3 was completely missed and the analyzed pressures were up 15 hPa too high.

Over the next day and a half the cyclone moved slowly northward as shown by Figure 1. Its position away from the Ross Ice Shelf was fixed by thermal infrared satellite images. This trajectory is generally consistent with the movement of a low over the Ross Sea shown on the Australian 500-hPa charts; two McMurdo

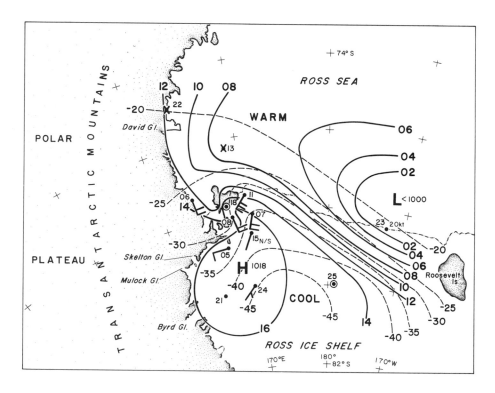

Figure 3. Cyclone generated near Franklin Island (Figure 2) has moved to the northeastern edge of the Ross Ice Shelf by 0900 UTC 12 May 1985. No directional information was available for Station 23, and the wind speed was missing for Station 15.

rawinsonde ascents are available to constrain the analyses. Throughout this period the subsynoptic cyclone contained mostly low clouds, thus showing that it was confined to the lower troposphere.

Between 2000 UTC 13 May and 1200 UTC 14 May the cyclone probably moved to the northeast at 16 m s^{-1} to be analyzed as a center on the Australian surface analysis at 70 °S 145 °W. This low was linked by a cold front to a cyclone 15 ° of latitude to the north-northeast, which was a remnant of the synoptic-scale depression that appeared to trigger the subsynoptic formation. The Southern Ocean low moved in a similar fashion to the Ross Sea cyclone, and both trajectories roughly followed the Australian 500-hPa geostrophic winds.

Around 2100 UTC 13 May a second subsynoptic-scale cyclone developed near Terra Nova Bay (Figure 1) in a trough tied to another strong maritime depression to the northwest of the Ross Sea. At 0000 UTC 14 May, the synoptic-scale cyclone was located near 67 °S 148 °E with a central pressure of 975 hPa, and caused winds of 40 m s^{-1} at Dumont D'Urville and 20 m s^{-1} at Leningradskaya. The Australian analysis also resolved the cyclone in the western Ross Sea, even though a thermal infrared DMSP image for 2114 UTC 13 May did not show a distinct cloud circulation. This image, as well as those for 0215 UTC and 0322 UTC on 14 May, contained the thermal signature of katabatic winds blowing into Terra Nova Bay (Parish and Bromwich, 1989; Bromwich, 1989b). Regional temperature readings suggest the presence of maritime air near Terra Nova Bay in conjunction with the synoptic-scale cyclone. It appears that the subsynoptic cyclone formed in the vicinity of a boundary layer baroclinic zone between relatively cold continental air from the ice sheet and warmer maritime air over the ocean.

The rawinsonde ascent from McMurdo Station at 0000 UTC 14 May was similar to the one just preceding the genesis of the first subsynoptic low. Here the low-level stability was much more marked with a sharp 7 °C inversion between 900 and 860 hPa. Also the atmosphere between the 900- and 500-hPa levels was much drier, consistent with the mostly clear DMSP image for 2114 UTC 13 May. The abrupt drying associated with the inversion together with 10–15 m s^{-1} northwest winds between 900 and 650 hPa suggests this air subsided from the polar plateau. Cyclonic vorticity generation by vertical stretching (Holton, 1979) in the low- to mid-troposphere may have accompanied subsynoptic cyclone formation on the boundary layer baroclinic zone.

Over the next 24 hr the subsynoptic-scale storm remained nearly stationary and steadily intensified. During the same period, the Australian 500-hPa maps showed a large, rapidly deepening vortex over the polar plateau to the west of the Ross Sea. This suggests the development of marked cyclonic vorticity above the subsynoptic cyclone and lower tropospheric ascent (Holton, 1979). Between 2100 UTC 14 May and 0600 UTC 15 May the surface cyclone weakened and remained stationary. Then it appeared to move northeastward to merge with the initiating synoptic-scale cyclone as it transited the northern edge of the Ross Sea. Between 0000 UTC and 1200 UTC on 15 May, the 500-hPa low weakened and moved away from the Ross Sea indicating that lower tropospheric ascent may have turned into descent, and caused the subsynoptic low to weaken. From 0000 UTC 14 May to 1200 UTC 15 May there were no rawinsonde ascents from McMurdo Station, thus adding a degree of uncertainty to the discussion in this paragraph. However, it seems that the large size of the upper low may offset this deficiency to a substantial extent.

3. SUBSYNOPTIC-SCALE CYCLOGENESES
NEAR BYRD GLACIER

This section discusses a series of three developments over the Ross Ice Shelf near Byrd Glacier (Figure 4) during late September and early October in 1985. This particular sequence was chosen because of the variety of fates experienced by the cyclones. The scheduled McMurdo rawinsonde ascents changed from once a day to twice daily on 28 September, and a complete record was obtained during this case study period.

At 0900 UTC 27 September a new center was analyzed adjacent to Byrd Glacier. The Australian sea level chart for 1200 UTC 27 September showed a deep low at 65°S 180°E with a central pressure of 955 hPa moving east-southeastward. The northern Ross Ice Shelf was barely affected by this synoptic-scale feature. On the corresponding 500-hPa chart a weak low was analyzed over the Ross Sea–Ross Ice Shelf area. As this circulation appeared during the previous 12 hr, cyclonic vorticity development and weak lower tropospheric vertical motion is suggested. A DMSP thermal infrared image for 0618 UTC 27 September showed katabatic signatures from Byrd, Mulock, and Skelton glaciers (Kurtz and Bromwich, 1985; Bromwich, 1989b) converging into the cloud-free area where the subsynoptic-scale cyclone was first analyzed 3 hr later. By 2016 UTC 27 September the cyclone had developed a low-level cloud field, and katabatic airflow from at least Mulock and Skelton glaciers continued to

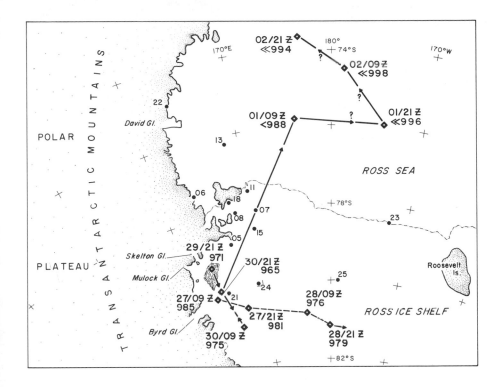

Figure 4. Surface trajectories at 12-hr intervals of subsynoptic cyclones that formed near Byrd Glacier, 27 September to 3 October 1985. Same notation as used in Figure 1. The shaded area was occupied by a weak, nearly stationary cyclone between 0900 UTC 2 October and 2100 UTC 3 October.

feed into the circulation (Figure 5). Bromwich and Parish (1988) described a similar katabatic-wind forced cyclogenesis near Terra Nova Bay in February 1988. The subsynoptic cyclone moved slowly eastward and by 0900 UTC 29 September could no longer be detected by observations from AWS 25. A DMSP image at 1955 UTC 28 September showed that the circulation was restricted to the lower part of the atmosphere and that the cloud field had grown substantially in size. Subsequent images (2051 UTC 29 September and 0151 UTC 30 September) suggest that this cyclone deepened dramatically over the western slopes of Marie Byrd Land (81°S 146°W) around 0000 UTC 30 September. Moisture for the well defined cloud vortex was probably supplied by the intense (approximately 945 hPa) synoptic-scale depression, which was mentioned previously and now

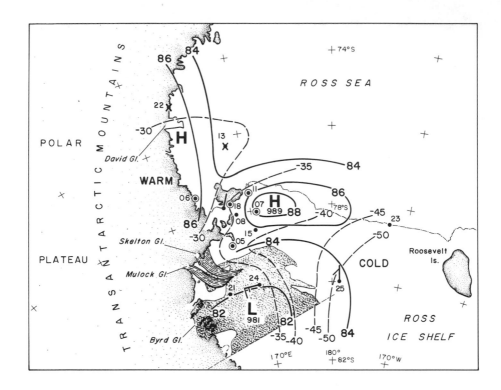

Figure 5. Subsynoptic-scale cyclogenesis near Byrd Glacier, 2100 UTC 27 September 1985. Same notation as Figure 2. Katabatic signatures (broken-lined) and low cloud (shaded) are taken from a DMSP infrared image at 2016 UTC 27 September 1985. Wind observation plotted at the southern tip of Ross Island is from Scott Base.

located by the Australian analysis (for 0000 UTC 30 September) just to the north-west of Russkaya at 73 °S 140 °W. The transformation of the subsynoptic cyclone to a synoptic-scale feature (around 0000 UTC 30 September) took place adjacent to a 500-hPa low resolved by the Australian analysis. At 1200 UTC 30 September this synoptic-scale low was analyzed for the first time on Australian sea level charts and placed over the interior of Marie Byrd Land near 78 °S 125 °W. The Australian analyses showed that this synoptic-scale cyclone continued to develop and moved northward along the west side of the Antarctic Peninsula bringing 25 m s^{-1} winds to Rothera at 1200 UTC 1 October. The cyclone had passed around the northern tip of the Antarctic Peninsula and into the Weddell Sea by 1200 UTC 2 October. In summary, a clear sequence of events links katabatic-wind forced subsynoptic-scale cyclogenesis near Byrd Glacier to strong winds along the west side of the Antarctic Peninsula some 4 days later and over 4000 km to the east; this only happened because of the right combination of large-scale conditions.

The second subsynoptic cyclone formed just to the north of Byrd Glacier around 2100 UTC 29 September. The thermal infrared DMSP image for 2051 UTC 29 September again showed katabatic airstreams primarily from Byrd Glacier, but also from Mulock and Skelton glaciers converging into the cloud-free region of cyclogenesis. As there were no significant changes on the modified Australian 500-hPa charts between 1200 UTC 29 September and 0000 UTC 30 September, no inferences can be made about the upper level support at the time of surface cyclogenesis. The subsynoptic cyclone slowly and steadily strengthened for the next 18 hr and did not move around very much. Over the following 6 hr (1500 UTC to 2100 UTC on 30 September), the storm deepened rapidly (see Figures 4 and 6). The Australian 500-hPa map for 0000 UTC 1 October, when modified to be consistent with a DMSP thermal image for 2054 UTC 30 September, showed that this deepening was probably associated with the rapid development of a narrow 500-hPa ridge. The ridge did not quite extend northward to Ross Island (at 2054 UTC), but was part of a huge high that stretched across the continent down 10 °W to the South Pole and then along 170 °E to near McMurdo. McMurdo rawinsonde ascents at 1200 UTC 30 September and 0000 UTC 1 October showed that this deepening was associated with marked warm advection in the 900- to 500-hPa layer. Presumably this warm advection generated lower tropospheric vertical motion and associated upper level divergence that deepened the surface cyclone (compare Holton, 1979, pp. 138–139). Orographic lifting of this comparatively warm air over the polar plateau could have helped to amplify the upper level ridge (Holton, 1979, pp. 90–91). It is very likely that this warm air was advected far to the south in a clockwise motion around the first cyclone (discussed in the previous paragraph) as it traversed Marie Byrd Land; then the air was advected to the northwest toward the Byrd Glacier area and caused the rapid subsynoptic deepening. After 0000 UTC 1 October the subsynoptic-scale storm moved rapidly northward as the 500-hPa low, which was located near Terra Nova Bay at 0000 UTC, moved to the northwest. The surface cyclone weakened substantially as it jumped across the 1.5 km-high obstacle of Ross Island between 0000 UTC and 0300 UTC on 1 October. The fate of this storm is uncertain because of the lack of surface data over the Ross Sea and because the satellite images are difficult to interpret; one scenario is presented in Figure 4. Around 0900 UTC 2 October the third cyclone in the sequence formed near Byrd Glacier beneath a weak 500-hPa

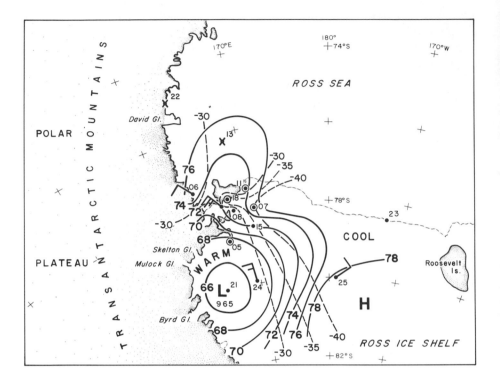

Figure 6. Same as Figure 5, but for 2100 UTC 30 September 1985.

low, which developed around the same time. Cloud cover precluded any evaluation of katabatic wind activity. The surface low was nearly stationary and showed little development before dissipating by 2100 UTC 3 October. Throughout this 36-hr period, the 500-hPa low was variable, and showed no consistent upper level support for the surface circulation.

4. DISCUSSION

Case studies of subsynoptic-scale cyclone development have been carried out to clarify the genesis mechanisms in areas affected by strong and persistent transport of cold boundary layer air from East Antarctica. In general, it appears that (katabatic-induced) boundary layer baroclinicity is a necessary condition. Another prerequisite seems to be a weak surface trough; this was the case for the Franklin Island sequence, but could not be verified for the Byrd Glacier area because of inadequate data. These boundary layer conditions accompanying subsynoptic-scale cyclogenesis in the Ross Sea sector are the same as those occurring in conjunction with polar low formation in the Arctic (Businger, 1987). Lower tropospheric vertical motion often appears to accompany surface cyclogenesis, and subsequent development is strongly influenced by the upper level support. The above discussion is substantially limited by the coarse and sometimes inaccurate 500-hPa analyses produced by the Australian Bureau of Meteorology. This situation underscores the need to acquire data about tropospheric dynamics on the same spatial scale as the subsynoptic cyclones. The satellite-based techniques described by Turner and Warren (1988) and Warren and Turner (1988) may supply much of the required information. The ongoing acquisition at McMurdo Station of satellite image and sounding data from the U.S. civilian and military polar-orbiting meteorological satellites should allow such data to be used in future quantitative diagnoses. The broad-scale significance of the very active subsynoptic-scale cyclogenesis regions in the Ross Sea sector has not been established. Both sequences showed that some cyclones can exit the Ross Sea area either as distinct synoptic-scale circulation features or to reinforce existing synoptic-scale lows. It is not known whether at later stages of their evolution these subsynoptic perturbations routinely amplify into major synoptic-scale cyclones, or dissipate and have negligible climatic impact.

ACKNOWLEDGMENTS

The DMSP images were obtained from the National Snow and Ice Data Center, University of Colorado, Campus Box 449, Boulder, Colorado 80309. Contribution 662 of Byrd Polar Research Center.

REFERENCES

Barry, R.G., and R.J. Chorley, 1987: *Atmosphere, Weather and Climate,* fifth edition, Metheun, London, 460 pp.

Bromwich, D.H., 1987: A case study of mesoscale cyclogenesis over the southwestern Ross Sea. *Antarct. J. U.S., 22(5),* 254–256.

Bromwich, D.H., 1989a: An extraordinary katabatic wind regime at Terra Nova Bay, Antarctica. *Mon. Wea. Rev., 117,* 688–695.

Bromwich, D.H., 1989b: Satellite analyses of Antarctic katabatic wind behavior. *Bull. Am. Meteor. Soc.* (accepted for publication).

Bromwich, D.H., and T.R. Parish, 1988: Mesoscale cyclone interactions with the surface windfield near Terra Nova Bay, Antarctica. *Antarct. J. U.S., 23(5)* (in press).

Businger, S. 1987: The synoptic climatology of polar-low outbreaks over the Gulf of Alaska and the Bering Sea. *Tellus, 39A,* 307–325.

Carleton, A.M., 1979: A Synoptic climatology of satellite-observed extratropical cyclone activity for the Southern Hemisphere winter. *Arch. Meteor. Geophys. Bioklim., B27,* 265–279.

Carleton, A.M., and D.A. Carpenter, 1989: Intermediate-scale sea ice-atmosphere interactions over high southern latitudes in winter. *GeoJournal, 18(1),* 87–101.

Carleton, A.M., and D.A. Carpenter, in review: Satellite climatology of "polar lows" and associations with climatic indices for the Southern Hemisphere. *J. Climatol.*

Holton, J.R., 1979: *An Introduction to Dynamic Meteorology,* second edition, Academic Press, New York, 391 pp.

Kurtz, D.D., and D.H. Bromwich, 1985: A recurring, atmospherically-forced polynya in Terra Nova Bay. In S.S. Jacobs (Ed.), *Oceanology of the Antarctic Continental Shelf, Antarct. Res. Ser., 43,* American Geophysical Union, Washington, D.C., 177–201.

Parish, T.R., and D.H. Bromwich, 1987: The surface windfield over the Antarctic Ice Sheets. *Nature, 328,* 51–54.

Parish, T.R., and D.H. Bromwich, 1989: Instrumented aircraft observations of the katabatic wind regime near Terra Nova Bay. *Mon. Wea. Rev.* (accepted for publication).

Sievers, M.F., G.A. Weidner, and C.R. Stearns, 1986: *Antarctic Automatic Weather Station Data for the Calendar Year 1985.* Department of Meteorology, University of Wisconsin, 254 pp. (Available from Dept. Meteor., 1225 West Dayton Street, Madison, WI 53706.)

Streten, N.A., and A.J. Troup, 1973: A synoptic climatology of satellite observed cloud vortices over the Southern Hemisphere. *Quart. J. Roy. Meteor. Soc., 99,* 56–72.

Turner, J., and D. Warren, 1988: The structure of subsynoptic-scale vortices in polar air-streams from AVHRR and TOVS data. Preprint, *Second Conference on Polar Meteorology and Oceanography,* American Meteorological Society, Boston, 126–128. (Available from Am. Meteor. Soc., 45 Beacon St., Boston, MA 02108.)

Uccellini, U., and P.J. Kocin, 1987: The interaction of jet streak circulations during heavy snow events along the East Coast of the United States. *Wea. Forecasting, 2,* 289–308.

Warren, D.E., and J. Turner, 1988: High latitude wind velocities derived from polar orbiting satellite imagery. Preprint, *Second Conference on Polar Meteorology and Oceanography,* American Meteorological Society, Boston, 119–121. (Available from Am. Meteor. Soc., 45 Beacon Street, Boston, MA 02108.)

MESOSCALE VORTICES IN THE BRITISH ANTARCTIC TERRITORY

J. Turner and M. Row

British Antarctic Survey, Natural Environment Research Council
Cambridge, United Kingdom

ABSTRACT

Satellite imagery and hand-drawn analyses prepared by the UK Meteorological Office have been examined to determine the occurrence and movement of cyclonic systems over the Weddell Sea and Antarctic Peninsula. Both forms of data indicated that two types of systems were occurring. First, eastward moving synoptic-scale depressions occurred in the latitude band of 65–70°S. Second, slow moving mesoscale vortices formed over the coastal area of the eastern Weddell Sea. It is proposed that this second type of circulation consists mainly of lee vortices forming in the easterly flow off the high Antarctic Plateau.

1. INTRODUCTION

High latitude mesoscale vortices have been studied for some years over the ocean areas of the Northern Hemisphere and are frequently referred to in the literature as polar lows. These systems form during the winter months in a few preferred areas of the world and never develop over the land or sea ice. The early work on polar lows was concerned with systems around Iceland and in the Barents and Norwegian Seas, where a large number of ship observations were available to help in the analysis. Recently, with the availability of high resolution satellite imagery, polar lows have also been identified in the Gulf of Alaska (Businger, 1987), however, no polar lows have ever been reported in the high latitude areas of the Southern Hemisphere. This could be due, in part, to the limited amount of research carried out into subsynoptic-scale systems in the Southern Hemisphere, but is more probably due to the differences in climatic conditions between the two hemispheres.

POLAR AND ARCTIC LOWS
Paul F. Twitchell, Erik A. Rasmussen,
and Kenneth L. Davidson (Eds.)

347

For many years polar lows were thought of primarily as thermal instability phenomena that developed during cold air outbreaks over relatively warm ocean areas. These conditions are found in the Norwegian Sea, where the North Atlantic Drift brings water with a surface temperature as high as $-10\,°C$ into high latitude areas where air with a temperature as low as $-40\,°C$ is flowing off the sea ice. Under such conditions very intense convection takes place in the extremely unstable lower layers of the troposphere. In the Southern Hemisphere the ocean circulation is much more zonal and there is no such dramatic poleward transport of warm water towards the Antarctic. What are often referred to as "classic" polar lows, where convection plays a major role, therefore probably do not occur in the Southern Hemisphere. Nevertheless, mesoscale vortices, with some of the characteristics of polar lows, do occur in the Southern Hemisphere and this paper is concerned with the occurrence of such systems and the possible mechanisms behind their formation.

In the Northern Hemisphere strong horizontal temperature gradients exist close to the edge of the sea ice and small vortices are often noted in satellite imagery in these areas. The majority of these vortices are short-lived phenomena but a small number develop into very active polar lows with strong winds and heavy precipitation. Harrold and Browning (1969) were the first to propose that baroclinic instability could be responsible for the formation of polar lows and subsequent theoretical studies (Duncan, 1977; Mansfield, 1974) have shown that this mechanism may indeed play a role in many developments. Over recent years no one theory for the development of polar lows has gained universal acceptance, and from the many diagnostic case studies carried out it does appear that a wide spectrum of systems occur in the Northern Hemisphere ranging from the largely convective system (e.g., Rasmussen, 1985) to the more baroclinic disturbance (e.g., Shapiro et al., 1987).

Around the Antarctic strong baroclinic zones are often found at the boundary between the very cold air from the centre of the continent and the milder air advected from lower latitudes. Small vortices often occur around the edge of the continent and occasionally develop into major circulations. Without the very warm sea surface temperatures found in the Northern Hemisphere it is unlikely that convective heating plays a significant role in the amplification of these vortices and that baroclinic instability is of more importance.

The following sections describe a number of investigations taking place to examine the development and structure of the Antarctic mesoscale vortices observed on meteorological satellite imagery. Preliminary results are presented from studies using meteorological analyses, satellite imagery received at the Antarctic bases of the British Antarctic Survey (BAS), and high resolution atmospheric sounder data from polar-orbiting satellites.

The locations referred to in the text are shown in Figure 1.

2. ANALYSIS OF SURFACE CHARTS

Producing a synoptic climatology of mesoscale vortices around the Antarctic is extremely difficult due to the limited number of meteorological observations and the problems associated with creating accurate analyses. Nevertheless, the Southern Hemispheric surface pressure analyses prepared by the major forecasting centres do provide an easily accessible means of investigating the

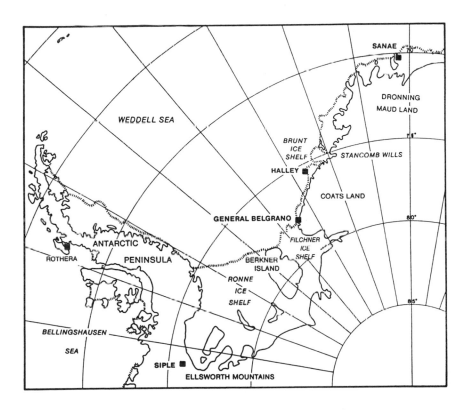

Figure 1. A section of the Antarctic indicating places referred to in the text.

frequency, distribution, and track of mesoscale vortices. One year of surface pressure charts, as drawn by the UK Meteorological Office, were used therefore to try to determine the number of mesoscale vortices occurring in one particular part of the Antarctic. The area chosen roughly corresponded to the British Antarctic Territory (BAT), which includes the Weddell Sea and Antarctic Peninsula. This area has a reasonable number of surface and upper air observing sites whose observations were useful for validation purposes.

Operational numerical analyses can often correctly represent small cyclones since they make use of all available surface pressure and wind observations provided by buoys, ships and land bases, and the upper air data from satellite sounding instruments and radiosonde ascents. However, the analysis systems cannot make use directly of the information contained in satellite imagery from the polar-orbiting satellites, which is one of the most powerful tools in the polar regions. For this reason it was decided to base this study on the Southern Hemispheric hand-drawn surface analyses of the UK Meteorological Office, which have been prepared at 0000 and 1200 GMT each day since 1982. The analysts who prepare these charts have as input all the observations on the Global Telecommunications System (GTS), the analyses from the UK 15-level operational numerical prediction model, the visible and infrared imagery from the TIROS-N/NOAA polar-orbiting satellites, and the analyses from other meteorological centres around the world. These charts should therefore represent as good an analysis as possible considering the limited amounts of in situ data available.

The study is based on one year of data from September 1983 to August 1984 inclusive. Data were extracted from the 1200 GMT charts for a latitude/longitude sector of the Antarctic bounded by the South Pole, 60°S 10°W, and 70°W. Data were not extracted from the 0000 GMT analyses, but these charts were frequently consulted to help maintain continuity. The following information was extracted from each 1200 GMT chart: (1) the location of each low pressure centre, (2) the location of cold, warm, and occluded fronts within the area, and (3) the synoptic conditions under which each depression developed. For systems not developing in the sector the track into the area was also noted.

During the year a total of 170 depressions were identified within this area. The distribution of these systems by area, and grouped according to the four seasons, autumn (March–May), winter (June–August), spring (September–November), and summer (December–February) is shown in Table 1.

First, the statistics for all vortices occurring in the area during the year are considered. As would be expected, the greatest number of depressions occurred

TABLE 1. DEPRESSIONS OVER THE WEDDELL SEA AND ANTARCTIC PENINSULA—SEPTEMBER 1983–AUGUST 1984

	Spring	*Summer*	*Autumn*	*Winter*	*Total*
All Cyclones	50	31	48	41	170
Systems Developing in Area	23	16	26	11	76
Systems Entering Area	27	15	22	30	94
Eastern Weddell Sea Coastal Systems	6	1	9	2	18
Developments in Lee of Peninsula	6	3	6	6	21

in the spring and autumn with a marked decrease in the number of lows in the area during the summer months. This is consistent with the peak in cyclonic activity found in the midlatitude areas of both hemispheres in the spring and autumn. Second, the number of depressions developing within the sector and moving in from outside were about equal in all seasons, except during the winter when the latter category predominated. The majority of systems entering the area were mature depressions moving towards the peninsula from the west, with a smaller number moving south from lower latitudes. During the one year analyzed no depressions at all entered the area from the east. Most of the mature depressions tracked eastwards across the Weddell Sea, while a smaller number veered southeastwards towards the coast of the Antarctic near Halley. Some of these systems then moved westwards towards the Ronne Ice Shelf in the prevailing easterly flow over Coats Land. Third, plots of the depression tracks for the four seasons clearly showed the seasonal migration of the lows, and the mean latitudes of the tracks were about 65 °S in summer and 70 °S in winter. The charts examined managed to resolve a number of the lee lows that form to the east of the Antarctic Peninsula. The number identified in each season remained constant at about 10–12% of the total number of depressions found.

The plots of depression tracks indicated that two forms of low pressure systems were occurring in the Weddell Sea: First, the eastward moving mature depressions located across the peninsula and central Weddell Sea, and second, relatively short-lived coastal lows developing around the edge of the Ronne Ice Shelf and at the eastern edge of the Weddell Sea. This second type of vortex rarely existed for more than 2 days and usually moved westward before dissipating. The number of such systems was relatively small compared to the number of mature depressions. A total of 18 mesoscale cyclones were found during the year and almost all occurred during the spring and autumn seasons. An analysis was carried out of the conditions under which each of these cyclones developed using the available surface and upper air analysis charts. The vast majority of the cyclones formed very close to Halley at the eastern side of the Weddell Sea. In almost all of the cases the synoptic situation prior to cyclogenesis had an area of low pressure off the coast of the Antarctic near SANAE, with a trough of low pressure extending poleward down through Coats Land or along the coastal zone over Halley. The low pressure centre then appeared on the surface charts about 12 or 24 hr after the establishment of the trough. The remainder of the mesoscale cyclones either formed around Halley or on the Ronne Ice Shelf from troughs extending from lower latitude areas of the Weddell Sea or the Bellingshausen Sea, respectively.

3. SATELLITE IMAGERY

Satellite imagery is a particularly valuable tool to the synoptic meteorologist concerned with the Southern Hemisphere, since few in situ observations are available. However, the usefulness of the imagery available at the main meteorological centres is often limited by communications problems and the relatively small area of the continent visible on each satellite pass. Although Southern Hemispheric polar stereographic composite images are prepared by NOAA/NESDIS and distributed via the GOES geostationary satellite these are not always available quickly enough to help in the analysis process. Hence, looking for mesoscale systems near the coastal margin from the operational data available on a real-time basis is extremely difficult, considering the small scale of some of the features.

The installation in recent years of receivers for meteorological satellite imagery at the BAS bases of Rothera and Halley has enabled the base meteorologists to observe local weather patterns from the imagery of the NOAA and METEOR series polar-orbiting satellites. Because of the proximity of these bases to the pole up to nine passes of these satellites can be received each day. At Halley (75°36′S, 26°42′W), a Feedback Instruments WSR513 receiver was installed at the beginning of 1986. This has proved useful for local forecasting and monitoring of sea ice conditions.

Routine monitoring of the imagery at Halley soon showed that the area tended to be affected by two distinct types of depressions: First, the large-scale systems, which moved mainly east or southeast through the Weddell Sea, having developed near or rounded the top of the Antarctic Peninsula, and second, smaller scale lows with a diameter of less than 500 km that had a much more local origin. The larger systems affecting Halley brought up to a week of heavy blizzard and snowfall and were usually associated with wind speeds in excess of 40 knots. However, the smaller depressions brought only a few days of bad weather and winds rarely to gale force. Such small systems could be observed at most times of the year, but were most common outside of the winter months. The following sections discuss three mesoscale lows that developed close to Halley.

3.1 10–13 February 1986

This case occurred towards the end of the Austral summer and brought moderate snowfall and winds of about 20 knots to Halley. The afternoon TIROS-N/NOAA imagery on 10 February gave the first indication of the system, when a tight swirl of cloud was apparent at the edge of the Filchner Ice Shelf, approximately 200 miles southwest of Halley. It was subsequently steered up the coast by a 15-knot west-southwest wind at midlevels and became slow moving and decayed to the northeast of Halley. Snowfall was mainly light and intermittent with surface easterly winds veering to the southwest as the disturbance passed. Although marked as a shallow enclosed depression on the UK Meteorological Office analysis charts, the cloud comma, as visible on the satellite imagery, meant that this may have been a rather more active system than indicated. The passage of the system was associated with a fall in 500-mb thickness over Halley and subsequent rise behind it. The 500-mb height of 516 gpdm and 700-mb wet-bulb potential temperature of $-1.8\,°C$ at 1200 GMT on 10 February were the lowest of the month. A small cut-off in the 1000- to 500-mb thickness field was indicated just to the north of Halley as the cloud structure broke up on 13 February. The cloud spiral on this occasion was never more than 500 km in diameter and consisted mainly of thick bands of lower cloud, although there was evidence from visible imagery that there was some cirriform cloud in its circulation. From the radiosonde ascent on 11 February, the profile was seen to be comparatively moist indicating cloud from near the surface to the tropopause, which had fallen to 350 mb.

3.2 6–8 March 1986

The events of this case were broadly similar to the above example with a small circulation once again observed over the Halley area in the satellite imagery.

The UK Meteorological Office analysis in this case placed a low inland of Halley, but surface winds of 20 knots from the east and satellite evidence would indicate that the system had in fact moved to the north of the station. A cold cut-off pool had formed to the north of Halley in association with a sudden intensification of a trough in the 500-mb height field between 1200 GMT 6 February and 0000 GMT 7 February. The winds were generally from a northeasterly point on this occasion, steering the system from this direction. There was a fall in thickness values to near their lowest values for the month, quickly recovering later as the cloud mass was engulfed by the approach of a much larger system from the north. This large-scale depression brought winds in excess of 50 knots and considerable snowfall. The contrast in thermal structure between the synoptic and mesoscale systems was dramatic and highlighted the differences between the two types of system encountered at Halley. The smaller systems had comparatively low temperatures throughout their evolution, while the major storm brought an injection of much warmer air from the north, which saw surface temperatures rise to near $-2\,°C$ and thickness values increase to 524 gpdm.

3.3 3–6 January 1986

This system was noted on the NOAA-9 imagery for 5 January 1986 as a 600-km diameter spiral of cloud just off the coast of Coats Land, close to Halley. The Halley observations for this day reported gale force winds and significant precipitation. As in the cases described above this vortex formed from a surface trough lying over Coats Land, but developed rapidly when it crossed the coast of the continent near Halley. Although the imagery and Halley observations usefully documented the evolution of the vortex and cloud comma they provided only limited information on the broader scale environment in which the development took place. For this reason the case was studied using satellite sounder data from the polar-orbiting satellites.

High resolution satellite sounder data have not been used extensively yet in polar meteorology but promise to be a powerful tool over the coming decades as satellite instrumentation and retrieval techniques are improved. The TIROS Operational Vertical Sounder (TOVS) on the TIROS-N/NOAA series of polar-orbiting satellites allows the production of temperature and humidity profiles at a horizontal resolution of 40 km over the whole globe every 12 hr. A low resolution form of these data are distributed over the GTS as SATEM messages for use in the numerical prediction system, however, the full resolution data are of great value for research in data sparse regions. Steffensen and Rasmussen

(1986) were the first to apply TOVS temperature profiles to polar lows research and used these data in a number of case studies of lows in the Barents Sea. Their work showed that mesoscale temperature structure could be studied with the TOVS retrievals and that they could even resolve the warm cores often found at the heart of polar lows.

The application of TOVS data to this case has recently been described by Turner and Warren (1988). Here the TOVS data showed that the amplification of the vortex occurred when, simultaneously, an upper cold trough moved over the vortex and the system moved into a strong baroclinic zone over Halley at the boundary of cold air off the Antarctic Plateau and warmer air advected down the western Weddell Sea from lower latitudes. The TOVS temperature fields provided the detailed thermal structure of the vortex every few hours and, when used in conjunction with the imagery, allowed the creation of a conceptual model of the system in its mature phase.

4. DISCUSSION

The analysis of satellite imagery and surface analyses has shown that mesoscale vortices frequently form over the coastal area of the eastern Weddell Sea. The vortices usually develop well to the south of the main track of mature depressions and evolve in the very cold air associated with the Antarctic Continent. Many form from troughs aligned northeast-southwest down the coast of Coats Land from a quasi-stationary area of low pressure that is often located off the coast near SANAE.

The large number of developments in the Halley area tends to suggest that they are caused by some local feature of the environment in this part of the Weddell Sea. One possibility is that the area of open water close to Halley could cause a sufficiently large flux of heat and moisture into the cold air crossing the area to trigger the developments. This ice-free area is present throughout the year in the prevailing easterly wind, however, the very short track the air has over the open water would probably not provide sufficient heat for the developments. The more likely cause of so many vortices in this location is the high topography of the Antarctic Plateau inland of Halley. As virtually all of the developments occurred when the wind direction was easterly and when the flow was already cyclonic around troughs lying over the coastal area, it appears likely that the systems were formed due to forcing by the high topography. Lee cyclogenesis occurs downwind of many mountain ranges, and over the western Weddell Sea small vortices are frequently found in the lee of the peninsula.

Future work will attempt to gain greater insight into the cyclogenesis taking place in this sector of the Antarctic. Further detailed case studies are required to understand the evolution of the cyclones along with numerical studies to determine the details of the processes in operation.

REFERENCES

Businger, S., 1987: The synoptic climatology of polar-low outbreaks over the Gulf of Alaska and the Bering Sea. *Tellus, 39A,* 307–325.

Duncan, C. N., 1977: A numerical investigation of polar lows. *Quart. J. Roy. Meteor. Soc., 103,* 255–267.

Harrold, T.W., and K.A. Browning, 1969: The polar low as a baroclinic disturbance. *Quart. J. Roy. Meteor. Soc., 95,* 710–723.

Mansfield, D.A., 1974: Polar lows: The development of baroclinic disturbances in cold air outbreaks. *Quart. J. Roy. Meteor. Soc., 100,* 541–554.

Rasmussen, E., 1985: A case study of a polar low development over the Barents Sea. *Tellus, 37A,* 407–418.

Shapiro, M.A., L.S. Fedor, and T. Hampel, 1987: Research aircraft measurements of a polar low over the Norwegian Sea. *Tellus, 39A,* 272–306.

Steffensen, M., and E. Rasmussen, 1986: *An Investigation of the Use of TOVS Data in Polar Low Research.* Tech. Rep. of the Polar Lows Project, 25 (available from the Norwegian Meteorological Institute).

Turner, J., and D.E. Warren, 1988: The structure of subsynoptic-scale vortices in polar airstreams from AVHRR and TOVS data. *Proceedings of the Second International Conference on Polar Meteorology and Oceanography,* Madison, Wisconsin.

CHAPTER 5 – OBSERVATIONS AND CLIMATOLOGIES

Introduction

Polar or arctic lows occur in regions where conventional meteorological observations are sparse and these potentially destructive storms develop rapidly between standard synoptic observation times. For these reasons high latitude storms can go undetected by the conventional spatial and temporal observation system. Often the first reports on the existence of these storms occur when the winds, precipitation, or ocean waves impact ships, drilling platforms, or coastal areas. The life cycles of the Greenland, Barents, and Norwegian Sea systems were analyzed years after their occurrences from a variety of synoptic network information. The resulting climatology has been published by Wilhelmsen in Tellus.

For more than a decade satellite-derived data were employed for scientific studies and for forecasting, with a focus on early detection of these short-lived but intense storms. Throughout this book there are numerous depictions of storms as observed by satellite and infrared imagery of cloud structure. An excellent example is the storm shown on the frontispiece. At the Copenhagen Workshop, in August 1984, and at the International Conference on Polar Lows, held in Oslo in May 1986, the trend was toward using satellite-derived data to understand the life cycles of these storms and toward developing prediction models that incorporate satellite data.

At present in the late 1980s there are several polar orbiting environmental satellites providing data on clouds, atmospheric structure, ice (including edge, leads and polynyas), and ocean surface (waves and temperature), all of which are related to the life of storms that meteorologists call polar or arctic lows. Poleward of about 70° (N or S) polar orbiting environmental satellites are capable of providing observations at any location about once an hour. Scientists and operational weather forecasters have this tool; the question is how effectively have environmental satellite data been used in the past 20 plus years and what can be done to improve use of these data. In this chapter the paper by Gloersen et al. is an example of how observations of sea surface waves can be used to detect the existence of a storm and estimate its strength. The paper by Claud et al. is another innovative example of how existing satellite-borne instruments can be used to extract information useful for predicting a storm's motions and intensity changes. The Warren and Turner paper describes the potential use of satellite imagery and satellite-derived sounding information to improve detection and understanding of high-latitude storm systems.

The final paper by Carleton and Carpenter is an example of multi-applications of satellite data toward improved understanding of high latitude vortices occurring in data sparse regions.

OBSERVATIONS OF ARCTIC POLAR LOWS WITH THE NIMBUS–7 SCANNING MULTICHANNEL MICROWAVE RADIOMETER[1]

Per Gloersen and Erik Mollo-Christensen
Laboratory for Oceans
NASA Goddard Space Flight Center
Greenbelt, Maryland, U.S.A.

Paul Hubanks
Research and Data Sytems Corporation
Lanham, Maryland, U.S.A.

ABSTRACT

A technique for estimating scalar oceanic winds and cloud water content in polar regions from multispectral microwave radiances has been developed and subjected to comparisons with a limited set of direct surface observations. The utilization of microwave frequencies of 18 GHz and 37 GHz permit these estimates as close as 50 km to land and sea ice boundaries. The limited comparisons described yield estimates with average errors of 8–9 knots with a comparison data set that was available in 5-knot bins.

1. INTRODUCTION

Polar lows are intense, small-scale cyclones that form in polar air streams behind or north of the polar front. They are a result of air-sea interaction where the temperature contrast between the cold atmosphere and the warmer ocean can feed energy into a developing cyclone. Cyclogenesis and explosive development of polar lows can occur in a time scale as short as a few hours. The regions

[1]Support for this work was provided by the Office of Naval Research under reimbursable document No. N0001488WM22006.

POLAR AND ARCTIC LOWS
Paul F. Twitchell, Erik A. Rasmussen,
and Kenneth L. Davidson (Eds.)

in which they form have few observation stations, and a polar low can remain undetected and develop to significant strength before it passes over a station.

Polar lows play an important role in air-sea interaction in terms of heat transfer; they affect water mass formation, ice motions, and air mass modification. Their effect on operations in the Arctic is also significant. It is thus highly desirable to be able to detect arctic lows with more certainty, and to follow their development and motion. This is both for operational reasons and for the purpose of understanding more quantitatively the formation of deep water in the Arctic Ocean, to mention one of several research purposes for wanting better knowledge and census of arctic lows.

We present here a method of observing surface winds over the ocean at high latitudes, using passive satellite radiometry. The method is applicable at high latitudes because the air is relatively dry, and the corrections for cloudiness and rain are possible without incurring excessive errors.

The Scanning Multichannel Microwave Radiometer (SMMR) (Gloersen and Barath, 1977) on board the Nimbus-7 SMMR was launched in October 1978 for the purpose of determining the utility and precision of obtaining on a global basis information on sea ice coverage, near surface oceanic winds, sea surface temperatures, atmospheric water vapor, and cloudiness (Gloersen et al., 1984). Also, snow water equivalence and soil moisture were some of the terrain parameters to be studied. All of the algorithms for estimating these parameters made use of the multispectral, dual-polarized radiances measured by SMMR, ten channels in all at the five wavelengths 0.8, 1.4, 1.7, 2.8, and 4.6 cm. In particular, a prelaunch algorithm was developed for obtaining near-surface scalar winds on a global basis (Wilheit et al., 1984). This algorithm makes use of four of the ten SMMR channels, including the 2.8-cm channels with nominal footprint sizes of about 100 km. While this algorithm works satisfactorily at distances of 300 km or more from land or sea ice, it is poorly suited for estimating winds in the polar regions near land or the ice edge, and tends to wash out the details of polar low events that are smaller in scale.

In this paper, we describe an alternative algorithm specifically for the polar regions that utilizes ratios of a different combination of four channels, at 0.8 and 1.7 cm, and subsequently yields valid estimates closer to land and sea ice boundaries. Restricting the estimates to polar regions minimizes the interference from atmospheric water vapor fluctuations. At this point in time, the algorithm has been tested against a limited data set. This paper represents a status report on the ongoing algorithm development.

2. PHYSICAL BASIS FOR THE ALGORITHM

Radiation emanating from the ocean surface in the microwave regime can be described by the Rayleigh-Jeans approximation to the Planck radiation law, i.e., the radiated power is linearly dependent on the sensible temperature of the radiating layer. Thus, microwave radiance is commonly given in units of degrees kelvin and referred to as brightness temperature. The radiance also depends linearly on the emissivity of the radiating surface. For calm seas, the emissivities are typically about 0.5 and 0.2 for the vertically and horizontally polarized channels, respectively (Gloersen and Barath, 1977). When subjected to wind stress, the reflectivity of the oceanic surface decreases from its specular, calm seas value due to the formation of waves and whitecaps. Therefore, the emissivity of a wind-swept ocean increases, resulting in a nonlinear increase in radiance from the ocean with increasing near-surface wind (Hollinger, 1971; Webster et al., 1976). The horizontally polarized radiance increases at a higher rate than the vertically polarized (Gloersen and Barath, 1977).

Estimation of scalar near-surface oceanic winds from observations of microwave radiances is complicated by atmospheric interference arising from fluctuations in atmospheric water vapor and cloud water content. It is further complicated by variations in the ocean surface temperature in two ways, by the direct dependence mentioned earlier and by a temperature dependence on the onset of whitecapping (Monahan and O'Muircheartaigh, 1986). All of these complications are minimized by restricting these estimations to polar waters, where the ranges of sea surface temperature and the atmospheric water vapor are small. Variation in cloud water content is also generally smaller in the polar regions, but must be taken into account when estimating winds.

3. FUNCTIONAL FORM OF THE ALGORITHM

In order to minimize further the dependence of wind estimation from microwave radiance on oceanic surface temperature, the microwave polarization, defined as the ratio of the difference of the vertically and horizontally polarized radiances at a given wavelength and their sum, is used as the independent variable. In this way the effect of sea surface temperature is eliminated to first order. As we shall show below, this approach has the additional advantage of a linear relationship between the oceanic scalar winds and the observed polarization. Either the polarization at the 0.8- or 1.7-cm wavelength could be used for these estimates, but the longer of the two is less subject to interference from clouds and is therefore selected. Since the sensitivity of polarization to wind is about the same at either wavelength, but not the same for cloud water, the difference in the polarizations at the two wavelengths is used to detect cloud water amount.

Defining the polarization at 1.7 cm as PR and the difference in the polarizations at 0.8 and 1.7 cm as DP, the dependence of these input parameters to near-surface scalar winds, W, and cloud liquid water amount, L, is assumed to be linear as follows:

$$PR = W_0 + w_1{}^*W + w_2{}^*L \tag{1}$$

$$DP = L_0 + l_1{}^*W + l_2{}^*L \tag{2}$$

If our previously stated hypotheses are correct, then the coefficients W_2 and L_1 in Eqs. (1) and (2) are relatively small. The estimation algorithm is the simultaneous solution of Eqs. (1) and (2):

$$W = W_1{}^*(PR - W_0) + W_2{}^*(DP - L_0) \tag{3}$$

$$L = L_1{}^*(PR - W_0) + L_2{}^*(DP - L_0) \tag{4}$$

where W_i, L_i are functions of w_i, l_i.

4. DETERMINATION OF THE ALGORITHM COEFFICIENTS

Model calculations for the microwave radiance emanating from atmospheric cloud water can be made with good confidence (Gaut and Reifenstein, 1970). This is fortunate, since wide-area measurements of cloud water amount suitable for comparison within the appropriate SMMR footprint (about 50 km) are not available. Thus, model calculations were used to determine the coefficients w_2 and c_2 in Eqs. (1) and (2).

While various models and observations (Hollinger, 1971; Webster et al., 1976) have provided insight as to the functional form of the dependence of oceanic microwave emissivity on wind, neither is sufficiently accurate for calculating the coefficients w_1 and c_1 in Eqs. (1) and (2). Instead, we have chosen to determine these coefficients directly by means of linear regression analysis of PR and DP against various observations of near-surface winds.

The comparison data sets utilized so far consist of 16 National Weather Service surface maps for 1984 selected for severe weather situations, five similar maps obtained by one of us (Mollo-Christensen) during the 1984 Marginal Ice Zone Experiment (Johannessen, 1987) from the Norwegian Weather Service at Tromsφ, nine surface maps of polar lows events nearly coinciding with SMMR overpasses during 1979–1982 extracted from a polar lows data set provided by Wilhelmsen (1981; also, personal communication, Kari Wilhelmsen, June 1987, Det Norske Meteorologiske Institutt, Tromsφ, Norway), and dropsonde data obtained during the 27 February 1984 flight of the National Oceanographic and Atmospheric Administration P3 aircraft during the 1984 Polar Lows Experiment (Shapiro et al., 1987). The study area was confined to the Norwegian, Greenland, and Barents Seas. In all cases, point measurements of winds appearing on the records rather than surface analyses were used for comparison. It should be recognized at the outset that comparisons are being made between point observations of scalar winds and broad footprint (about 50 km) observations of PR and DP from the SMMR within 4 hr of each other. Due to the rapid temporal and spatial variations in oceanic surface winds, such comparisons are subject to appreciable errors, as evidenced in the scatter in the comparisons. A total number of 131 cases were selected from these sources.

The results of the model calculations and the linear regression fit of observed winds to PR and DP are summarized in Table 1. In Table 1, the coefficients correspond to near-surface scalar winds in knots and cloud water in the column in centimeters. Despite the poor goodness of fit parameter for L_0 and l_2, use of the non-zero values in Table 1 resulted in better observed-estimated wind comparisons than their neglect.

TABLE 1. COEFFICIENTS FOR THE SCALAR WIND ESTIMATION ALGORITHM

Coefficient	Value	Goodness of Fit (R^2)	Source
W_0	0.242	0.71	Empirical
w_1	−1.15	N.A.	Model
w_2	−1.128e-03	0.71	Empirical
L_0	0.056	0.06	Empirical
l_1	1.5	N.A.	Model
l_2	−1.460e-04	0.06	Empirical
W_1	−806.4	------Matrix Inversion------	
W_2	−618.3	"	
L_1	−0.217	"	
L_2	0.499	"	

The regression fit for *PR* vs wind is also shown in Figure 1, along with model calculations of the effect of cloud liquid water. The curve labeled Cloud water = 0 corresponds to the regression fit. The one labeled $L = 0.02$ cm would correspond to very heavy nonprecipitating clouds in the polar regions, but this value could be exceeded during storms with rain. The observations fall into columns because of the binning of the surface observations into 5-knot intervals. The scatter of the points is attributed mostly to the temporal and spatial windows (4 hr and 50 km, respectively) allowed for the comparisons.

An illustration of how the algorithm operates is shown in Figure 2, where the abscissa and ordinate are *PR* and *DP,* respectively. Equations (1) and (2) and the coefficients in Table 1 were used to calculate "model" effects of wind and cloud water in Figure 2. For the most part, the observations fall inside the parallelopiped described by the algorithm. Those that fall below result from the offsets determined by the linear regression fit of the observations, and may be partly due to the neglect of water vapor fluctuations in the algorithm.

5. COMPARISON OF OBSERVATIONS AND SCALAR WIND ESTIMATES FROM THE ALGORITHM

In order to determine how well Eq. (3) performs as an interpolation algorithm, winds estimated by it were compared with the corresponding surface

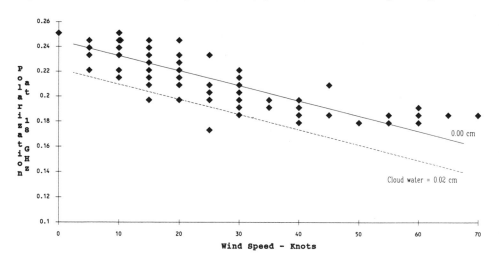

Figure 1. Observed and calculated polarization at 18 GHz vs wind speed.

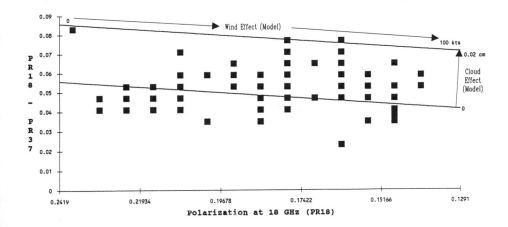

Figure 2. Difference of polarizations at 18 and 37 GHz vs polarization at 18 GHz (model and observed) for microwave emission from the sea surface.

observations again by linear regression. For this purpose, the SMMR wind estimates were also binned into 5-knot intervals, and the number of matches recorded. The results are illustrated in Figure 3, where the number of observations in each 5×5 predicted-observed wind bin is shown. A blank or "."

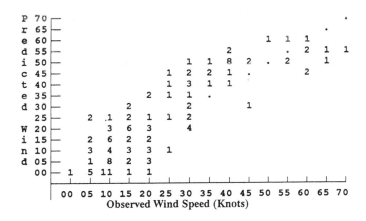

Figure 3. Predicted vs observed winds (5 knot bins).

corresponds to no observations. The slope of this regression fit is, of course unity, and is indicated at the high end by the dots. The goodness of fit parameter for this comparison (R^2) is 0.73, the sigma is 10, and the average error is 8.5 knots, not an unreasonable error in light of the 5-knot bins of the surface observations.

6. ILLUSTRATIONS OF SURFACE WIND AND CLOUD WATER ESTIMATES FROM SMMR

In Figures 4–7, we show grid-print maps of near-surface oceanic scalar winds and cloud liquid water content in the Norwegian, Greenland, and Barents Seas for the descending and ascending nodal passes of the SMMR on 27 February

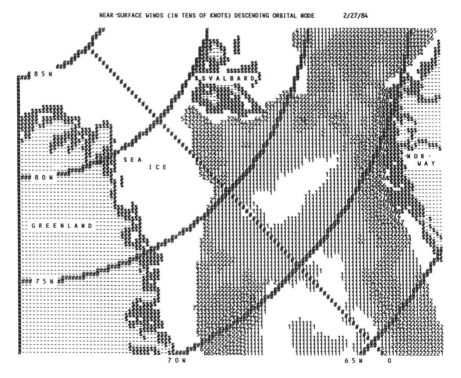

Figure 4. Near-surface wind estimates from the Nimbus-7 SMMR. The winds are indicated in 10-knot intervals, starting with "0"=0–10 knots. Point "0" indicates 90–100 knots. The time of passage over 0°E, 70°N was about 0100 GMT.

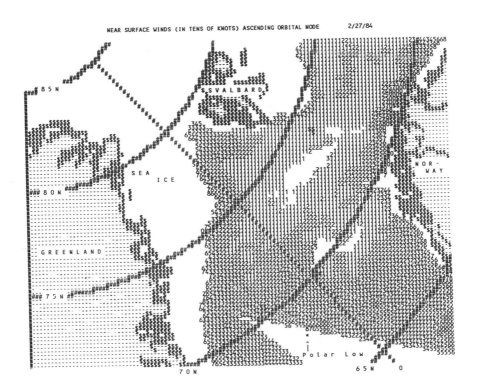

NEAR SURFACE WINDS (IN TENS OF KNOTS) ASCENDING ORBITAL NODE   2/27/84

Figure 5. Near-surface wind estimates from the Nimbus-7 SMMR. The winds are indicated in 10-knot intervals, starting with "1"=0–10 knots. Point "0" indicates 90–100 knots. The time of passage over 0°E, 70°N was about 1000 GMT.

1984, which occurred at approximately 0100 and 1000 GMT, respectively. The polar low observed at 69°N and 3°W by Shapiro et al. (1987) can be seen as well in Figures 5 and 7 at the same location. In Figure 5, the "*" points to the center of the storm, located near the edge of the SMMR orbital swath. (The wedge-shaped data gap is the space between adjacent swaths.) Winds ranging from 30–70 knots can be seen in this vicinity. An area of strong wind extends all the way from the sea ice edge near Greenland to the coast of Norway. This is a marked change from the situation 9 hr earlier (Figure 4) when the winds were generally weaker. The cloud patterns (Figure 7) are in the form of circular bands to the east of the storm center, with one band just off the coast of Norway

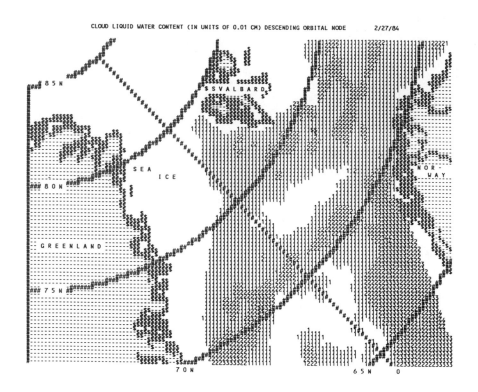

Figure 6. Estimates of liquid water content of clouds from the Nimbus-7 SMMR. The water is indicated in 0.01-cm intervals, starting with "1"=0–0.01 cm. Point "0" indicates 0.09=0.10 cm. The time of passage over 0°E, 70°N was about 0100 GMT.

and another about 300 km to the west. They are approximately centered on the polar low. Again, 9 hr earlier (Figure 6), the clouds appear generally heavier.

Shapiro's local scale surface analysis chart, corresponding to the time 1340 GMT, overlays the wind patterns of the 0800 GMT SMMR pass very well. However, other than the location of the center of the polar low, there is little correspondence between the smooth 1000-mb height contours on the larger scale European Center for Medium Range Weather Forecasts (ECMWF) surface analysis for 1200 GMT and the detailed patterns shown in the SMMR images of scalar winds and cloud water amounts.

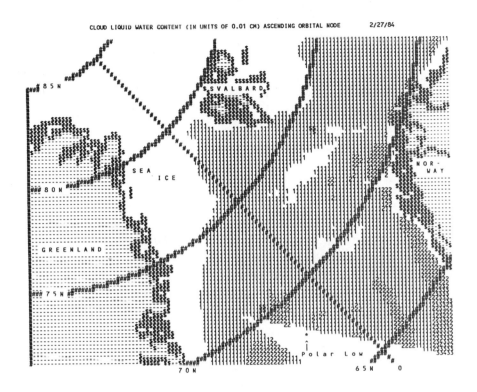

Figure 7. Estimates of liquid water content of clouds from the Nimbus-7 SMMR. The water is indicated in 0.01-cm intervals, starting with "1"=0–0.01 cm. Point "0" indicates 0.09=0.10 cm. The time of passage over 0°E, 70°N was about 1000 GMT.

The SMMR images in Figures 4–7 show a number of interesting features that demonstrate the utility of passive microwave observations of polar storms. As would be expected as a result of using them in the determination of the algorithm coefficients, the scalar winds in the vicinity of the polar low correspond to those observed with the dropsondes deployed by the NOAA P-3 aircraft (Shapiro et al., 1987). The SMMR data in this instance serve to provide a detailed interpolation between the sparse dropsonde data, and extrapolation away from the actual P-3 flight path. Thus, a detailed two-dimensional field can be developed from the combination of SMMR data and a one-dimensional set of surface observations.

In addition to the polar low, a severe storm can be seen near the northern coast of Norway in Figures 4 and 5. Its existence is confirmed by high-level winds also observed in that area by the NOAA P-3 on its transit flight from Bodo, Norway to the polar low event. Although the accuracy of SMMR wind estimates within 50–100 km of land is limited due to the footprint size of 50 km, the existence of such storms can be distinguished at this proximity. As can be seen in Figure 6, winds in the 80–90-knot range are indicated close to shore; the actual winds are probably lower due to land contamination in the SMMR footprint.

7. SUMMARY

Multispectral microwave radiances obtained from the SMMR on board the Nimbus-7 spacecraft have been used to estimate near-surface scalar oceanic winds and columnar cloud water amounts. A specific example calculated for 27 February 1984, coincident with a NOAA P-3 aircraft profiling of a polar low event demonstrates that the estimates from SMMR data are in broad agreement with the surface analysis of that storm made from the surface data, and that the SMMR estimates show considerably greater detail in the scalar wind distributions than can be inferred from the smooth 1000-mb contours shown on the surface analysis of the aircraft data. The detail is even greater on the larger scale surface analysis maps provided by ECMWF in general and for this particular event. It is clear that the more detailed wind information available from multispectral passive microwave sensors such as SMMR are essential for high-resolution air-sea inter-action models for calculating, for example, sea ice motion and formation, and deep water formation.

The coefficients in the semi-empirical estimation algorithm were determined by calculating cloud effects with a radiative transfer model and obtaining wind effects from linear regression analysis between SMMR data and a limited number of surface wind observations. Ratios of SMMR radiances were used as input to the estimation algorithm to eliminate ocean surface temperature effects to first order. Within the limited tuning data set, which reports the scalar winds in 5-knot increments, the self-consistent average error of estimation is 8.5 knots. A specific polar low event, which occurred on 27 February 1984 and for which there are detailed airborne meteorological observations, is used to illustrate the utility of multispectral passive microwave observations of polar storms from space. In particular, it appears possible to improve the detail in large-scale surface weather maps with the use of such data.

ACKNOWLEDGMENT

The authors wish to thank Dr. Paul F. Twitchell for useful discussions and encouragement.

REFERENCES

Gaut, N.E., and E.C. Reifenstein III, 1970: Interaction of microwave energy with the atmosphere. Paper No. 70-197 presented at *The AIAA Earth Resources Observations and Information Systems Conference,* Annapolis, Md.

Gloersen, P., and F.T. Barath, 1977: A scanning multichannel microwave radiometer for Nimbus-G and SEASAT-A. *IEEE J. Oceanic Eng., OE-2,* 172–178.

Gloersen, P., D.J. Cavalieri, A.T.C. Chang, T.T. Wilheit, W.J. Campbell, O.M. Johannessen, K.B. Katsoros, K.F. Kunzi, D.B. Ross, D. Staelin, E.P.L. Windsor, F.T. Barath, P. Gudmandsen, E. Lanham, and R.O. Ramseier, 1984: A summary of results from the first Nimbus-7 SMMR observations. *J. Geophys. Res., 89,* 5335–5344.

Hollinger, P., 1971: Passive microwave measurements of sea surface roughness. *IEEE Trans. Geosci. Electron., GE-9,* 165–169.

Johannessen, O.M., 1987: Introduction: Summer marginal ice zone experiments during 1983 and 1984 in Fram Strait and the Greenland Sea. *J. Geophys Res., 92,* 6716–6718.

Monohan, E.C., and I.G. O'Muircheartaigh, 1986: Whitecaps and the passive remote sensing of the ocean surface. *Int. J. Remote Sensing, 7,* 627–642.

Shapiro, M.A., L.S. Fedor, and T. Hampel, 1987: Research aircraft measurements of a polar low over the Norwegian Sea. *Tellus, 39A,* 272–306.

Webster, W.J., T.T. Wilheit, D.B. Ross, and P. Gloersen, 1976: Spectral characteristics of the microwave emission from a wind-driven, foam-covered sea. *J. Geophys. Res., 81,* 3095–3099.

Wilheit. T.T., J. Greaves, D. Han, B.M. Krupp, and A.S. Milman. 1984: Retrieval of ocean surface parameters from the scanning multichannel microwave radiometer on the Nimbus-7 satellite. *IEEE Trans. GRS, GE-22,* 133–143.

Wilhelmsen, K., 1981: *The Polar Low Near the Norwegian Coast.* NMI Tech. Rep. No. 55, Det Norske Meteorologiske Institutt, Oslo, Norway, 31 pp.

A STUDY OF AN ARCTIC DEPRESSION THROUGH RETRIEVAL OF MESOSCALE METEOROLOGICAL PARAMETERS FROM NOAA-9 FOR THREE CONSECUTIVE DAYS OF JUNE 1986

C. Claud, A. Chedin, and N.A. Scott
Laboratoire de Météorologie Dynamique, Ecole Polytechnique
Palaiseau Cedex, France.

J.C. Gascard
Laboratoire d'Océanographie Dynamique et de Climatologie
Paris, France

G.J. Prangsma
KNMI, de Bilt, Netherlands

ABSTRACT

The Improved Initialization Inversion (3I) method has been designed for retrieving meteorological parameters from observations of the satellites of the TIROS-N series. This procedure has been applied to high latitude situations and among them to a sequence of 3 consecutive days (6–8 June 1986) corresponding to the development of a depression in arctic waters. Although this system is not a polar low, the qualities displayed by the method for this kind of study (identification of cyclonic conditions, additional information in comparison with conventional analyses) indicate that it should be of great interest for the study of polar low cases.

1. INTRODUCTION

Observations from the satellites of the TIROS-N series in polar regions, where the in situ data are very scarce, are helpful for the study of mesoscale meteorological phenomena through the retrieval of atmospheric and surface parameters. The Improved Initialization Inversion (3I) method (Chedin and Scott,

POLAR AND ARCTIC LOWS
Paul F. Twitchell, Erik A. Rasmussen,
and Kenneth L. Davidson (Eds.)

373

1984, 1985), developed over the past few years at the Laboratoire de Météorologie Dynamique, has shown the ability to retrieve temperature profiles with good accuracy through inversion of the radiative transfer equation. The Marginal Ice Zone of the European ARCTIC (ARCTEMIZ) campaign, conducted since 1986 north and south of the Fram Strait, provided the opportunity to apply the retrieval algorithm when numerous high latitude in situ observations were available, and therefore allowing comparisons of in situ data with the retrieved products.

The TIROS-N Operational Vertical Sounder (TOVS) data have been analyzed using the 3I method for 15 consecutive days of June 1986. Satellite-retrieved atmospheric parameters are presented here for a sequence of 3 consecutive days (6–8 June at respectively, 1128, 1117, and 1107 UT), during which period a small-scale depression developed and decayed over Norwegian arctic waters. These parameters are then compared with conventional products in order to show the capability of this procedure to provide high quality results and additional information of particular interest in the case of mesoscale developments.

2. THE 3I RETRIEVAL METHOD

The 3I procedure is a physicostatistical method for retrieval of atmospheric parameters from satellite data. It is a physical type method in that it uses atmospheric transmittances obtained through computations of the full radiative transfer equation, transmittances that directly enter the inversion process itself. It is a statistical type method since it relies upon a priori knowledge of the observations (the brightness temperatures in each channel of the vertical sounder) as well as of atmospheric structure parameters. This a priori knowledge is contained in a large precomputed data set, the TOVS initial guess retrieval (TIGR) data set. Provided the observed brightness temperatures correspond to clear areas, or have been properly "cleared" (decontaminated from clouds) using the so-called psi method (Chedin and Scott, 1984), the 3I procedure follows two principal steps: search of the "best" initial guess solution using the TIGR data set, and retrieval of the "exact" solution by a maximum probability estimation procedure using the Jacobian associated with the initial guess, in the TIGR data set. This Jacobian links variations in the atmospheric profile to variations in the observed brightness temperatures.

At present, the spatial resolution of the 3I algorithm is a compromise between the spatial resolutions of the two major sounders of TOVS: High Resolution

Infrared Radiation Sounder (HIRS-2) and Microwave Sounding Unit (MSU). Retrievals are made for fields of view (or boxes) of 3 HIRS-2 spots along the suborbital track by 4 to 2 HIRS-2 spots along the scan line according to the viewing angle. Such boxes approximately represent a surface of 100×100 km^2.

Due to the particularities of polar regions, refinements have been incorporated into the method. One important point concerns the cloud detection algorithm and the problem of discriminating sea ice from extended cloudiness since they may have similar radiative properties. To overcome this difficulty, two different methods that discriminate sea ice from open water directly from the satellite observations, using data from the MSU instrument (four microwave channels, relatively insensitive to clouds), have been developed. We have shown that this discrimination significantly improves cloud detection (Claud et al., 1989).

The physical parameters retrieved by the 3I method over the area covered by the satellite passes are: air mass types (Moine et al., 1987), temperature profiles, geopotential thicknesses, thermal winds, temperature of the lower stratosphere, cloud heights, cloud amounts (equivalent of black clouds), total water vapor content, and relative humidities for three layers delimited by the levels 1000 mb, 800 mb, 500 mb, and 300 mb.

3. RESULTS

3.1 Synoptic Description of the Situations

As an example of the results obtained by the 3I method, a synoptic development is studied on a day-by-day basis (6, 7, and 8 June). From observation of surface analysis charts, the synoptic situation can be described as follows:

At 1200 UT 6 June (Figure 1a), a frontal system with a wave is developing near the Lofoten Islands, a system visible in the high resolution imagery (not shown). The cloud structure associated with the wave is double: a cirrus layer overhangs low clouds (stratocumulus). The precipitation zone is very small. The region around Spitsbergen is covered with fog.

At 1200 UT 7 June (Figure 1b), the wave is now completely developed and occluded (see also Figure 2, AVHRR channel 2 image). The system has moved

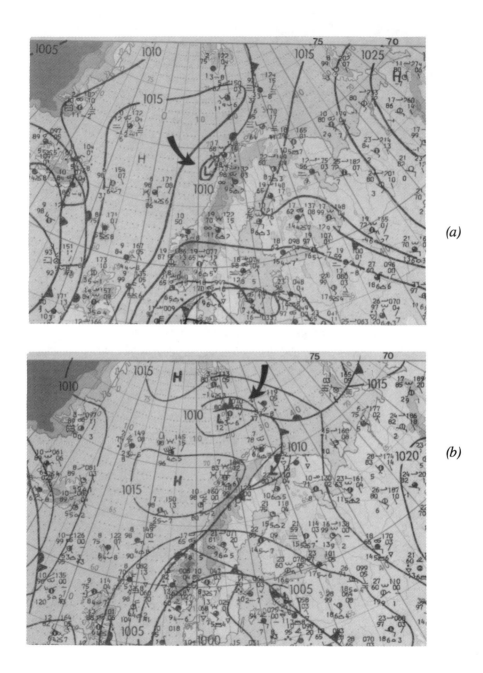

(a)

(b)

Figure 1. Surface charts provided at 1200 UT by the Deutscher Wetterdienst.
(a) 6 June 1986. (b) 7 June 1986.

Figure 2. AVHRR picture (channel 2), 1117 UT 7 June 1986.

over the Norwegian Sea and lies south of Spitzbergen. Clouds are low with tops below the 400-mb level; the precipitation zone is still very limited.

At 1200 UT 8 June, the center of the depression has moved eastward into the Barents Sea and the depression is decaying. The clouds are low with almost no precipitation.

For all 3 days the depression is depicted by the 1010-mb isobar.

3.2 Results of the Inversion

The 3I method results are classified as follows:

(1) Sea ice/cloud detection

An essential point of the 3I method is cloud detection, which requires knowledge of the sea ice edge. Figure 3 illustrates the results of cloud detection for 7 June. Note the good agreement with Figure 2. In Figure 3, the dashed line indicates the sea ice edge as obtained from our calculations. Similar results are obtained for 6 and 8 June.

(2) Identification of cyclonic conditions

Directly determined from the 3I procedure, the temperature of the lower stratosphere (Chedin et al., 1985) can be considered as an indication of cyclonic conditions that are characterized by a lower tropopause level and thus a higher tropopause temperature. Figures 4a, 4b, and

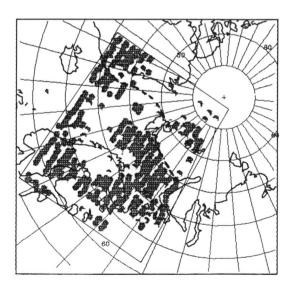

Figure 3. Cloud detection in 3I boxes for 1117 UT 7 June 1986. Gray areas represent cloudy fields of view; the dashed line indicates the sea ice edge.

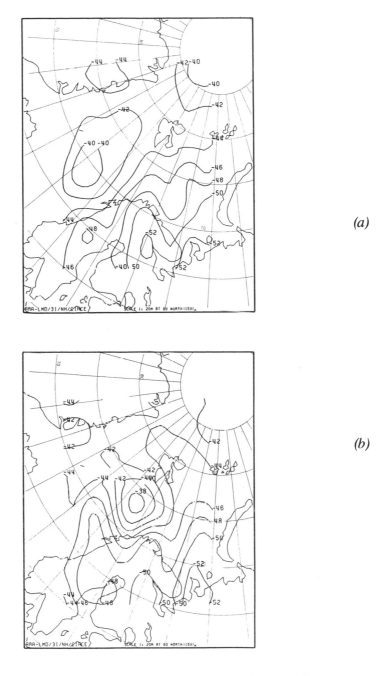

Figure 4. Values of the lower stratospheric temperature determined by the 3I method (in °C). (a) 1128 UT 6 June; (b) 1117 UT 7 June.

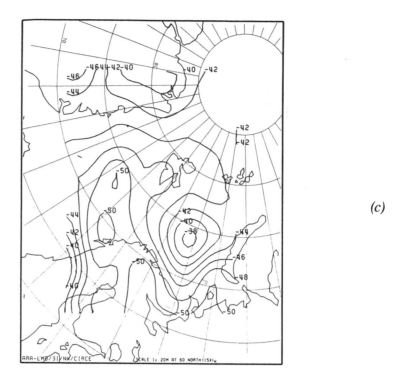

(c)

Figure 4. (Concluded) (c) 1107 UT 8 June.

4c, which display this product for the 3 days, show the displacement
of the system: cyclonic conditions are clearly identified in the Norwegian
Sea on 6 June, between Spitsbergen and Scandinavia on the next day,
and in the Barents Sea on 8 June. Moreover, other cyclonic centers,
which appear in these figures, are corroborated by the charts provided
by the European Meteorological Bulletin (Figure 1).

(3) Geopotential thicknesses: Comparisons with numerical analyses

From the 3I-retrieved geopotential thicknesses, maps are produced
for each day and for different layers. We have compared them to upper
air analyses provided by The Norwegian Meteorological Institute
(DNMI) using results of a limited area model, LAM150. This model
gives, among other things (with a spatial resolution of 150 km),

geopotential thicknesses at 1200 UT, which is in good coincidence with satellite-derived data.

For 6 June, Figures 5a and 5b display geopotential thicknesses between 1000 and 500 mb retrieved through 3I and obtained by the LAM150 respectively; the strong temperature gradients in Figure 5a coincide nicely with the frontal position. The cold tongue west of the wave is more clearly visible in satellite data than in synoptic analysis. Furthermore, the thermal vorticity (i.e., the vorticity of the thermal wind field) displays a maximum in the zone of wave development (not shown), in line with synoptic development theory (Prangsma et al., 1987) and thermal vorticity advection into the centre of wave activity. Comparison with the analysis (Figure 5b) shows fair agreement; the interesting point is, however, that 3I maps show more detail and mesoscale activity than the corresponding synoptic charts.

For 7 June, the different air mass movements can easily be seen in Figure 6a, which represents the retrieved geopotential thicknesses between 850 and 500 mb. Colder air coming from Greenland is tracking eastward over the Norwegian Sea, whereas warmer air is invading the Spitzbergen region, traveling westward toward north of Greenland. The gradient is not strong. The analysis map (Figure 6b) gives similar results.

Then, on 8 June, only large-scale dynamical structures with little mesoscale structure left in the area of interest can be seen (no important gradient in the depression zone) in Figure 7a, which displays 1000–500 mb retrieved geopotential thicknesses. This indicates that the depression is decaying; corroborated by the corresponding analysis (Figure 7b).

For each day, the 500- to 300-mb charts (not shown) display homogeneous air in the depression area, indicating that the vertical extent of the depression is limited to the atmosphere below 500 mb.

This comparison brings into evidence two conclusions:

(i) There is good agreement between these two types of products for large-scale structure.

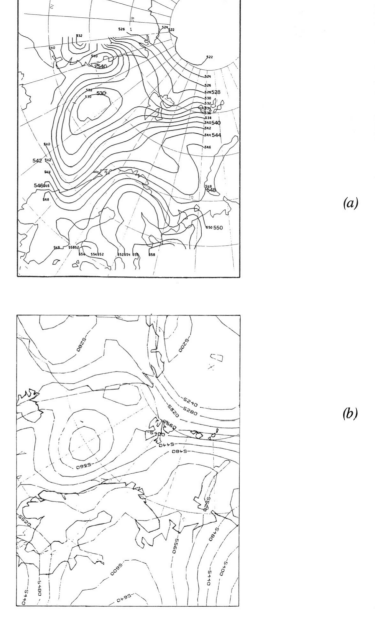

(a)

(b)

Figure 5. Geopotential thicknesses between 1000 and 500 mb for 1128 UT 6 June. (a) retrieved from the 3I algorithm, in decameters (contouring interval: 2 dam); (b) from LAM150, in meters (contouring interval: 40 m) (from The Norwegian Meteorological Institute).

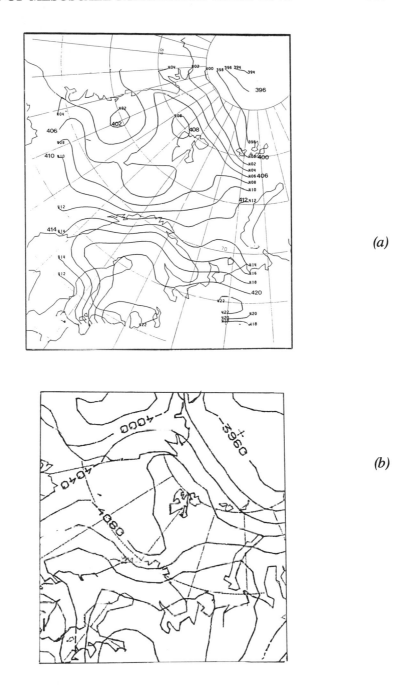

Figure 6. Same as Figure 5 but between 850 and 500 mb and for 1117 UT 7 June.

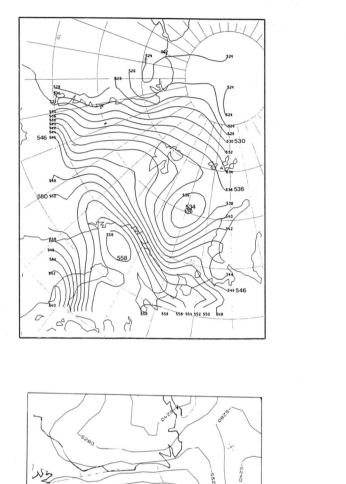

(a)

(b)

Figure 7. Same as Figure 5 but for 1107 UT 8 June.

(ii) Some additional information is provided, particularly concerning the mesoscale character of the wave on 6 June, which is more pronounced on the 3I products than inferred from synoptic analysis, thus in better coincidence with the later development.

(4) Comparisons with co-located radiosondes

The 3I retrieved temperature profiles have also been co-located with radiosondes to assess the quality of the retrievals. Radiosondes are at 1200 UT, and the maximum distance to 3I boxes is 100 km. The comparison includes 25 cases. Results in terms of mean and standard deviation values for the differences between 3I and radiosonde temperatures are presented in Figure 8. The results are very satisfactory (i.e., the accuracy is similar to the accuracy found for other latitudes or for situations that do not display such cyclonic activity). The standard deviation values are less than 2 K except for the 1000- to 850-mb layer; this is due to the remaining surface problems on the one hand and to the limited vertical resolution of the radiometers aboard the TIROS-N satellites on the other hand. Moreover similar accuracies are found for cloudy and clear situations, and areas characterized by large temperature

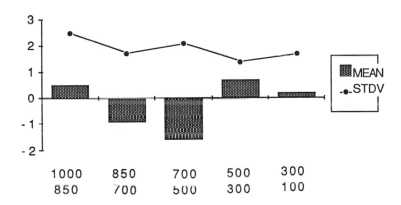

Figure 8. Results of the comparison between 3I profiles and co-located radiosondes for the 3 days (25 items in terms of mean and standard deviation values (in K); the layers considered are 1000–850 mb; 850–700 mb; 700–500 mb; 500–300 mb; 300–100 mb.

gradients do not display particular problems, as suggested in Figure 9, displaying comparisons between a radiosonde ascent and a nearby 3I retrieved temperature profile. The first case is Jan Mayen on 6 June (i.e., in an area with important thermal gradient), the second one Bear Island on 7 June (cloudy area): although the 3I profiles appear smoother, due to the vertical resolution of the instruments, the profiles are very close. In other words, the comparison confirms that the 3I method is able to produce atmospheric profiles of similar accuracies for clear and overcast conditions as well as for active meteorological systems (developing occlusions and depressions, frontal areas, etc.).

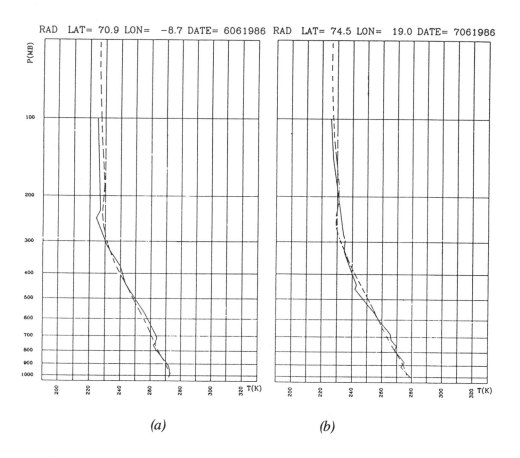

(a)   (b)

Figure 9. Comparison between radiosonde ascents and nearby 3I temperature profiles. Radiosondes are at 1200 UT, the distance between radiosounding and 3I box is less than 100 km. (a) Jan Mayen, on 6 June. (b) Bear Island, on 7 June.

4. CONCLUSION

This study case demonstrates the capability of the 3I method to delineate meteorological synoptic and subsynoptic features, clearly identified on high-resolution satellite images. Although the spatial scale of this depression is larger than those of a polar low, the quality of the results indicates that this procedure should be powerful for the analysis of polar lows in the future.

ACKNOWLEDGMENTS

Help and support from the Tromso Telemetry Station (Norway) and the CMS–Lannion are gratefully acknowledged. The processing of the data has been done on the computers of CIRCE (Centre Inter Regional de Calcul Electronique). Thanks are due to DNMI for providing us with radiosoundings and results of the LAM150.

REFERENCES

Chedin, A., and N.A. Scott, 1984: Improved initialization inversion procedure. *Proceedings of the First International TOVS Study Conference,* Igls, Austria, August 1983, 14 79.

Chedin, A., and N.A. Scott, 1985: Initialization of the radiative transfer equation inversion problem from a pattern recognition type approach: Applications to the satellites of the TIROS-N series. In A. Deepak, H.E. Fleming, and M.T. Chahine (Eds.), *Advances in Remote Sensing Retrieval Methods,* A. Deepak Publishing, Hampton, Virginia, 495–515.

Chedin A., N.A. Scott, C. Wahiche, and P. Moulinier, 1985: The improved initialization inversion method: A high resolution physical method for temperature retrievals from the TIROS-N series. *J. Clim. Appl. Meteor.,* 24, 124–143.

Claud, C., A. Chedin, N.A. Scott, and J.C. Gascard, 1989: Retrieval of mesoscale meteorological parameters for polar latitudes (MIZEX and ARCTEMIZ campaigns). *Annales Geophysicae, 7,* 3 (to be published).

Moine, P., A. Chedin, and N.A. Scott, 1987: Automatic classification of air mass type from satellite vertical sounding data: Application to NOAA-7 observations. *Ocean-Air Interaction, 1,* 95–108.

Prangsma, G.J., N.A. Scott, A. Chedin, N. Husson, J. Quere, and G. Rochard, 1987: Observation of the development of mesoscale systems by operational meteorological satellites. *Proceedings of the IUGG Conference,* Vancouver, Canada.

STUDIES OF ANTARCTIC MESOSCALE SYSTEMS USING SATELLITE IMAGERY AND SOUNDER DATA

D.E. Warren and J. Turner

British Antarctic Survey, Natural Environment Research Council
Cambridge, United Kingdom

ABSTRACT

Satellite images and temperature sounding data are of immense value in unraveling the structure of mesoscale meteorological systems at high latitudes where conventional surfaces and upper air observations are made only on a very widely spaced network. Two techniques and examples of their application are presented.

The use of high-resolution temperature soundings derived from the TIROS Operational Vertical Sounder is illustrated in the study of the thermal structure of a small cyclone in the southern Weddell Sea.

The large overlap of images from successive passes of the Advanced Very High Resolution Radiometer at latitudes poleward of about 60° can be exploited to estimate wind vectors by comparing the positions of cloud features in a sequence of three images. The technique for deriving the wind vectors, the quality control methods, likely errors and limitations, and the results from a single case study are compared with conventional observations.

1. INTRODUCTION

In recent years the use of global numerical weather prediction models in operational and research applications has increased dramatically and interest has been focused on the ability of such models to represent the range of weather systems found in the polar regions. One difficulty, which has been encountered, is that few detailed, observational studies have been carried out on high-latitude

POLAR AND ARCTIC LOWS
Paul F. Twitchell, Erik A. Rasmussen,
and Kenneth L. Davidson (Eds.)

synoptic and mesoscale weather systems, and little is known of the mechanisms that are important in their formation and development. The available surface and upper air observations from polar regions are very sparse and unable to provide detailed information on the structure of such systems, and, although data from meteorological polar orbiting satellites have been available for 20 years, only now this area of enquiry is beginning to be exploited to the fullest.

In what follows we describe investigations that are being carried out at the British Antarctic Survey (BAS) into the structure of high-latitude mesoscale atmospheric systems in which extensive use is being made of satellite data to supplement conventional surface observations. Data from the TIROS Operational Vertical Sounder (TOVS) provide measurements of the temperature and humidity fields and sequences of images from the Advanced Very High Resolution Radiometer (AVHRR) are used to estimate wind speeds and directions by comparing the positions of cloud features over periods of several hours. Brief descriptions of both the TOVS processing scheme and the wind derivation algorithm are followed by case studies that illustrate both the strengths and limitations of these two techniques.

2. THE TOVS AND ITS APPLICATION TO POLAR RESEARCH

The TOVS package includes the High Resolution Infrared Sounder and the Microwave Sounding Unit. These instruments can resolve the structure of the atmosphere at a horizontal scale of a few tens of kilometres. This is of tremendous value in areas where few radiosonde ascents are available. However, the derived temperature profiles are relatively smooth in the vertical and give poor resolution of the tropopause and boundary layer.

Temperature profiles derived from TOVS data are distributed over the Global Telecommunications System and are used extensively as input for operational numerical weather prediction but more rarely in research studies because of their coarse 250-km, horizontal resolution. The processing scheme in use at BAS converts the raw radiance measurements into atmospheric profiles over a swath approximately 2,000 km wide with a horizontal resolution of 80 km at nadir and is based on the local area sounding system, which is run routinely at the UK Meteorological Office (Eyre and Jerrett, 1982). It relies on a multiple linear regression algorithm to calculate atmospheric profiles from cloud-cleared, limb-corrected brightness temperatures. Regression coefficients for the Weddell Sea area have been derived using Halley radiosonde data and the cloud clearing is carried out with the "sequential estimation" algorithm in use at the UK Meteorological Office (Eyre and Watts, 1987).

Experience gained in the operational use of soundings of this type has shown that integrated quantities, such as the thicknesses of atmospheric layers, are more useful than level values, the 500-mb temperature for example, and are capable of providing reliable information on strong baroclinic layers, such as those found associated with midlatitude cyclones.

Recent studies of North Atlantic polar lows (Rasmussen, 1985; Shapiro et al., 1987) confirm that a knowledge of the three-dimensional temperature structure is vital if the mechanisms behind their formation are to be understood. With the limited amounts of radiosonde data available it is very difficult to track small upper level cold pools and shallow baroclinic layers that are likely to be important in triggering their development. Satellite sounding data allow these mesoscale thermal features to be tracked over long periods since the Arctic and Antarctic regions (Figure 1) are frequently covered by passes of the polar-orbiting satellites.

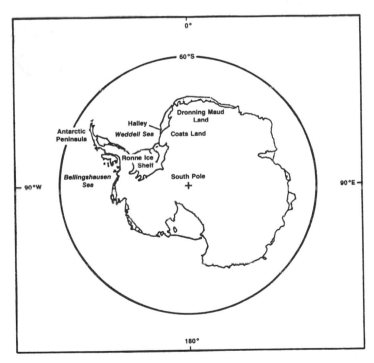

Figure 1. The Antarctic continent showing the places mentioned in the text.

3. THE TEMPERATURE STRUCTURE OF A MESOSCALE VORTEX OVER THE WEDDELL

Figure 2 shows the channel 4 (11-μm) AVHRR image of the area around Halley at 1650 GMT 5 January 1986 when a subsynoptic-scale vortex of approximately 700-km diameter was located just off the coast with associated cloud covering much of the southern Weddell Sea. The Halley surface meteorological records show that there was intermittent snow for much of the day and easterly winds of between 25 and 30 knots. The calibrated AVHRR imagery indicated cloud top temperatures of around $-15\,°C$ in the low cloud surrounding the system and temperatues as low as $-35\,°C$ in the band of higher cloud running along the coast. Comparison of these temperatures with the 1200 GMT Halley radiosonde ascent implies that the cloud tops were at approximately 800 and 500 mb, respectively.

Figure 2. NOAA-7 AVHRR channel 4 image, 1650 GMT 5 January 1986. The location of Halley is marked and the estimated position of the centre of the developing vortex is indicated with a white star.

From an inspection of the sequence of AVHRR passes over the area the formation of the vortex could be traced back to 3 January when a small circulation was visible between Halley and the Stancomb-Wills Glacier Tongue. Figure 3 shows the visible image taken at 1710 GMT when the system was first apparent. This would have been a preferred area for development because of the low-level convergence caused by the flow off the ice shelves coupled with the destabilization of the atmosphere over the relatively warm ocean surface. The 1000- to 500-mb thickness fields over this period, which were constructed from the TOVS soundings, show a pool of cold air over Dronning Maud Land, which deepened from around 512 deka geopotential metres (dagpm) to less than 500 dagpm on 3 January. At the same time a warm ridge of over 528 dagpm extended northeastward across the Ronne Ice Shelf, creating a strong barocolinic zone across Halley Bay on 3 January. An analysis of the TOVS 1000- to 500-mb thickness field at 1710 GMT is shown in Figure 4. It was within this deep baroclinic zone, which, according to the subsequent analyses of TOVS data, persisted throughout 4 January, that the vortex developed. During 5 January the

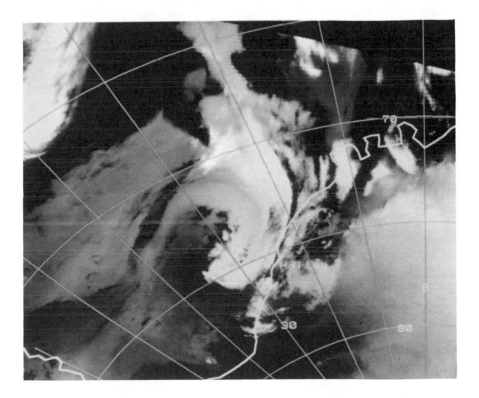

Figure 3. NOAA-7 AVHRR channel 2 image, 1710 GMT 3 January 1986.

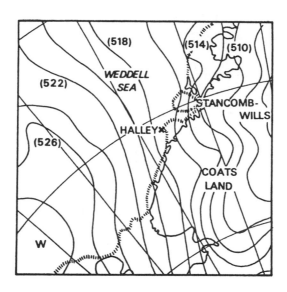

Figure 4. TOVS 1000- to 500-mb thickness analysis, 1710 GMT 3 January 1986. The contours are plotted at intervals of 2 dagpm; the height of every other contour is printed in parentheses.

system started to draw significant amounts of cold air from the continent and a low-level cold trough extended over Halley and into the southern Weddell Sea. This trough can be seen in the 1000- to 500-mb thickness analysis for 1650 GMT shown in Figure 5.

The southward extension of cold air weakened the temperature gradient over the region, and when a synoptic-scale disturbance started to travel eastward across the Weddell Sea on 6 January the vortex moved over the continent close to Halley. Its movement inland was apparent in the three hourly Halley surface observations as a pressure minimum at around midday on 6 January, with continuous moderate snow and an increase in wind speed to almost 30 knots. Once away from the relatively warm ocean surface the system rapidly declined and late on that day the imagery shows only an area of unorganised cloud inland of Halley.

In summary then, the available AVHRR and TOVS data show that this system was a relatively shallow disturbance that formed within an intense baroclinic zone extending from the eastern Ronne Ice Shelf into the southern Weddell Sea. The TOVS analyses highlight the absence of any source of warm air over the Weddell Sea and clearly identify the inflow of very cold, continental air, which, coupled with the relatively high sea surface temperatures, decreased the stability of the atmosphere and enabled the system to persist.

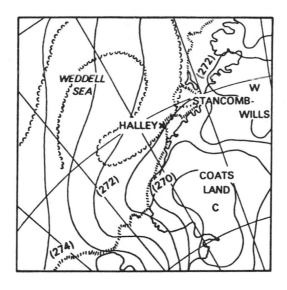

Figure 5. TOVS 1000- to 700-mb thickness analysis, 1650 GMT 5 January 1986. The contours are plotted at intervals of 1 dagpm; the height of every other contour is printed in parentheses.

4. ESTIMATING WIND VELOCITIES FROM AVHRR IMAGES

We now turn our attention to the problem of measuring wind velocities from satellite imagery. Leese et al. (1971) describe a well-established technique, variants of which are used routinely by the operators of the geostationary meteorological satellites. Wind velocities are derived by noting the positions of individual cloud features, which can be identified on each of three successive images, and calculating their respective latitudes, longitudes, and their displacements between images. Great care has to be taken in selecting features that are likely to behave as passive tracers and move with the wind rather than developing features whose displacements are the result of both advection and development.

The viewing geometry of geostationary satellites limits this technique to latitudes equatorward of 50°. However the orbital height of the TIROS-N/NOAA polar-orbiting satellites and the width of the swath of the AVHRR mean that at high latitudes there is considerable overlap between images from successive passes. In fact, points poleward of 70° latitude are bound to fall within the instrument's field of view on three consecutive passes at least twice a day. Data from these overlapping images can be resampled onto a common map projection and then used to derive winds in the same way as is done with the geostationary imagery. This technique has already been successfully applied to estimating the motion of pack ice (Ninnis et al., 1986); however, the orbital period of the TIROS-N satellites is approximately 100 min and useful wind velocities can only be derived if suitably long-lived cloud features exist in sufficient numbers.

To test how useful this technique might be we have obtained from NOAA/NESDIS, several sequences, each of three consecutive high latitude NOAA-7 AVHRR images. The data have a nominal resolution of 4 km at nadir and come with Earth location information appended. Each image of the sequence is resampled onto a polar stereographic map projection, using the NOAA Earth location data and a nearest neighbour interpolation scheme. We have only used data from the thermal infrared channel (channel 4), which apart from remaining insensitive to changes of solar elevation, contains information from which estimates of cloud height can be made. Only a very coarse height assignment is used. Each derived wind is classified as either high (above 450 mb), medium (450 to 700 mb) or low (below 700 mb) by comparing the brightness temperature of the tracer with a radiosonde temperature profile from a nearby upper air station.

The process of estimating wind velocities involves two distinct steps—identification and tracking of likely tracers, followed by a thorough quality control process. Three strategies have been tried to identify and track tracers. The simplest is a purely manual technique. The reprojected image sequence is displayed on an interactive image processing system, and a latitude-longitude grid and coastline are overlaid. After the operator has defined a likely track using a trackerball controlled cursor, the host computer calculates the latitude and longitude of the tracer in each of the three images, its displacement from one image to the next, and the magnitude and direction of the implied velocity vectors. The results are then written to a file for future reference.

Manual identification of many tracers over an image is a time consuming affair. Careful examination of a sequence of three 512×512 element images can take as long as 30 min. In order to reduce the time taken a fully automatic scheme has also been implemented. This works by segmenting the first image of the sequence into non-overlapping 7×7 pixel templates. To avoid uniform, featureless areas of the image the variance of the pixel values over the template is tested and the subsequent search is only undertaken if the standard deviation is greater than 0.3 K. Corresponding areas on the second and third images are searched for within a 100×100 pixel area centred on the undisplaced template. A spiral search pattern is followed and a match is assumed at the first local maximum in the cross-correlation function that has a value greater than 0.7. This step takes considerable time, but use of the difference, rather than the cross correlation, between the two images and a more sophisticated search algorithm should reduce this in the future.

The automatic procedure generates a large number of tracks, many of them spurious. As a result a comprehensive set of quality control tests have to be applied. First, each member of every pair of estimated wind velocities must agree to within 50% in speed and 30° in direction. Second, the brightness temperatures of the tracer in each image must not vary by more than 5 K. Last, primarily to avoid the difficulties associated with developing features, a manual quality control step allows an operator to delete those suspect winds that remain.

A third option, which combines the desirable features of both the manual and fully automatic procedures, has also been developed. In this case the operator uses the cursor to define a rectangular template in the first image and a search area in the second and third. The exact match is then found by finding the displacements that minimize the pixel-by-pixel difference between the template in each pair of images. This relieves the operator of considerable detailed examination of the screen but avoids the computational overheads involved in the automatic process.

5. THE WIND FIELD OVER THE BELLINGSHAUSEN SEA

Figure 6 shows the reprojected data from a NOAA-7 pass over the Bellingshausen Sea on 1 July 1984 at 0437 GMT. This image was the first in the sequence of three, the second, taken at 0627 GMT, is shown in Figure 7.

The analysis issued by the UK Meteorological Office for 0000 GMT shows that there was a region of high pressure centred over the Antarctic Peninsula, and the surface winds over the area were mostly light southwesterlies with speeds between 10 and 20 knots. The derived wind velocities, both manual and automatic, from the central part of the image are shown, in conventional form, in Figure 8. These were derived by tracking what appears to be high-level cirrus, giving wind speeds in the range of 10 to 15 knots from the northwest.

Several sources of possible error exist and great care has to be taken to reduce their effects if the derived winds are to be of any value. The most difficult errors

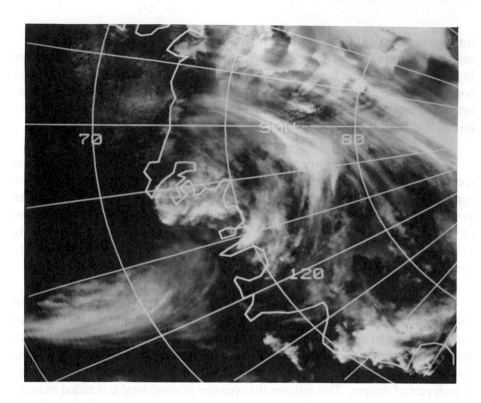

Figure 6. NOAA-7 AVHRR channel 4 image, 0437 GMT 1 July 1984. The data were resampled onto a polar stereographic projection and a latitude-longitude grid and coastline overlaid. The meridians are drawn every 10° and the parallels every 5°. The centre of the image is at approximately 100°W 74°S.

to eliminate are those that arise as a result of tracking a feature whose displacement arises from local development as much as from advection by the wind. Certain formations are more likely to fall into this category than others, and with experience we hope to be able to identify and avoid them. One obvious example is provided by cloud band edges at fronts—prominent features that may be moving with a very different velocity from the local wind.

The automatic scheme is prone to mismatch tracers, particularly if relatively featureless templates are not excluded, but consistency checks on pairs of winds were introduced to eliminate such mismatches. The results obtained so far from the automatic and manual schemes agree well with one another, although wind speeds derived by the former are somewhat smaller, and this gives us confidence that the quality control is performing well.

Figure 7. NOAA-7 AVHRR channel 4 image, 0627 GMT 1 July 1984. This image, the second in the sequence of three used to estimate the wind vectors, is an enlargement covering the central area of Figure 6. The wind velocities are plotted in the usual meteorological convention.

Finally there are constraints on the accuracy of the results that arise from the limitations of the initial data:

—errors in the navigation of the images
—effects of the finite spatial resolution of the imagery
—parallax errors
—errors in height assignment.

The total effect of these has yet to be evaluated. Opportunities for comparison of the cloud track winds with conventional data are rather few in the Antarctic because of the very limited number of upper air stations in the region. There are likely to be more in the Northern Hemisphere, where the observing network is less sparse. An interesting possibility is to use cloud track winds derived from Meteosat imagery. Zick (1986) has successfully derived wind velocities from Meteosat imagery at latitudes as high as 70°N. Obviously there are great differences in the spatial resolution of the two observing systems at these latitudes, but the similar global coverage is a great advantage compared to ground-based observations.

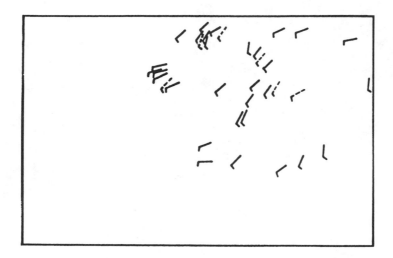

Figure 8. Cloud track winds taken from Figure 7 and redrawn for clarity. Those derived by the automatic process are shown by solid arrows, those from the manual scheme have been added in and drawn as dashed arrows.

REFERENCES

Eyre, J.R., and D. Jerrett, 1982: Local-area atmospheric sounding from satellites. *Weather, 37,* 314–322.

Eyre, J.R., and P.D. Watts, 1987: A sequential estimation approach to cloud-clearing for satellite temperature soundings. *Quart. J. Roy. Meteor. Soc., 113,* 1369–1376.

Leese, J.A., C.S. Novak, and B. Clarke, 1971: An automated technique for obtaining cloud motion from geosynchronous satellite data using cross correlation. *J. Appl. Meteor., 10,* 110–132.

Ninnis, R.M., W.J. Emery, and M.J. Collins, 1986: Automated extraction of pack ice motion from AVHRR imagery. *J. Geophys. Res., 95,* 10725–10734.

Rasmussen, E., 1985: A case study of a polar low development over the Barents Sea. *Tellus, 37A,* 407–418.

Shapiro, M.A., L.S. Fedor, and T. Hampel, 1987: Research aircraft measurements of a polar low over the Norwegian Sea. *Tellus, 39A,* 272–306.

Zick, C., 1986: The use of Meteosat image sequences for perception and analysis of weather in the area of the Norwegian Sea. *Proceedings, Sixth Meteosat Scientific Users Meeting,* EUMETSAT, D-6100 Darmstadt, FRG.

SATELLITE CLIMATOLOGY OF "POLAR AIR" VORTICES FOR THE SOUTHERN HEMISPHERE WINTER[1]

Andrew M. Carleton and Duane A. Carpenter
Indiana University
Bloomington, Indiana, U.S.A.

ABSTRACT

A climatology of polar air cloud systems ("polar lows") is developed for the Southern Hemisphere winter (June/July/August/September—JJAS) from Defense Meteorological Satellite Program (DMSP) imagery for the 7 years 1977–1983. Both comma cloud and spiraliform polar lows are identified. The former, which is the more numerous, occurs over a wide range of ocean latitudes, while the latter predominates over the seasonal sea ice zone. For the hemisphere poleward of 30°S, the ratio of spirals to commas is about 1:10; however, this increases to about 1:3 over sub-Antarctic latitudes in midwinter. Within-season and interannual variations in the preferred longitudes of polar low occurrences reflect changes in the hemispheric long waves and, accordingly, the incidence of cold air outbreaks to lower latitudes. Between June and September, on average, the maximum frequencies of polar lows shift westward into the southeast Indian Ocean, in connection with the semiannual oscillation. Interannually, a signal of the El Nino Southern Oscillation (ENSO) is apparent, and involves maximum frequencies of comma clouds in the Indian Ocean (southwest Pacific) for the winter of year 1 (year 0) of a "warm" event, represented by 1981 (1982).

Seasonal and interannual relationships of polar lows with the Antarctic sea ice extent are identified.

[1]This research was supported by NSF grant SES–8603470.

POLAR AND ARCTIC LOWS
Paul F. Twitchell, Erik A. Rasmussen, and Kenneth L. Davidson (Eds.)

401

1. INTRODUCTION

Previous satellite-based studies of synoptic-scale cyclones for the Southern Hemisphere suggest that the greatest incidence of "polar air" (primarily comma cloud) systems occurs in the cold season, as is the tendency in the Northern Hemisphere. There, polar lows are the subject of intensive diagnostic research. However, in the Southern Hemisphere, an even more basic need—that of their synoptic climatology—remains to be filled since the few available studies of polar lows are restricted to either case study or composite analyses for limited longitude sectors (e.g., Lyons, 1983; Auer, 1986). Further, an assessment of the incidence of the spiraliform ("arctic hurricane") type of polar low in relation to the more commonly observed comma cloud system has not hitherto been made for the Southern Hemisphere. Studies for the Northern Hemisphere (e.g., Forbes and Lottes, 1985; Businger, 1985; Carleton, 1985) suggest that polar lows indicate anomalous atmospheric circulation conditions and that they exhibit some relationship with regional sea ice extent. Accordingly, the present study examines these relationships for the Southern Hemisphere on seasonal and interannual time scales for the winters of 1977–1983. This period was characterized by large interannual variations of atmospheric circulation and climate in the Southern Hemisphere.

2. DEFENSE METEOROLOGICAL SATELLITE PROGRAM IMAGERY AND ANALYSIS

The wintertime climatology of polar lows utilizes twice-daily medium resolution (5.4 km) thermal infrared (IR) (8–13 μm) mosaics acquired by the polar-orbiting DMSP system. The large size of the DMSP mosaics and the long data record, coupled with their relatively high spatial resolution and high latitude coverage, optimize the detection of mesoscale cyclogenesis events over the Southern Hemisphere for the winter season.

The analysis of polar air cloud vortices appearing on the DMSP imagery makes use of subjective pattern recognition similar to that employed elsewhere in both case study and climatological analyses (e.g., Reed, 1979; Rasmussen, 1981; Forbes and Lottes, 1985; Carleton, 1985, 1987). Three main cyclogenetic signature types are identified on the imagery: the polar low, the frontal wave, and the cyclogenesis resulting from the merger of these two incipient formations (the "instant occlusion"). For the present study, polar air systems are categorized either as comma clouds developing in enhanced cumulus fields to the rear of frontal cloud bands (Figure 1a), as smaller spiraliform ("Antarctic") polar lows (Figure 1b), or as systems remaining from decayed frontal cyclones. The last is described in some detail for the North Atlantic by Zick (1983). Further DMSP examples of these systems and full details of the classification system used appear in Carleton and Carpenter (1989).

3. CLIMATOLOGY

3.1 Polar Air Cyclogenesis

Figure 2a and Figure 2b show genesis (first-observed) locations of polar air vortices (comma, spiral) for the seven June and August months of 1977–1983. The comma systems are by far the more numerous for the hemisphere, and occur over a wide range of ocean latitudes. For the hemisphere poleward of 30 °S and for polar air vortices in all stages of development, the spiraliform systems comprise only about 10 % of the total population of polar lows for much of the winter, and considerably less in September (not shown). However, in the June through August period, and over latitudes close to and south of the sea ice margin (about 60 °–70 °S), the spirals comprise about 20 % of the total for that zone and, in July, exceed 40 %. Thus, the situation with regard to the relative frequencies of spiral to comma cloud systems over the southern oceans in winter seems to resemble the North Atlantic Ocean (Forbes and Lottes, 1985; Carleton, 1985).

Figure 3 shows composite latitude variations of polar lows by winter month (all longitudes). In all four months, the frequencies of polar air cyclogenesis increase dramatically equatorward of the Antarctic continent, but the biggest intraseasonal changes occur over ocean latitudes between about 45 °–60 °S. There is the suggestion (Figure 3) that polar low frequencies expand equatorward between June and September, possibly in association with the seasonal movement of the sea ice and sea surface temperatures (Budd, 1982). However, this pattern is not a simple one, and the abrupt increase in comma clouds in the 35 °–40 °S zone between August and September seems unlikely to be due solely to these surface influences. The effect is probably also partly the result of substantial interannual and longitudinal variations in polar low occurrences occurring over the seven-winter period.

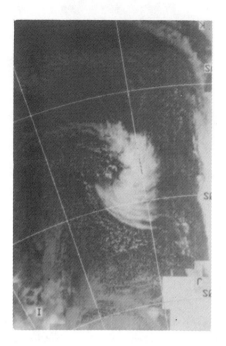

(a)

(b)

Figure 1.   DMSP IR examples for the Southern Hemisphere of (a) comma cloud polar low that formed in a field of enhanced cumulus between two large frontal wave cyclones, and (b) spiraliform polar low located over the pack ice zone (arrowed, image center). The vortex is comprised dominantly of stratiform clouds (from Carleton and Carpenter, 1989).

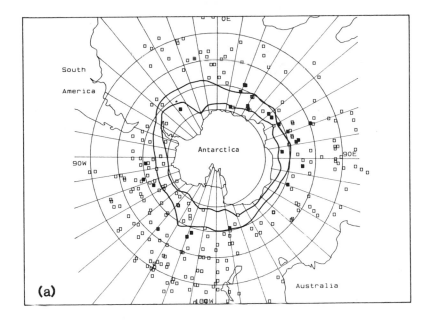

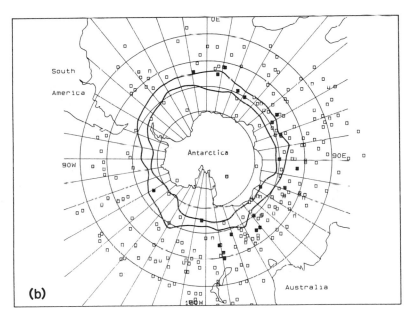

Figure 2. Polar low locations for 1977–1983 in (a) June, and (b) August. Commas (spirals) are shown by open (filled) squares. Maximum and minimum sea ice limits in each month are also shown.

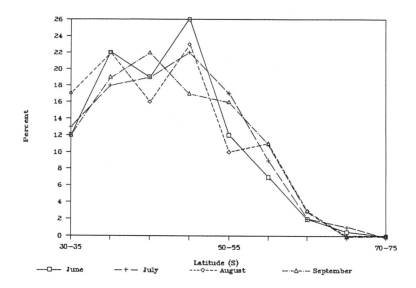

*Figure 3. Monthly polar low frequency variations by latitude. Values in each
5° zone are expressed as the percentage of the total number south of 30°S
(from Carleton and Carpenter, 1989).*

3.2 Longitudinal Variations

Examination of Figure 2 suggests a marked westward shift in the maximum
frequency of polar lows from the South Pacific (90°–180°W) to the southeast
Indian Ocean (180°–90°E) sectors between early and late winter. This impres-
sion is confirmed in Figure 4, which shows the longitude variations of polar
air cyclogenesis by winter month. Maximum frequencies of polar lows occur
in the western Pacific and Indian Ocean, and these are also the longitudes of
largest interannual variability for these systems (below). Frequencies considerably
below the hemispheric mean characterize the South American and western South
Atlantic sectors, and imply a reduced incidence of deep cold air outbreaks in
those longitudes. The sectors of peak polar low occurrence, such as between
60°–100°E and in longitudes of New Zealand (about 170°–180°E), coincide
with or are displaced slightly to the west of, the longitudes of peak occurrence
of frontal lows in most months (not shown). This is to be expected when considering that the majority of polar lows are comma clouds that develop in cold
air outbreaks behind major frontal systems (e.g., Reed, 1979). The marked
increases (decreases) noted for polar lows in the Indian Ocean (western Pacific)
between early and late winter are less evident for frontal cyclogenesis.

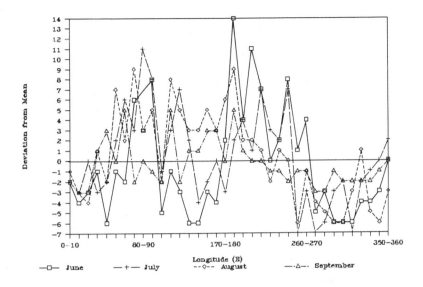

Figure 4. Monthly polar low variations by longitude. Values in each 10° zone are shown as the departure from the hemispheric mean in that month.

3.3 Relationships With Hemispheric Long Waves

A major feature of the spatial and longitudinal patterns of polar lows (Figures 2 and 4) is the statistically significant westward shift in activity into the southeast Indian Ocean between early and late winter. Since the extent of the intraseasonal change of the Antarctic sea ice between June and September is generally no more marked in the south Indian Ocean than it is, on average, in the southwest Pacific (Ropelewski, 1983), the source for such a large-scale shift in polar low activity must be sought in the atmospheric long waves and associated changes in the longitudes of cold air outbreaks.

Table 1 shows the average monthly values of an index of sea level pressure (SLP) wavenumber one in the Southern Hemisphere; the Trans-Polar Index (TPI), or SLP anomaly difference Hobart minus Stanley (Pittock, 1984) for the winters 1977–1981. SLP data were not available for Stanley for the last 2 years of the analysis (1982 and 1983). There is a progressive change from positive to quite strongly negative TPI between June and September (Table 1).

TABLE 1. WITHIN SEASON CIRCULATION CHANGES

	Trans-Polar Index:ZHobart minus ZStanley SLP (1977–1981)				MI Index:Hobart minus Chatham Is. (1977–1983)			
	June	July	August	Sept	June	July	August	Sept
$\bar{x}(mb)$	+3.00	−2.23	−0.65	−3.38	+4.1	+4.0	+1.4	+1.7
$\hat{s}$	5.17	2.36	0.60	3.78	5.3	4.0	3.8	6.5
	Sept–June: −6.38 mb				Sept–June: −2.43 mb			

The mean change between June and September is −6.4 mb, however, the interannual variability (standard deviations) is large. These results confirm the amplification (weakening) of the trough between early and late winter months in the Australia (South America) sectors. This change results in a reduction of cold air advection and observed frequencies of polar lows over the South Pacific (90°–180°W) after early winter as winds take on more northerly components. The opposite occurs for the southeast Indian Ocean. Regional-scale pressure indices for the Australia/New Zealand region [e.g., Trenberth's (1976) meridional index, MI—Table 1] confirm these changes in the waves and their influence on polar low genesis regions. Further details of the indices used are given in Carleton (1989).

4. INTERANNUAL VARIABILITY

4.1 Polar Low Variations, 1977–1983, and FGGE 1979)

Table 2 lists winter frequencies of polar lows for the Southern Hemisphere by year. The greatest number of polar lows occurred in the 1979 winter, coincident with the intensive observing period of FGGE (First GARP Global Experiment). The 1979 FGGE year was highly anomalous over middle and higher southern latitudes in terms of atmospheric circulation (strong westerlies, deep synoptic systems) and Antarctic sea ice extent, although the latter was most apparent for the preceding fall. Comparison of the spatial distribution of polar lows for the 1979 winter (not shown) with the departures from the long-term means of SLP for June, July, and August 1979 presented in van Loon and Rogers (1981—their Figure 5), shows that most of the polar lows occur in association

TABLE 2. HEMISPHERIC TOTALS OF POLAR LOWS,
1977–1983

Year	Comma Type	Spiral Type	Totals
1977	82	4	86
1978	119	27	146
1979 (FGGE)	186	17	203
1980	78	11	89
1981 (ENSO yr−1)	118	11	129
1982 (ENSO yr 0)	170	10	180
1983	115	8	123
$\bar{x}$	124	13	137
$\hat{s}$	41	7	44

(from Carleton and Carpenter, 1989)

with the strong gradient between negative (positive) pressure anomalies over sub-Antarctic (middle) latitudes. These are at a maximum in the Eastern Hemisphere between 0° eastward through about 130°W and result in the anomalously strong westerlies noted for that season.

Similarly, examination of the broadscale pressure indices for the 1979 winter (Table 3) reveals minimal eccentricity of wavenumber 1 (TPI) for the season (mean=0.3 mb) (Carleton and Carpenter, 1989). These results indicate the broad scale zonality of the FGGE winter, and as revealed by a lack of longitudinal differentiation in polar low occurrences.

TABLE 3. CIRCULATION INDICES, 1977–1983

	TPI (mb)		MI (mb)	
	Sept–June	x*	Sept–June	x*
1977	−3.2	1.7	5.4	4.1
1978	−14.0	−1.7	−19.0	1.9
1979	−7.7	0.3	3.1	0.1
1980	−11.6	−1.2	−4.3	3.9
1981	4.6	−3.3	11.0	2.9
1982	—	—	−8.7	4.3
1983	—	—	−4.5	2.3

*Represents the winter season mean, JJAS.

4.2 Relationships With ENSO

The occurrence of the major "warm" event of 1982–1983 during our period of study offers the opportunity to study polar low relationships with ENSO. The winters of 1981 and 1982 correspond, respectively, to the year (-1) and year (0) of that event. A comparison of the occurrences of polar lows for those winters (Figures 5 and 6) shows considerably greater numbers of polar lows in the 1982 winter than in 1981 (also Table 2), and distinct changes in their longitudes of occurrence. There is a marked peak in activity around the 90°E meridian in 1981, but relatively little activity in the New Zealand–west Pacific regions at that time. The reverse is the case for 1982, when high frequencies of polar lows occur in the latter region (Figure 6). The intraseasonal change in the wave in the Tasman Sea, given by MI, is strongly positive ($+11.0$ mb) in 1981 (yr-1) but strongly negative (-8.7 mb) in 1982 (yr 0). Thus, the reduced (increased) occurrences of polar lows in the New Zealand sector in 1981 (1982) are consistent with a suppressed (enhanced) annual cycle of the wave, as described by van Loon (1984).

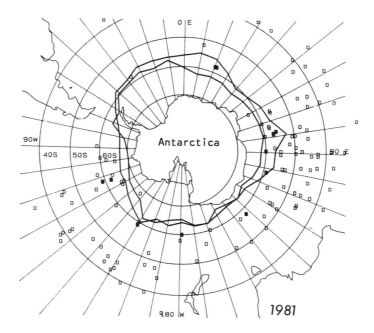

Figure 5. Polar low locations and maximum/minimum sea ice extent for winter 1981 (commas: open squares; spirals: filled squares) (from Carleton and Carpenter, 1989).

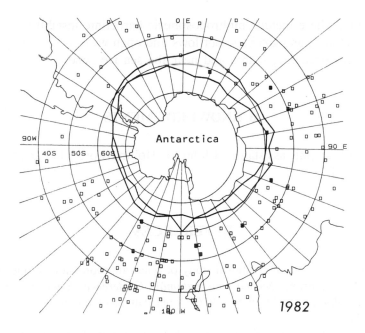

Figure 6. Similar to Figure 5, except for winter 1982 (from Carleton and Carpenter, 1989).

4.3 Polar Low–Sea Ice Relationships

The foregoing results indicate that increased (decreased) frequencies of polar lows, in particular longitudes, coincide with enhanced (reduced) southerly cold air outflows and reduced (enhanced) static stability. Similarly, seasonal and inter-annual changes in the longitudes of frequent cold air outlows show some positive correspondence with the regional extent of the Antarctic sea ice, since equator-ward ice advance is favored to the west of major cyclones. Thus, with reference to Figures 5 and 6, it is noted that polar lows increase (decrease) in association with greater (reduced) sea ice in 1981 (1982) over the southern Indian Ocean between about 60°–100°E. A similar relationship holds for different sectors in some of the other winters (not shown), however, this relationship is not always apparent. Further, the sea ice–cold air outflow–polar low feedback is more evi-dent on a seasonally averaged basis than on a case-by-case (daily to weekly) basis (Carleton and Carpenter, 1989). Future work should examine the role of candidate contributory processes in polar low development for the Southern

Hemisphere. These include midtropospheric baroclinic instability in the case of comma clouds, and katabatic outflows and open water areas for spiraliform developments occurring within and near the sea ice zone.

ACKNOWLEDGMENTS

The cartographic assistance of John M. Hollingsworth (Indiana University) is appreciated.

REFERENCES

Auer, A.H., 1986: An observational study of polar air depressions in the Australian region. Preprint, *Second International Conference Southern Hemisphere Meteorology,* 1–5 December 1986, American Meteorological Society, Boston, 46–49.

Budd, W.F., 1982: The role of Antarctica in Southern Hemisphere weather and climate. *Austral. Meteor. Mag., 30,* 265–272.

Businger, S., 1985: The synoptic climatology of polar low outbreaks. *Tellus, 37A,* 419–432.

Carleton, A.M., 1985: Satellite climatological aspects of the "polar low" and "instant occlusion." *Tellus, 37A,* 433–450.

Carleton, A.M., 1987: Satellite-derived attributes of cloud vortex systems and their application to climate studies. *Remote Sensing Envt., 22,* 271–296.

Carleton, A.M., 1989: Antarctic sea-ice relationships with indices of the atmospheric circulation of the Southern Hemisphere. *Climate Dynamics, 4* (in press).

Carleton, A.M., and D.A. Carpenter, 1989: Intermediate-scale sea ice-atmosphere interactions over high southern latitudes in winter. *GeoJournal, 18,* 87–101.

Forbes, G.S., and W.D. Lottes, 1985: Classification of mesoscale vortices in polar airstreams and the influence of the large-scale environment on their evolutions. *Tellus, 37A,* 132–155.

Lyons, S.W., 1983: Characteristics of intense Antarctic depressions near the Drake Passage. Preprint, *First International Conference Southern Hemisphere Meteorology,* 31 July–6 August 1983, American Meteorological Society, Boston, 283–240.

Pittock, A.B., 1984: On the reality, stability, and usefulness of Southern Hemisphere teleconnections. *Austral. Meteor. Mag., 32,* 75–82.

Rasmussen, E., 1981: An investigation of a polar low with a spiral cloud structure. *J.Atmos. Sci., 38,* 1785–1792.

Reed, R.J., 1979: Cyclogenesis in polar air streams. *Mon Wea. Rev., 107,* 38–52.

Ropelewski, C.F., 1983: Spatial and temporal variations in Antarctic sea ice (1973–1982). *J. Clim. Appl. Meteor.*, *22*, 470–473.

Trenberth, K.E., 1976: Fluctuations and trends in indices of the Southern Hemisphere circulation. *Quart. J. Roy. Meteor. Soc.*, *102*, 65–75.

van Loon, H., 1984: The Southern Oscillation, Part III: Associations with the trades and with the trough in the westerlies of the South Pacific Ocean. *Mon. Wea. Rev.*, *112*, 947–954.

van Loon, H., and J.C. Rogers, 1981: Remarks on the circulation over the Southern Hemisphere in FGGE and on its relation to the phases of the Southern Oscillation. *Mon. Wea. Rev.*, *109*, 2255–2259.

Zick, C., 1983: Method and results of an analysis of comma cloud developments by means of vorticity fields from upper tropospheric satellite wind data. *Meteor. Rundsch.*, *36*, 69–84.

AUTHOR INDEX